Geotechnical, Geological and Earthquake Engineering

Volume 47

Series Editor
Atilla Ansal, School of Engineering, Özyeğin University, Istanbul, Turkey

Editorial Advisory Boards
Julian Bommer, Imperial College London, U.K.
Jonathan D. Bray, University of California, Berkeley, U.S.A.
Kyriazis Pitilakis, Aristotle University of Thessaloniki, Greece
Susumu Yasuda, Tokyo Denki University, Japan

More information about this series at http://www.springer.com/series/6011

Rajesh Rupakhety • Simon Olafsson
Bjarni Bessason
Editors

Proceedings of the International Conference on Earthquake Engineering and Structural Dynamics

Editors
Rajesh Rupakhety
Faculty of Civil and Environmental
Engineering, School of Engineering
and Natural Sciences
University of Iceland
Reykjavik, Iceland

Director of Research, Earthquake
Engineering Research Centre (EERC)
University of Iceland
Selfoss, Iceland

Bjarni Bessason
Faculty of Civil and Environmental
Engineering, School of Engineering
and Natural Sciences
University of Iceland
Reykjavik, Iceland

Simon Olafsson
Earthquake Engineering Research
Centre (EERC)
University of Iceland
Selfoss, Iceland

International Conference on Earthquake Engineering and Structural Dynamics ICESD2017
Reykjavík, Iceland 12–14 June 2017 https://icesd.hi.is/

ISSN 1573-6059   ISSN 1872-4671   (electronic)
Geotechnical, Geological and Earthquake Engineering
ISBN 978-3-319-78186-0   ISBN 978-3-319-78187-7   (eBook)
https://doi.org/10.1007/978-3-319-78187-7

Library of Congress Control Number: 2018945552

© Springer International Publishing AG, part of Springer Nature 2019
This work is subject to copyright. All rights are reserved by the Publisher, whether the whole or part of the material is concerned, specifically the rights of translation, reprinting, reuse of illustrations, recitation, broadcasting, reproduction on microfilms or in any other physical way, and transmission or information storage and retrieval, electronic adaptation, computer software, or by similar or dissimilar methodology now known or hereafter developed.
The use of general descriptive names, registered names, trademarks, service marks, etc. in this publication does not imply, even in the absence of a specific statement, that such names are exempt from the relevant protective laws and regulations and therefore free for general use.
The publisher, the authors and the editors are safe to assume that the advice and information in this book are believed to be true and accurate at the date of publication. Neither the publisher nor the authors or the editors give a warranty, express or implied, with respect to the material contained herein or for any errors or omissions that may have been made. The publisher remains neutral with regard to jurisdictional claims in published maps and institutional affiliations.

Printed on acid-free paper

This Springer imprint is published by the registered company Springer International Publishing AG part of Springer Nature.
The registered company address is: Gewerbestrasse 11, 6330 Cham, Switzerland

We dedicate this work to Ragnar Sigbjörnsson, our teacher and mentor

Foreword

The *International Conference on Earthquake Engineering and Structural Dynamics* (ICESD) was organized by the Earthquake Engineering Research Centre of the University of Iceland in honour of late Professor Ragnar Sigbjörnsson. The conference was held in Reykjavik, Iceland, from 12 to 14 June 2017. There were 11 keynote lectures and 46 oral presentations related to the different topics in earthquake engineering and structural dynamics. Some of the keynote lectures and invited papers were published in a separate book entitled *Earthquake Engineering and Structural Dynamics in Memory of Ragnar Sigbjörnsson* with R. Rupakhety and S. Ólafsson as editors in the Springer Book Series on *Geotechnical, Geological and Earthquake Engineering* as Volume 44.

Ragnar was a very dear and kind friend who always supported the European Association for Earthquake Engineering (EAEE) as the Delegate of Iceland and the journal *Bulletin of Earthquake Engineering* as an Editorial Board Member whole heartedly. He will always be remembered for his enthusiasm.

This book contains a very sound collection of 37 contributions that were submitted to the ICESD. They represent new ideas and methodologies in the fields of engineering seismology, soil dynamics and site effects, earthquake disaster prevention and mitigation, lessons from past earthquakes, seismic analysis and design, earthquake engineering, and structural dynamics. It will always enable us to remember our very kind and considerate friend Ragnar from Iceland.

School of Engineering, Özyeğin
University, Istanbul, Turkey

Atilla Ansal

Preface

This proceedings is a collection of papers presented at the International Conference on Earthquake Engineering and Structural Dynamics (ICESD). The ICESD (www.icesd.hi.is) was held in Reykjavík, Iceland, from 12–14 June 2017. It was held in memory and honour of late Professor Ragnar Sigbjörnsson, who was an active researcher and inspiring teacher in the fields of earthquake engineering and structural dynamics. The 90 participants, of whom many were students, presented and discussed state of the art and new developments in earthquake engineering and structural dynamics, making the event a pleasant success. Researchers and students from 25 countries participated in the conference, making it a truly international event.

Each of the papers submitted to the conference was peer reviewed by two experts, and 46 papers were accepted for presentation. Peer review was managed by the editors with assistance from the international scientific committee of ICESD and other experts. Apart from these 46 papers, 11 keynote lectures were delivered by some of the most distinguished researchers and scientists in the fields of earthquake engineering, engineering seismology, and structural dynamics. Some of the keynote lectures along with selected conference papers have been published as book chapters in *Earthquake Engineering and Structural Dynamics in Memory of Ragnar Sigbjörnsson: Selected Topics*, edited by R. Rupakhety and S. Ólafsson, in the *Geotechnical, Geological and Earthquake Engineering* book series by Springer. This proceedings contains another 37 papers presented at the conference. One of the papers was presented as keynote lecture by Professor Andreas J. Kappos.

The two main themes of the conference were earthquake engineering and structural dynamics. Papers included in this volume cover a wide range of topics such as engineering seismology; seismic analysis, design and retrofit; non-structural damage and risk; seismic risk management and communication; and experience from past earthquakes. Some papers address the dynamics of special structures such as long-span suspension bridges and end-supported pontoon bridges.

We are grateful to the participants of the conference for making it a lively and enjoyable event. We wish to extend our gratitude to all the authors of the papers included in this volume. Special thanks are due to Professor Atilla Ansal, who is also the series editor of this book series, for his instrumental role in the success of the conference. We also thank all the keynote speakers for their valuable contributions to the conference, showcasing excellent research in diverse subjects. The conference was attended by many practising engineers from Iceland, to whom we are grateful for their support and enthusiasm. The international scientific committee and many other academics were instrumental in reviewing the submitted papers and maintaining their academic standard. We are grateful for their contribution. We thank the local organizing committee for successfully managing and organizing the conference. Special thanks are due to Petra van Steenbergen, Springer executive editor of Earth Sciences, Geography and Environment, for her continuous support in preparing and publishing this volume.

Selfoss, Iceland Rajesh Rupakhety
Selfoss, Iceland Símon Ólafsson
Reykjavik, Iceland Bjarni Bessason

Contents

1 **The Ranges of Uncertainty Among the Use of NGA-West1 and NGA-West 2 Ground Motion Prediction Equations** 1
Teraphan Ornthammarath and Pennung Warnitchai

2 **On the Manifestation of Ground Motion Model Differences on Seismic Hazard Sensitivity in North Iceland** 11
Milad Kowsari, Benedikt Halldorsson, and Jónas Thor Snæbjörnsson

3 **Bayesian Hierarchical Model of Peak Ground Acceleration for the Icelandic Strong-Motion Arrays** 25
Sahar Rahpeyma, Benedikt Halldorsson, and Birgir Hrafnkelsson

4 **Towards an Automated Kappa Measurement Procedure** 39
Tim Sonnemann and Benedikt Halldorsson

5 **Application of MASW in the South Iceland Seismic Zone** 53
Elín Ásta Ólafsdóttir, Bjarni Bessason, and Sigurður Erlingsson

6 **Experimental Study of All-Steel Buckling-Restrained Braces Under Cyclic Loading** 67
Ahmad Fayeq Ghowsi and Dipti Ranjan Sahoo

7 **Seismic Design Procedure for Staggered Steel Plate Shear Wall** 81
Abhishek Verma and Dipti Ranjan Sahoo

8 **Passive-Hybrid System of Base-Isolated Bridge with Tuned Mass Absorbers** 95
Said Elias and Vasant Matsagar

9	Monitoring and Damage Detection of a 70-Year-Old Suspension Bridge – Ölfusá Bridge in Selfoss, Case Study 111
	Gudmundur Valur Gudmundsson, Einar Thor Ingólfsson, Kristján Uni Óskarsson, Bjarni Bessason, Baldvin Einarsson, and Aron Bjarnason
10	Performance of Base Isolated Bridges in Recent South Iceland Earthquakes 123
	Bjarni Bessason, Einar Hafliðason, and Guðmundur Valur Guðmundsson
11	Cyclic Capacity of Dowel Connections 137
	Tatjana Isakovic, B. Zoubek, and M. Fischinger
12	Ductile Knee-Braced Frames for Seismic Applications 149
	Sutat Leelataviwat, P. Doung, E. Junda, and W. Chan-anan
13	Seismic Capacity Reduction Factors for a RC Beam and Two RC Columns 159
	Pablo Mariano Barlek Mendoza, Daniela Micaela Scotta and Enrique Emilio Galíndez
14	Single-Degree-of-Freedom Analytical Predictive Models for Seismic Isolators 173
	Todor Zhelyazov
15	The Evaluation of Nonlinear Seismic Demands of RC Shear Wall Buildings Using a Modified Response Spectrum Analysis Procedure 185
	Fawad Ahmed Najam and Pennung Warnitchai
16	Seismic Fragility Assessment of Reinforced Concrete High-Rise Buildings Using the Uncoupled Modal Response History Analysis (UMRHA) 201
	Muhammad Zain, Naveed Anwar, Fawad Ahmed Najam, and Tahir Mehmood
17	Ambient Vibration Testing of a Three-Storey Substandard RC Building at Different Levels of Structural Seismic Damage 219
	Pinar Inci, Caglar Goksu, Ugur Demir, and Alper Ilki
18	System Identification of a Residential Building in Kathmandu Using Aftershocks of 2015 Gorkha Earthquake and Triggered Noise Data 233
	Yoshio Sawaki, Rajesh Rupakhety, Simon Ólafsson, and Dipendra Gautam
19	Damage Observations Following the M_w 7.8 2016 Kaikoura Earthquake 249
	Dmytro Dizhur, Marta Giaretton, and Jason M. Ingham

20	**Seismic Rehabilitation of Masonry Heritage Structures with Base-Isolation and FRP Laminates – A Case Study Comparison**... Simon Petrovčič and Vojko Kilar	263
21	**From Seismic Input to Damage Scenario: An Example for the Pilot Area of Mt. Etna Volcano (Italy) in the KnowRISK Project**... Raffaele Azzaro, Salvatore D'Amico, Horst Langer, Fabrizio Meroni, Thea Squarcina, Giuseppina Tusa, Tiziana Tuvè, and Rajesh Rupakhety	277
22	**Seismic Performance of Non-structural Elements Assessed Through Shake Table Tests: The KnowRISK Room Set-Up**... Paulo Candeias, Marta Vicente, Rajesh Rupakhety, Mário Lopes, Mónica Amaral Ferreira, and Carlos Sousa Oliveira	293
23	**KnowRISK Practical Guide for Mitigation of Seismic Risk Due to Non-structural Components**... Hugo O'Neill, Mónica Amaral Ferreira, Carlos Sousa Oliveira, Mário Lopes, Stefano Solarino, Gemma Musacchio, Paulo Candeias, Marta Vicente, and Delta Sousa Silva	309
24	**A Study of Rigid Blocks Rocking Against Rigid Walls**... Gudmundur Örn Sigurdsson, Rajesh Rupakhety, and Símon Ólafsson	323
25	**Finite Element Model Updating of a Long Span Suspension Bridge**... Øyvind Wiig Petersen and Ole Øiseth	335
26	**Characterization of the Wave Field Around an Existing End-Supported Pontoon Bridge from Simulated Data**... Knut Andreas Kvåle and Ole Øiseth	345
27	**Identification of Rational Functions with a Forced Vibration Technique Using Random Motion Histories**... Bartosz Siedziako and Ole Øiseth	361
28	**The Dynamic Intelligent Bridge: A New Concept in Bridge Dynamics**... Andreas J. Kappos	373
29	**Systematic Methodology for Planning and Evaluation of a Multi-source Geohazard Monitoring System. Application of a Reusable Template**... Fjóla G. Sigtryggsdóttir and Jónas Th. Snæbjörnsson	385

30 How to Survive Earthquakes: The Example of Norcia............ 403
    Mário Lopes, Francisco Mota de Sá, Mónica Amaral Ferreira,
    Carlos Sousa Oliveira, Cristina F. Oliveira, Fabrizio Meroni,
    Thea Squarcina, and Gemma Musacchio

31 KnowRISK on Seismic Risk Communication: The Set-Up of a
    Participatory Strategy- Italy Case Study...................... 413
    Gemma Musacchio, Susanna Falsaperla, Stefano Solarino,
    Giovanna Lucia Piangiamore, Massimo Crescimbene,
    Nicola Alessandro Pino, Elena Eva, Danilo Reitano,
    Federica Manzoli, Michele Fabbri, Mariangela Butturi,
    and Mariasilvia Accardo

32 Seismic Risk Communication: How to Assess It?
    The Case of Lisbon Pilot-Area............................. 429
    Delta Sousa e Silva, A. Pereira, Marta Vicente, R. Bernardo,
    Monica Amarel Ferreira, Mario Lopes, and Carlos Sousa Oliveira

33 Shaping Favorable Beliefs Towards Seismic Protection Through
    Risk Communication: A Pilot-Experience in Two Lisbon
    Schools (Portugal)...................................... 445
    Delta Sousa e Silva, Marta Vicente, A. Pereira, R. Bernardo,
    Paulo Candeias, Monica Amarel Ferreira, Mario Lopes,
    Carlos Sousa Oliveira, and P. Henriques

34 The KnowRISK Action for Schools: A Case Study in Italy....... 459
    Gemma Musacchio, Elena Eva, and Giovanna Lucia Piangiamore

35 Risk Perception and Knowledge: The Construction of the
    Italian Questionnaire to Assess the Effectiveness of the
    KnowRISK Project Actions................................ 471
    Massimo Crescimbene, Nicola Alessandro Pino,
    and Gemma Musacchio

36 Awareness on Seismic Risk: How Can Augmented
    Reality Help?.. 485
    Danilo Reitano, Susanna Falsaperla, Gemma Musacchio,
    and Riccardo Merenda

37 Development of a Common (European) Tool to Assess
    Earthquake Risk Communication........................... 493
    Stephen Platt, Gemma Musacchio, Massimo Crescimbene,
    Nicola Alessandro Pino, Delta S. Silva, Mónica A. Ferreira,
    Carlos S. Oliveira, Mário Lopes, and Rajesh Rupakhety

Chapter 1
The Ranges of Uncertainty Among the Use of NGA-West1 and NGA-West 2 Ground Motion Prediction Equations

Teraphan Ornthammarath and Pennung Warnitchai

Abstract In this study, a comparison of the use of NGA-West1 and NGA-West2 ground motion prediction equations (GMPEs) to estimate peak ground acceleration (PGA) and spectral acceleration (SA) at 1.0 s for moderate to high seismic hazard area have been presented. This paper focuses on updated estimated ground motion due to the use of NGA-West2, and their impact on the hazard map related to those estimated by NGA-West1 for two cities in South East Asia with different level of seismic hazard. In addition, comparison of the range of epistemic uncertainty between NGA-West1 and NGA-West2 have also been determined. In general, the combined effects of lower medians and increased standard deviations in the new GMPEs have caused only small changes, within 5–20%, in the probabilistic ground motions for considered sites compared to the previous results. In addition, the results illustrate that the variation in seismic hazard due to GMPEs seems to be lower for NGA-West2 comparing to NGA-West1 for area with controlling earthquake magnitude of 6.0–6.5. However, for area with controlling earthquakes of small magnitudes in the range 5.5–6.0 or very strong earthquakes (M > 7.5), the variations in seismic hazard seems to be similar for both NGA-West1 and NGA-West2.

Keywords NGA-West1 · NGA-West2 · Seismic hazard · Epistemic Uncertainty

T. Ornthammarath (✉)
Department of Civil and Environmental Engineering, Faculty of Engineering, Mahidol University, Salaya, Thailand
e-mail: teraphan.orn@mahidol.edu

P. Warnitchai
School of Engineering and Technology, Asian Institute of Technology, Khlong Nueng, Pathumthani, Thailand

1.1 Introduction

The development of seismic hazard models and their characteristics are obviously the combination of a wide range of possible outcomes and their uncertainties. Many past studies emphasize and address uncertainties in PSHA (e.g. McGuire and Shedlock 1981; Senior Seismic Hazard Analysis Committee SSHAC 1997). There are two types of variability that are formalized and included in PSHA. Epistemic uncertainty, or modeling uncertainty, is derived from the recognition that diverse alternative models could describe specific phenomena equally well. This uncertainty is due to insufficient knowledge about the validity of alternative assumptions, mathematical models, and values of the parameters of each model. Aleatory variability, or randomness, is uncertainty in the data used in an analysis and generally accounts for randomness associated with the prediction of a parameter from a specific model, assuming that the model is correct. The standard deviation (σ) of an individual GMPE is a representation of aleatory variability.

In PSHA, the epistemic uncertainty is mostly considered by applying logic trees that handle the use of alternative models and parameter values of each model. Contrary to aleatory variability, the epistemic uncertainty might be reduced by acquiring a better understanding – that is, by acquiring additional data and improved information. Recognition of the two kinds of uncertainty is initially useful when selecting and combining inputs. Hazard evaluators need to be aware of the sources of uncertainties (e.g., limitations of available data) so that they can make informed assessments of the validity of alternative hypotheses, the accuracy of alternative models, and the value of data and then transmit these uncertainties to the end users. For example, epistemic uncertainty would generally be much greater for the assessment of seismic hazard in regions where there are relatively few ground-motion records and undetermined locations of slow-slip-rate faults to constrain the selection of appropriate models. The calculated mean seismic hazard from different source models (area or fault sources) could give similar results irrespective of different source models; however, the fractile hazard curves that represent epistemic uncertainty would be differing greatly.

For many regions, where a limited number of strong-motion records are available, one solution to this limitation in order to perform seismic hazard analysis is to assume that some existing GMPEs developed for other regions with similar seismotectonic characteristics can adequately represent ground-motion scaling in this region. For Thailand and the rest of South East Asia (SE Asia), Next Generation Attenuation West 1 (NGA-West 1) models developed for shallow crustal earthquakes in the western United States has generally been used in the past few years. However, with recent development of the NGA West 2 models, some improvement from the NGA-West 1 equations involved adding data at small-to moderate magnitudes, the richer database available for NGA-West 2 allows NGA West 2 developers to improve on prior work by considering additional variables that could not previously be adequately resolved.

In this study, a comparison of the use of NGA-West1 and NGA-West2 ground motion prediction equations (GMPEs) to estimate peak ground acceleration (PGA) and spectral acceleration (SA) at 1.0 at different soil classes for moderate to high seismic hazard area have been presented. This paper focuses on updated estimated ground motion due to the use of NGA-West2, and their impact on the hazard map related to those estimated by NGA-West1 for two cities in SE Asia with different level of seismic hazard. In addition, comparison of the range of epistemic uncertainty between NGA-West1 and NGA-West2 have also been determined and discussed.

1.2 Selected Study Area

In this study, two cities in SE Asia with different level of seismic hazard have been selected. The PGAs and SA at 1 s at 5% of critical damping had been determined for the mean, median, and 16th and 84th percentile for 475 and 2475 year return periods based on the model developed by Ornthammarath et al. (2011). The model is a mixture of background smooth seismicity, crustal faults, and subduction area. The background seismicity model represents random earthquakes in the whole study region except the subduction zones. The model accounts for all earthquakes in areas with no mapped seismic faults and for smaller earthquakes in areas with mapped faults. The magnitude-dependent characteristic of the seismicity rate in each background seismicity zone is modelled by a truncated exponential model (Gutenberg-Richter model). The obtained regional b-value is 0.90, and The a-value varies from place to place within each grid. In the truncated Gutenberg-Richter models of both background smooth seismicity inside Thailand (BG-I) and outside Thailand (BG-II), the minimum earthquake magnitude is set equal to 4.5 because earthquakes with smaller magnitude than this are judged not to cause damage to buildings and structures (Bommer et al. 2001). The maximum (upper bound) magnitude is set to 6.5 for BG-I and 7.5 for BG-II.

Table 1.1 shows PGA and SA for 1 s at 5% critical damping for 10-percent and 2-percent probability of exceedance in 50 years (475 and 2475-year, respectively) based on selected NGA West1. These selected NGA-West1 models were Boore and Atkinson (2008), Campbell and Bozorgnia (2008), and Chiou and Youngs (2008). Equal probabilities (i.e. 1/3) have been assigned to each of these three models in the logic tree analysis. Of the selected cities in SE Asia, Yangon has by far is one of the greatest seismic hazard, primarily due to observed seismicity and its proximity to the

Table 1.1 Probabilistic ground motions for selected cities by using NAG-West1 based on Ornthammarath et al. (2011)

City	PGA (g)		SA (T = 1 s)	
	T = 475	T = 2475	T = 475	T = 2475
Chiang Mai	0.16	0.25	0.07	0.10
Yangon	0.30	0.50	0.30	0.45

Table 1.2 Controlling earthquake scenarios for PGA and SA at 1 s for Chiang Mai and Yangon

City	Soil Class	PGA (g)				SA (T = 1 s)			
		T = 475		T = 2475		T = 475		T = 2475	
		M	Rrup (km)	M	Rrup (km)	M	Rrup (km)	M	Rrup (km)
Chiang Mai	B	5.68	13	5.79	9.9	5.9	23	6.4	58
	D	5.66	14	5.79	10.5	6.0	21	6.1	27
Yangon	B	7.1	53	7.9	50	7.1	53	7.9	50
	D	6.9	55	7.9	50	6.9	55	7.9	50

potential large earthquake fault, Sagaing fault. The estimated PGA value of 50% g for 2% in 50 years is comparable to the seismically active regions of the intermountain west in the United States. For Chiang Mai, the observed seismicity contributes mainly to moderate hazard with PGA value of 25% g for 2% in 50 years.

1.3 Deaggregation

Furthermore, deaggregation analysis is applied for selected sites to assess the combined effect of all magnitudes and distances on the probability of exceeding a given ground motion level. Considering a return period of 475 and 2475 year and PGA and SA at 1 s for Chiang Mai and Yangon, the deaggregation was computed and the controlling earthquakes are shown in Table 1.2 in terms of mean moment magnitude and mean rupture distance, Rrup. The deaggregation results are found to be similar for both soil classes. Observed local seismicity in and around Chiang Mai seems to govern the hazard for both short and long structural periods at considered return periods; however, for Yangon, the deaggregation shows a large contribution for the controlling earthquake scenarios from Sagaing fault. It can also be noticed that there is similar controlling earthquake scenarios between different site classes. For deaggregation of PGA at a 2475-year return period, the controlling earthquake scenario for Chiang Mai leads to larger earthquake size at a closer distance. While the deaggregation results of Yangon again show a large contribution from large earthquake magnitudes at 50 km distance.

1.4 Comparison Between NGA-WEST1 and NGA-WEST2

In this section, seismic hazard analysis has been computed for two selected cities based on selected NGA-West2. These selected NGA-West2 models were Boore et al. (2014), Campbell and Bozorgnia (2014), and Chiou and Youngs (2014). Equal probabilities (i.e. 1/3) have been assigned to each of these three models in the logic tree analysis. Figure 1.1 displays hazard results and the range of uncertainty

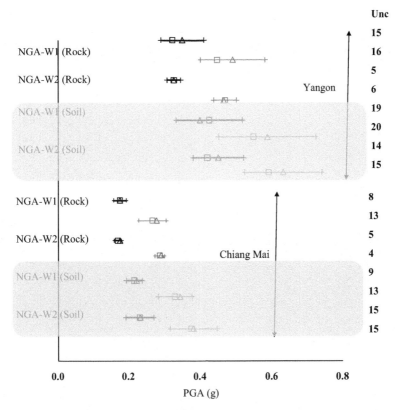

Fig. 1.1 Comparison of expected PGAs (AFEs of 1 = 475 (black) and 1 = 2475 (gray) return periods) from different probabilistic seismic-hazard assessments (PSHAs) (bars, 15th–85th fractiles; square, medians; and triangles, means) for moderate seismic hazard area (Chiang Mai) and high seismic hazard area (Yangon), the uncertainty metric 100 log(SA_{85}/SA_{15})

observed from the use of NGA-West1 and NGA-West2 for both considered sites for rock and soft soil condition at 475- and 2475-year return periods. It can be noticed that the hazard results among the use of NGA-West1 and NGA-West2 are comparable for all considered cases, within 5–10%, in the probabilistic ground motions for considered sites compared to the previous results. For area of moderate hazard in Chiang Mai, the observed epistemic uncertainty for NGA-West2 at rock site condition in the range between 0.02 and 0.026 g are much smaller than that for NGA-West1 in the range between 0.03 and 0.08 g especially where the controlling earthquake magnitude is in the range between 5.5 and 6.5 from 10 to 60 km distance. However, for the case of soft soil condition, both NGA-West1 and NGA-West2 show similar dispersion in the range between 0.04 and 0.13 g.

To more objectively compare epistemic uncertainty among NGA-West1 and NGA-West2, the range of uncertainty in Yangon, where the high level of seismic hazard is observed with large controlling earthquake magnitude (M > 7.5) at 50 km

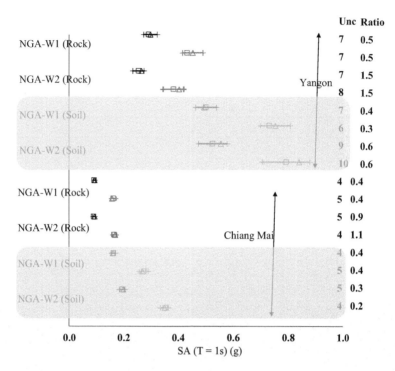

Fig. 1.2 Comparison of expected SAs (T = 1.0 s) (AFEs of 1 = 475 (black) and 1 = 2475 (gray) return periods) from different probabilistic seismic-hazard assessments (PSHAs) (bars, 15th–85th fractiles; square, medians; and triangles, means) for moderate seismic hazard area (Chiang Mai) and high seismic hazard area (Yangon), the uncertainty metric 100 log(SA_{85}/ SA_{15}) and the ratio of the uncertainty metrics for SA(1 s) and PGA

distance, has been computed. The distribution of predicted ground motion at 475- and 2475-year return period at rock site for Yangon by NGA-West2 in the range between 0.03 and 0.06 g shows lower uncertainty than the previous analysis with NGA-West1 in the range between 0.11 and 0.18 g, where the controlling earthquake magnitude is in the range of 6.9–7.9 at 50 km distance. However, the observed uncertainty is still higher than that observed in the area of moderate controlling earthquake magnitude (5.5 < M < 5.8) as previously observed in Chiang Mai. In addition, the uncertainty for soft site at high level hazard are larger than that observed in moderate hazard for both NGA-West1 and NGA West2 models in the range between 0.14 and 0.27 g.

Moreover, dispersion in GMPEs appear to be only weakly dependent on structural period, it would be expected that the uncertainties in the PSHA would also not show strong period dependency. This is examined by comparing Fig. 1.1 (for PGA) and Fig. 1.2 (for SA (1 s)) and particularly by examining the range of uncertainty. In general, the hazard results for both selected sites at all considered return periods and soil conditions displays similar results, within 5–12%, in the probabilistic ground motions. However, large changes within 22–29%, in the case for soft soil site could be observed in moderate seismic hazard area. For the uncertainties, in general the expected SA (1 s) values are show similar dispersion pattern as it is observed in the expected PGA. For moderate hazard area, the uncertainty is lower than 0.02 g and for both NGA-West1 and NGA-West2 and for both rock and soft soil conditions. For high hazard area, the dispersion is higher than that of moderate hazard area for both NGA-West1 and NGA West2 models at rock condition in the range between 0.04 and 0.07 g and for soft soil condition the uncertainty is in the range between 0.07 and 0.17 g for both NGA-West1 and NGA-West2.

1.5 Discussion and Conclusion

Since Kulkarni et al. (1984) first introduced the logic tree approach, it has been regularly employed as a means to account for epistemic uncertainty in seismic hazard analysis. The use of logic trees provides a convenient framework for the explicit treatment of model (i.e. epistemic) uncertainty. Apart from alternative models, e.g. different GMPEs, the logic tree approach could also consider ranges of parameters required for them and their associated uncertainties, which results in a set of hazard curves from a logic tree that can be represented using a set of percentile (or fractile) curves and a mean curve (i.e. mean with respect to the epistemic uncertainties). In practice, the mean hazard curve is generally employed. A characteristic of the mean hazard curve that is significantly different from the median curve is that the hazard distribution at very low annual rates of exceedance has a positive skewness toward larger values – that is, there is a long upper tail to the distribution. Abrahamson and Bommer (2005) also question the applicability of the mean hazard curve at long return periods (e.g. $\nu \geq 10^{-7}$ mean annual rate of exceedance), since the mean hazard curve at very long return period is strongly influenced by excessively large fractile hazard curves, and the mean hazard curve tends to climb across the fractile curves and can result in very high design ground motions.

Considering all sources of epistemic uncertainty can lead to very complex logic trees, making the final hazard calculations difficult to implement by reviewers or anyone not part of the project, thus resulting in a serious lack of transparency of the results. As the complexity of the logic trees grows, the number of possible combinations of their different branches becomes enormous, on the order of 1020 in some past studies, (Abrahamson et al. 2002). In the past the requirement has been to fully sample the logic tree, which is a huge computational effort. Hence, sensitivity analysis is sometimes introduced to discriminate which parameters contribute the most to the hazard and its uncertainty and can be used as a preliminary step for the construction of logic trees focusing efforts on the parameters found to be most important (Rabinowitz et al. 1998; Scherbaum et al. 2005).

Peruš and Fajfar (2009) use a non-parametric approach (CAE method) to investigate the possible reasons for difference among NGA-West1 equations. The results suggest that the predictions depend substantially on the selection of the effective database and on the adopted functional forms. These observed variations (i.e. the epistemic uncertainty) inevitably influence the application of these well-constrained GMPEs, especially in PSHA. Computed hazard curves from well-constrained GMPEs do not necessarily produce lower epistemic uncertainty. This could be due to the adopted functional form, selected database, and the determination of sigma.

Current analysis shows that despite the fact that the development of NGA-West2 models is a collaborative effort with many interaction and exchange of ideas among the developers the similar level of epistemic uncertainties still could be observed from both NGA-West1 and NGA-West2. The large differences are observed for the areas in which the NGA-West2 database is sparse, such as for large (M > 7.9) earthquakes at close distances and larger total standard deviation values for the smaller earthquakes comparting to those of moderate earthquake magnitudes. Based on this information, additional data collection or analysis should be undertaken to reduce this uncertainty in the future.

Acknowledgements This study was sponsored by Thailand Research Fund and Faculty of Engineering, Mahidol University under contract No. TRG5780243 & MRG5980243.

References

Abrahamson NA, Bommer JJ (2005) Opinion papers: probability and uncertainty in seismic hazard analysis. Earthquake Spectra 21:603–607
Abrahamson NA, Birkhauser P, Koller M, Mayer-Rosa D, Smit P, Sprecher C, Tinic S, Graf R (2002) PEGASOS – comprehensive probabilistic seismic hazard assessment for nuclear power plants in Switzerland. In: Proceedings of the Twelfth European Conference on Earthquake Engineering, Paper no. 633, London
Boore DM, Atkinson GM (2008) Ground-motion prediction equations for the average horizontal component of PGA, PGV, and 5%-damped PSA at spectral periods between 0.01 s and 10.0 s. Earthquake Spectra 24:99–138
Bommer JJ, Georgallides G, Tromans I (2001) Is there a near-field for small-to-moderate magnitude earthquakes? J Earthq Eng 5(3):395–423
Boore DM, Stewart JP, Seyhan E, Atkinson GM (2014) NGA-West2 equations for predicting PGA, PGV, and 5% damped PSA for shallow crustal earthquakes. Earthquake Spectra 30:1057–1085
Campbell KW, Bozorgnia Y (2008) NGA ground motion model for the geometric mean horizontal component of PGA, PGV, PGD and 5% damped linear elastic response spectra for periods ranging from 0.01 to 10 s. Earthquake Spectra 24:139–172
Campbell KW, Bozorgnia Y (2014) NGA-West2 ground motion model for the average horizontal components of PGA, PGV, and 5% damped linear acceleration response spectra. Earthquake Spectra 30:1087–1115
Chiou BSJ, Youngs RR (2008) Chiou-Youngs NGA ground motion relations for the geometric mean horizontal component of peak and spectral ground motion parameters. Earthquake Spectra 24:173–216
Chiou BSJ, Youngs RR (2014) Update of the Chiou and Youngs NGA model for the average horizontal component of peak ground motion and response spectra. Earthquake Spectra 30:1117–1153
Kulkarni RB, Youngs RR, Coppersmith KJ (1984) Assessment of confidence intervals for results of seismic hazard analysis. In: Proceedings, Eighth World conference on Earthquake Engineering, vol 1, San Francisco, pp 263–270

McGuire RK, Shedlock KM (1981) Statistical uncertainties in seismic hazard evaluations in the United States. Bull Seismol Soc Am 71:1287–1308

Ornthammarath T, Warnitchai P, Worakanchana K, Zaman S, Sigbjörnsson R, Lai CG (2011) Probabilistic seismic hazard assessment for Thailand. Bull Earthq Eng. https://doi.org/10.1007/s10518-010-9197-3

Peruš I, Fajfar P (2009) How reliable are the ground motion prediction equations? In: Proceedings, 20th International conference on Structural Mechanics in Reactor Technology (SMiRT 20), Paper No. 1662, Espoo, Finland

Rabinowitz N, Steinberg DM, Leonard G (1998) Logic trees, sensitivity analyses, and data reduction in probabilistic seismic hazard assessment. Earthquake Spectra 14:189–201

Scherbaum F, Bommer JJ, Bungum H, Cotton F, Abrahamson NA (2005) Composite ground-motion models and logic trees: methodology, sensitivities, and uncertainties. Bull Seismol Soc Am 95(5):1575–1593

SSHAC (Senior Seismic Hazard Assessment Committee) (1997) Recommendations for PSHA: Guidanceon uncertainty and use of experts. Report NUREG/CR-6372. U.S. Nuclear Regulatory Commission, Washington, DC

Chapter 2
On the Manifestation of Ground Motion Model Differences on Seismic Hazard Sensitivity in North Iceland

Milad Kowsari, Benedikt Halldorsson, and Jónas Thor Snæbjörnsson

Abstract In this study, probabilistic seismic hazard assessment (PSHA) for North Iceland is explored in terms of its sensitivity to one of its key elements, the selected ground-motion models (GMMs). The GMMs in previous PSHA studies for Iceland are reviewed and in some cases recalibrated to the Icelandic dataset using a Markov Chain Monte Carlo (MCMC) algorithm which is useful in regions where the earthquake records are scarce. To show the ground motion model variability as it is manifested in PSHA uncertainties, the hazard maps of standard deviation and coefficient of variation (CV) of PGA at two hazard levels for GMMs before and after recalibrating are shown. The results indicate that the recalibrated models are promising candidates to be applied for future hazard studies in Iceland, but more importantly they show how to what extent and how the epistemic uncertainty of the GMMs contribute to patches of heightened hazard uncertainties, especially at near- and far-fault distances where there is a particular lack of data.

Keywords Earthquake hazard · PSHA · GMMs · Epistemic uncertainty

---

M. Kowsari
Earthquake Engineering Research Centre, and Faculty of Civil and Environmental Engineering, School of Engineering and Natural Sciences, University of Iceland, Selfoss, Iceland
e-mail: milad@hi.is

B. Halldorsson (✉)
Earthquake Engineering Research Centre, and Faculty of Civil and Environmental Engineering, School of Engineering and Natural Sciences, University of Iceland, Selfoss, Iceland

Division of Processing and Research, Icelandic Meteorological Office, University of Iceland, Reykjavík, Iceland
e-mail: skykkur@hi.is

J. T. Snæbjörnsson
School of Science and Engineering, Reykjavík University, Reykjavík, Iceland
e-mail: jonasthor@ru.is

2.1 Introduction

Seismic hazard maps display the probability that a ground-motion parameter will exceed a certain level within a given unspecific time interval (usually at probabilities of exceedance of 10% or 2% in 50 years) for the area under study. In addition to seismic codes, these maps can be used in insurance rate structures, risk assessments and other public-policies. Since these hazard products are used in making public-policy decisions, they need to be updated regularly to keep pace with new scientific advancements (Petersen et al. 2007). Generally, hazard maps can be obtained through a probabilistic seismic hazard analysis (PSHA) which incorporates ground-motions and occurrence frequencies for all potentially dangerous earthquakes considering different source of uncertainties (McGuire 2004). The ground-motion intensity measures for such analysis are peak ground acceleration (PGA) and pseudo-spectral acceleration (PSA) which can also be used in seismic codes as the horizontal design loads.

In Iceland, the vast majority of reported strong earthquakes have occurred within the South Iceland Seismic Zone (SISZ) in the south-western lowlands and the Tjörnes Fracture Zone (TFZ) in north-eastern Iceland. One of the first Icelandic hazard maps was published by Halldórsson (1996) which was criticized for paying extreme attention to the Hvítársíða earthquake of 1974 and the Vatnafjöll earthquake of 1987, which would dominate parts of the hazard map (Sólnes 2016). An improved version of this map was then published in 2002 as the Icelandic hazard map for the National annex of Eurocode 8 prepared by Halldórsson and Sveinsson (2003). Nevertheless, Solnes et al. (2004) presented an alternative national hazard map for the 10%-in-50 years PGA based on past and future statistically generated earthquake events. They also used the theoretical attenuation relationship proposed by (Ólafsson and Sigbjörnsson 1999, 2002) for the near-, intermediate- and far-field spectra including an exponential term to account for anelastic attenuation. Another recent hazard map exists in which Iceland was divided into 26 sub-regions to cover all known earthquake (area) sources. The earthquake simulation method and the ground-motion models (GMMs) applied are in a similar manner as for the 2004 earthquake hazard map in Solnes et al. (2004). A key limitation is that none of the above hazard maps have been published in internationally accredited scientific journals and therefore, to our knowledge, not been subjected to peer-review as well.

The Icelandic seismic catalogue has been compiled in detail by Ambraseys and Sigbjörnsson (2000) and covers from 1896 to 1996 based on the teleseismic data obtained from station bulletins, books, periodicals, newspapers and public domain reports. The catalogue lists 422 events with surface-wave magnitudes, including 276 events with recalculated surface-wave magnitudes and maximum observed magnitude of 1910, the largest recorded earthquake in the Iceland area, reaching a magnitude of 7.2 (Sigbjörnsson and Ólafsson 2004). Due to uncertainties associated with this catalogue, it might not be appropriate for hazard assessment in North Iceland where many of the earthquakes occurred off-shore and may not have been mentioned in the historical annals. To overcome this problem, Monte-Carlo simulation can be used to generate and extend the earthquake catalog, thus reducing historical bias. The Monte-Carlo PSHA (MC-PSHA) involves taking a standard seismic source model and using it to generate a large number of synthetic catalogue representing possible future outcomes of regional seismicity in a period representing the lifetime of the structure being designed (Musson 2004). The solution provided is the same as the Cornell-McGuire approach (CM-PSHA) (Cornell 1968; McGuire 2004, 1976), which uses a numerical integration but there are some advantages to the MC-PSHA in terms of flexibility and transparency (Atkinson 2012).

The TFZ is a broad and complex region and cannot be associated with a single fault or clearly identified plate boundary. The historical events such as Ms. 7 in 11 September 1755, Ms. 6.5 in 12 June 1838, Ms. 6.5 in 18 April 1872 and Ms. 6.3 in 25 January 1885 indicate the high seismic activity of this region (Tryggvason 1973; Ambraseys and Sigbjörnsson 2000; Halldórsson 2005; Stefansson et al. 2008). However, most PSHA studies have been conducted by several research in the SISZ (Ólafsson et al. 2014; Sigbjörnsson et al. 1995; Snaebjornsson and Sigbjornsson 2008; Solnes et al. 1994, 2000; Solnes and Halldorsson 1995). So far, in North Iceland, there are just two PSHA studies by Snaebjornsson and Sigbjornsson (2007) and Sigbjörnsson and Snaebjornsson (2007), for four geothermal power plants (i.e., Krafla, Theistareykir, Gjástykki and Bjarnarflag) and an industrial site (i.e., Bakki) which is located in the Northern Volcanic Zone (NVZ) and TFZ, repectively. Snaebjornsson and Sigbjornsson (2007), based on the geological and geophysical findings, associate North Iceland seismicity with few hypothetical lines, visualised as seismic delineations which were used in the PSHA studies. The Grímsey lineament (A), the Húsavík–Flatey Fault (HFF) (3 segments: B1, B2 and B3), the Dalvík lineament (C) are three parallel WNW trending lines, the Krafla zone (D), Theistareykir zone (E), the Fremri-Námur zone (F) and the Askja zone (G) are four lines trending NNE represent the main fissure swarms of the NVZ which are shown in Fig. 2.1.

The selected GMM is undoubtedly one of the most important elements of any PSHA study. However, previous earthquake hazard estimates for North Iceland have relied on a single GMM, which indicates that uncertainties are incorporated in a limited way. Therefore, it is both timely and important, especially in light of the fast-growing heavy industry in the region, to revisit the earthquake hazard estimates for North Iceland. In this study, a Bayesian posterior inference by Markov Chain Monte Carlo (MCMC) simulation is used to recalibrate the selected GMMs to the Icelandic data. Then, both the original and recalibrated models are evaluated based on data-driven approaches to incorporate epistemic uncertainty. The uncertainty estimates as an inseparable part of the seismic hazard assessment enables the hazard map developers to describe the confidence in the mean hazard estimates (Cao et al. 2005). For this purpose, the coefficient of variation (CV) as the ratio of the standard deviation to the mean, effectively describing the amount of variability relative to the mean is used. As a result, the hazard maps in terms of PGA variability and CV at different hazard levels are presented to indicate the manifestation of GMM variability on the hazard in North Iceland.

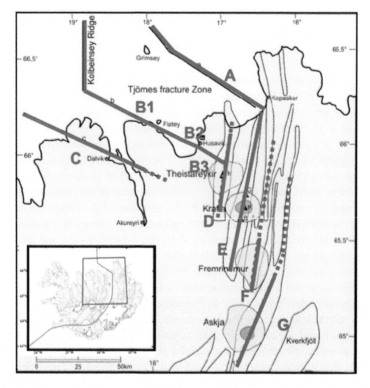

Fig. 2.1 The seismic source zones applied in probabilistic seismic hazard analysis for North Iceland (Sigbjörnsson and Snaebjornsson 2007). The small map inset at bottom left shows Iceland and the area under study. The solid red lines indicate seismic source zones producing earthquakes with magnitude greater than or equal to 4 and the dotted lines refer to source zones where event magnitude does not exceed 4

2.2 Selected GMMS for PSHA

In this study, GMMs are selected from local, regional and worldwide data all of which satisfy the minimum requirements proposed by Cotton et al. (2006) and Bommer et al. (2010). In Bommer et al. (2010), the exclusion criteria of Cotton et al. (2006) were updated to pre-select the GMMs. The updated exclusion criteria were addressed the problems regarding an inappropriate tectonic environment, the number of records and earthquakes, range of response periods, functional forms, regression methods, inappropriate definitions for explanatory variables, and the range of applicability. The selected GMMs are Rupakhety and Sigbjörnsson) 2009), RS09; Akkar and Bommer (2010), AB10; Ambraseys et al. (2005), Am05; Danciu and Tselentis (2007), DT07; Zhao et al. (2006), Zh06; Lin and Lee (2008), LL08. Apart from RS09, the other models had been proposed by Delavaud et al. (2012) in the SHARE project as the candidate GMMs for seismic hazard in Iceland.

The functional forms of each GMMs, region of origin and the magnitude and distance range of applicability are shown in Table 2.1.

2.3 Recalibration of GMMS Using MCMC

A Markov Chain Monte Carlo algorithm has recently been developed to recalibrate GMMs by means of updating the existing models with observations. The MCMC method that forms the backbone of modern Bayesian posterior inference is used to recalibrate the selected GMMs to the Icelandic ground-motions. The sampling is done sequentially, with the distribution of the sampled draws depending on the last value drawn, hence, the draws form a Markov chain (Gelman et al. 2014). The Metropolis-Hastings algorithm is used to sample from the joint posterior distribution. Figure 2.2 shows the decay of PGA for the selected GMMs before and after recalibrating at rock site and the circles are the recorded PGAs. A moment magnitude equal to the mean magnitude of the data (i.e., Mw = 6.1) is considered for comparison purposes in the plots. Of note, all GMMs are recalibrated based on the closest horizontal distance to the vertical projection of the rupture (Rjb). All the recalibrated models seem to fit the recorded data very well in the magnitude and distance range where data is available. The original models underestimate the PGA at short distances from the epicentre and overestimate it at distances farther from the epicentre, a behaviour that was already well known (Ólafsson and Sigbjörnsson 2004, 2002). The different behaviour at extreme near-fault distances indicates that GMMs with magnitude saturation, magnitude-dependent distance scaling and anelastic attenuation terms, as the important features in the functional form of GMMs for hazard analysis (Stewart et al. 2012, 2015).

In GMMs, the residuals are generally assumed to be normal with a mean zero and a standard deviation. The standard deviation of GMM as an aleatory uncertainty has a significant impact on the seismic hazard results. The recalibration has resulted in new standard deviations for earthquakes in Iceland which are shown in Table 2.2. As can be seen, this difference between the original and recalibrated sigmas, strongly affect the results of PSHA at low annual frequencies of exceedance (Bommer and Abrahamson 2006; Strasser et al. 2009). We note that the decreased uncertainty reflects primarily the better fit to the data, but may possibly also reflect a smaller dataset than used for the original models.

Moreover, the model-to-model variability in the median predictions is obtained for estimating the minimum epistemic uncertainty (Al Atik and Youngs 2014). The variability among the median ground-motion estimates of the original and recalibrated models for different magnitudes at rock site class is compared and shown in Fig. 2.3. The thick line and the gray shaded area represent the mean and the standard deviation, respectively. This figure indicates that at near and far distances where there is a lack of data, epistemic uncertainty is higher than mid-distance where the data is concentrated. The mean epistemic standard deviation is about 0.08–0.1 in base 10 logarithms for the original models and very small

Table 2.1 Description of the selected GMMs with different functional forms

GMM	Functional form	Mw range	R type	R range	Main region(s)
Am05	$\log_{10}(PGA) = C_1 + C_2 M_w$ $+ (C_3 + C_4 M_w) \cdot \log_{10} \sqrt{(C_5^2 + R^2)} + C_6 S + \sigma$	5.0–7.6	Rjb	0–100	Europe and Middle East
AB10	$\log_{10}(PGA) = C_1 + C_2 M_w + C_3 M_w^2$ $+ (C_4 + C_5 M_w) \log_{10} \sqrt{(C_6^2 + R^2)} + C_7 S + \sigma$	5.0–7.6	Rjb	0–100	Europe and Middle East
LL08	$\ln(PGA) = C_1 + C_2 M_w$ $+ C_3 \ln\left(R + C_4 e^{C_5 M_w}\right) + C_6 H + \sigma$	4.1–8.1	Rhyp	15–630	Northern Taiwan
Zh06	$\ln(PGA) = C_1 M_w + C_2 R - \ln\left(R + C_3 e^{C_4 M_w}\right)$ $+ C_5 (\min(\max(H, 15), 125) - 15) + C_6 S_I + C_7 S_{II} + \sigma$	5.0–8.3	Rrup	0–300	Japan
DT07	$\log_{10}(PGA) = C_1 + C_2 M_w$ $+ C_3 \log_{10} \sqrt{(C_4^2 + R^2)} + C_5 S + C_6 F + \sigma$	4.5–6.9	Repi	0–136	Greece
RS09	$\log_{10}(PGA) = C_1 + C_2 M_w$ $+ C_3 \log_{10} \sqrt{(C_4^2 + R^2)} + C_5 S + \sigma$	5.0–7.7	Rjb	1–97	Iceland, Greece, Turkey

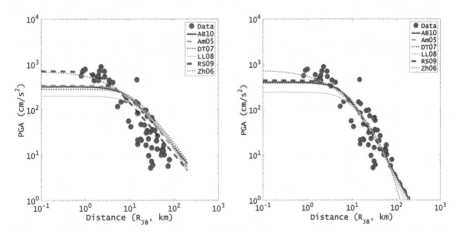

Fig. 2.2 Attenuation of PGA from a magnitude Mw = 6.1 earthquake with distance for the selected GMMs before (left) and after (right) recalibrating. The circles are the recorded PGAs on rock

Table 2.2 The standard deviation (in log-10 units) of the original and recalibrated GMMs

GMMs	AB10	Zh06	LL08	Am05	DT07	GK02[+]	RS09	CF08
Original σ	0.289	0.314	0.227	0.282[a]	0.290	0.243	0.287	0.344
Recalibrated σ	0.175	0.170	0.170	0.175	0.177	0.175	0.174	0.170

[a]The sigma in Am05 model is magnitude dependent. Here, the sigma is calculated for the mean magnitude of Icelandic data

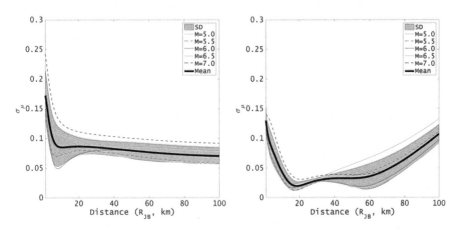

Fig. 2.3 Variability among the median ground-motion estimates of the original (left) and recalibrated (right) GMMs for rock site class (Kowsari et al. 2017). The thick line and the gray shaded area represent the mean and the standard deviation, respectively

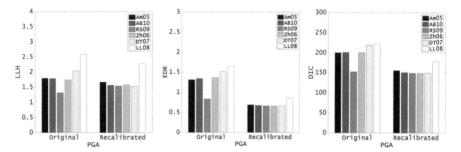

Fig. 2.4 The scores of the LLH (left), EDR (middle) and DIC (right) method in PGA for the candidate GMMs before and after recalibration

(smaller than 0.05 in most distances) for the recalibrated GMMs depending on the applied GMMs and quantity of the well-recorded ground-motions.

2.4 Data-Driven Methods to Select the GMMS

In PSHA, epistemic uncertainty has been modelled by the use of alternative equations in a logic tree framework which it is not necessarily the best way to deal with uncertainties (Atkinson 2011; Atkinson and Adams 2013; Bommer and Scherbaum 2008). Data-driven methods are able to decrease epistemic uncertainties by reducing subjectivity and also guiding the selection process in a quantitative way (Scherbaum et al. 2009). In this study, three data-driven methods, the likelihood-based (Scherbaum et al. 2009), the Euclidean distance-based ranking (Kale and Akkar 2013) and the deviance information criterion based (Kowsari et al. 2017) are used after the recalibration. Figure 2.4 shows the scores of the LLH, EDR and DIC ranking methods for the PGA of the selected GMMs before and after recalibration. The DIC method is able to optimize the selection of GMMs in an unbiased way through the Bayesian framework, but also overcomes the problems associated with the LLH and EDR methods (Kowsari et al. 2017). The results point out that the recalibrated GMMs perform better than the original methods, especially when the DIC is applied. In this study for both reliability and comparison purposes, however, both the original and recalibrated models are applied in PSHA to show how the GMM variability appears on seismic hazard estimates.

2.5 Results and Discussions

A Monte Carlo basis approach is used to provide probabilistic seismic hazard maps in North Iceland. Due to uncertainties in the Icelandic earthquake catalogue, especially in TFZ, the MC-PSHA that simulates synthetic catalogues based on the geophysical characteristics of the seismic zones, is preferred over classical CM-PSHA. In this study, the seismic source zones and associated seismicity parameters proposed in Sigbjörnsson and Snaebjornsson (2007) are used. GMMs have a major impact on seismic hazard estimates and should be carefully selected for such an analysis. In this vein, a MCMC algorithm is used to recalibrate the selected GMMs to the Icelandic data, shown in Fig. 2.2 how they improved the original models and fit the recorded data better. Furthermore, the results of three data driven methods to select the most appropriate GMMs are represented in Fig. 2.4. The use of only one GMM in hazard calculations is not recommended due to epistemic uncertainty, especially for regions where earthquake data is limited. To this end, multiple GMMs that are expected to apply for the region under study are used in a logic tree by assigning equal weights for each GMM.

To show how the GMM variability manifests as uncertainty of the earthquake hazard, the hazard has been calculated over a dense grid over North Iceland, where for each grid point the 10% and 2% probability of exceedance of PGA in 50 years is estimated for each GMM of the current study, assuming rock conditions. The mean hazard at each location was then calculated using equal weights for of the hazard maps based on a given GMM, along with the corresponding standard deviation at each grid point. Figure 2.5 shows the resulting maps of standard deviation of the hazard in terms of PGA at 10% (top row) and 2% (bottom row) probability of exceedance in 50 years for the original (left) and recalibrated GMMs (right). In all cases, the predicted ground-motions in terms of standard deviation at both hazard levels follow the spatial pattern of the faults and decreases when moving away from them. The standard deviation maps for recalibrated GMMs at both hazard levels are significantly reduced compared to those for the original models. This indicates the importance of recalibrated GMMs which was resulted in new standard deviation of residuals that expresses the variability of amplitudes about the median ground-motions for earthquakes in Iceland.

The coefficient of variation is the ratio of the standard deviation to the mean of the hazard, and acts as a useful statistical dimensionless measure indicating the relative variability of the hazard on a ratio scale. Effectively, maps of CV reflects where and to what extent our lack of knowledge (epistemic uncertainty) affects the hazard and as a result, highlights the areas where in the absence of new data modeling efforts should be directed at (Cramer 2001). Figure 2.6 shows the CV of PGA for a 10%-in-50 years (top) and 2%-in-50 years (bottom) based on the original (left) and recalibrated (right) GMMs, respectively. The background CV value for the hazard on the basis of the original models is around 0.4, indicating that the original models are less than ideal for PSHA in North Iceland. High CV values are found along the seismic sources which is compatible to the Fig. 2.3a (left figure) where the epistemic uncertainty of the original GMMs is higher at near-fault distances. Moreover, CV increases with decreasing annual probability of exceedance or in other words with increasing ground-motion i.e., in 2% probability of exceedance in 50 years. The CV maps corresponding to the recalibrated GMMs show much smaller background levels than those based on the original GMMs. Additionally, the CV shows minimum values is at 10–40 km distance where the epistemic uncertainty is the smallest

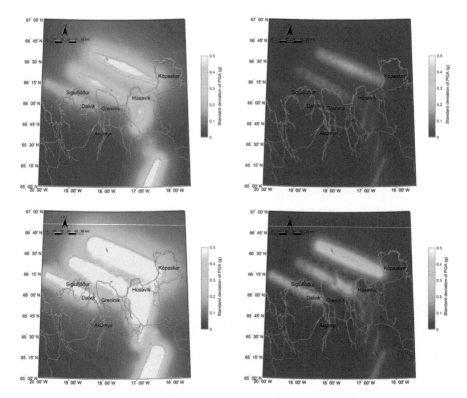

Fig. 2.5 Maps of standard deviation of the earthquake hazard in North Iceland calculated for PGA corresponding to 10% PE of in 50 years (top row) and 2% PE in 50 years (bottom row). The figures at left show the standard deviation of hazard when using the original recalibrated (right) GMMs

for the recalibrated GMMs (see Fig. 2.3b). Moreover, the CV in these cases increases to about 0.6 g away from areas of active seismicity which is expected since the epistemic uncertainty in GMMs grows with increasing distance. It should be kept in mind, that the epistemic uncertainties associated with the GMMs at very close (<10 km) and far distances (>50 km) from the seismic sources are high because of lack of strong ground motion data for such distances in Iceland, a characteristic pointed out in Fig. 2.3b (right figure).

2.6 Conclusions

Seismic hazard maps as one product of PSHA are used in seismic codes for design purposes and reduction of damages from earthquakes. This study builds on previous hazard studies for Iceland, specifically on the delineation of seismic sources, seismicity parameters and the GMMs. These assumptions need to be carefully analysed

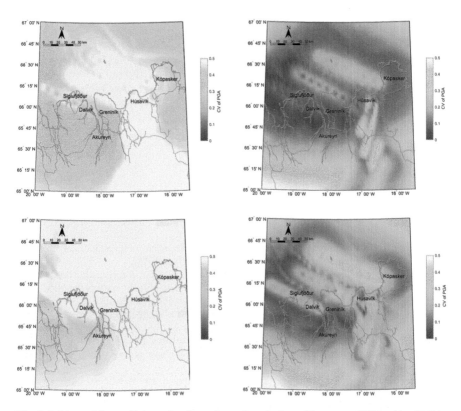

Fig. 2.6 Maps of the coefficient of variance for earthquake hazard in terms of PGA with a 10% in 50 years (top row) and 2% in 50 years (bottom row) in North Iceland. The figures at left and right show the CV based on the original and recalibrated GMMs, respectively

and possibly revised. In particular, selection of GMMs and the uncertainty associated with them tend to exert a great influence on the hazard results. In this study, we showed that some of the selected GMMs in previous PSHA studies in Iceland may not necessarily be appropriate due to the large uncertainties involved. Therefore, selected GMMs are recalibrated to the Icelandic earthquakes by a MCMC algorithm which is useful for regions like Iceland where the earthquake records are scarce. Despite the fact that the current data-driven methods for GMM selection indicate that some of them may in fact be appropriate models for use in Iceland, our results show that the recalibrated models are promising candidates to be applied in future hazard studies in Iceland. Furthermore, the epistemic uncertainty can have a large impact on the mean probabilities of exceedance (Douglas 2010) and it is vital that this is accounted for in PSHA. The results show how the GMMs contribute to large epistemic uncertainties at near and far distances where there is a lack of data. On the other hand, the recalibration has resulted in better fitting GMMs with lower standard deviation of residuals, which indicates that the aleatory uncertainty is handled in an appropriate way. We incorporate epistemic uncertainty into hazard calculation by considering multiple GMMs. The findings have important implications for the revision of hazard for North Iceland.

Acknowledgements This study was funded by Grant of Excellence (No. 141261-051/052/053) from the Icelandic centre for research, and partly by the Icelandic Catastrophe Insurance. The support is gratefully acknowledged.

References

Akkar S, Bommer JJ (2010) Empirical equations for the prediction of PGA, PGV, and spectral accelerations in Europe, the Mediterranean region, and the Middle East. Seismol Res Lett 81:195–206
Ambraseys NN, Sigbjörnsson R (2000) Re-appraisal of the seismicity of Iceland. Acta Polytechnica Scandinavica, Engineering Seismology Series 2000-003, pp 1–184
Ambraseys NN, Douglas J, Sarma SK, Smit PM (2005) Equations for the estimation of strong ground motions from shallow crustal earthquakes using data from Europe and the Middle East: horizontal peak ground acceleration and spectral acceleration. Bull Earthq Eng 3:1–53
Atik LA, Youngs RR (2014) Epistemic uncertainty for NGA-West2 models. Earthquake Spectra 30:1301–1318
Atkinson GM (2011) An empirical perspective on uncertainty in earthquake ground motion prediction 1 this paper is one of a selection of papers in this special issue in honour of professor davenport. Can J Civ Eng 38:1002–1015
Atkinson GM (2012) Integrating advances in ground-motion and seismic-hazard analysis. In: Proceedings of the 15th World conference on earthquake engineering
Atkinson GM, Adams J (2013) Ground motion prediction equations for application to the 2015 Canadian national seismic hazard maps. Can J Civ Eng 40:988–998
Bommer JJ, Abrahamson NA (2006) Why do modern probabilistic seismic-hazard analyses often lead to increased hazard estimates? Bull Seismol Soc Am 96:1967–1977
Bommer JJ, Scherbaum F (2008) The use and misuse of logic trees in probabilistic seismic hazard analysis. Earthquake Spectra 24:997–1009
Bommer JJ, Douglas J, Scherbaum F, Cotton F, Bungum H, Fäh D (2010) On the selection of ground-motion prediction equations for seismic hazard analysis. Seismol Res Lett 81:783–793
Cao T, Petersen MD, Frankel AD (2005) Model uncertainties of the 2002 update of California seismic hazard maps. Bull Seismol Soc Am 95:2040–2057
Cornell CA (1968) Engineering seismic risk analysis. Bull Seismol Soc Am 58:1583–1606
Cotton F, Scherbaum F, Bommer JJ, Bungum H (2006) Criteria for selecting and adjusting ground-motion models for specific target regions: application to Central Europe and rock sites. J Seismol 10:137–156
Cramer CH (2001) A seismic hazard uncertainty analysis for the New Madrid seismic zone. Eng Geol 62:251–266
Danciu L, Tselentis G-A (2007) Engineering ground-motion parameters attenuation relationships for Greece. Bull Seismol Soc Am 97:162–183
Delavaud E, Cotton F, Akkar S, Scherbaum F, Danciu L, Beauval C, Drouet S, Douglas J, Basili R, Sandikkaya MA (2012) Toward a ground-motion logic tree for probabilistic seismic hazard assessment in Europe. J Seismol 16:451–473
Douglas J (2010) Consistency of ground-motion predictions from the past four decades. Bull Earthq Eng 8:1515–1526
Gelman A, Carlin J, Stern HS, Rubin DB (2014) Bayesian data analysis. Chapman & Hall/CRC, London

Halldórsson P (1996) Seismic hazard assessment, Seismic and volcanic risk. In: Proceedings of the workshop on monitoring and research for Mitigating Seismic and Volcanic Risk. European Commission, Environment and climate Programme, pp 25–32

Halldórsson P (2005) Jarskjálftavirkni á Norurlandi-Earthquake activity in N-Iceland. Greinarger. Veurstofu Íslands, IMO report 5021, p 34

Halldórsson P, Sveinsson B (2003) Dvínun hröðunar á Íslandi. Greinargerð 030225, VÍJA 03. Veðurstofa Íslands, Reykjavík

Kale Ö, Akkar S (2013) A new procedure for selecting and ranking ground-motion prediction equations (GMPEs): the Euclidean distance-based ranking (EDR) method. Bull Seismol Soc Am 103:1069–1084

Kowsari M, Halldorsson B, Hrafnkelsson B, Snæbjörnsson JT, Ólafsson S, Rupakhety R (2017) On the selection of ground-motion prediction equations for seismic hazard assessment in the South Iceland Seismic Zone. In: 16th World Conference on Earthquake Engineering (16WCEE). Paper no. 2809, Santiago, Chile

Lin P-S, Lee C-T (2008) Ground-motion attenuation relationships for subduction-zone earthquakes in northeastern Taiwan. Bull Seismol Soc Am 98:220–240

McGuire RK (1976) FORTRAN computer program for seismic risk analysis. US Geological Survey

McGuire RK (2004) Seismic hazard and risk analysis. Earthquake Engineering Research Institute

Musson RMW (2004) Design earthquakes in the UK. Bull Earthq Eng 2:101–112

Ólafsson S, Sigbjörnsson R (1999) A theoretical attenuation model for earthquake-induced ground motion. J Earthq Eng 3:287–315

Ólafsson S, Sigbjörnsson R (2002) Attenuation of strong-motion in the South Iceland Earthquakes of June 2000. In: Proceedings of the 12th European conference on Earthquake Engineering. Elsevier, London

Ólafsson S, Sigbjörnsson R (2004) Attenuation of strong ground motion in shallow earthquakes. In: Proceedings of the 13th World conference on Earthquake Engineering, p 10

Ólafsson S, Rupakhety R, Sigbjörnsson R (2014) Earthquake design provisions for the extension of Búrfell Power Plant: basic parameters. Earthquake Engineering Research Centre, University of Iceland, Report No. 14002

Petersen MD, Cao T, Campbell KW, Frankel AD (2007) Time-independent and time-dependent seismic hazard assessment for the State of California: Uniform California earthquake rupture forecast model 1.0. Seismol Res Lett 78:99–109

Rupakhety R, Sigbjörnsson R (2009) Ground-motion prediction equations (GMPEs) for inelastic displacement and ductility demands of constant-strength SDOF systems. Bull Earthq Eng 7:661–679

Scherbaum F, Delavaud E, Riggelsen C (2009) Model selection in seismic hazard analysis: an information-theoretic perspective. Bull Seismol Soc Am 99:3234–3247

Sigbjörnsson R, Ólafsson S (2004) On the South Iceland earthquakes in June 2000: strong-motion effects and damage. Bollettino di Geofisica Teoricaed Applicata 45:131–152

Sigbjörnsson R, Snaebjornsson J (2007) Earthquake hazard – preliminary assessment for an industrial lot at Bakki near Húsavík. Earthquake Engineering Research Centre, University of Iceland

Sigbjörnsson R, Baldvinsson GI, Thrainsson H (1995) A stochastic simulation approach for assessment of seismic hazard maps in "European Seismic Design Practice." Balkema, Rotterdam

Snaebjornsson J, Sigbjornsson R (2007) Earthquake action in Geothermal projects in NE Iceland at Krafla, Bjarnarflag, Gjastykki and Theistareykir: assessment of geohazards affecting energy production and transmission systems emphasizing structural design criteria and mitigation of risk. Theistareykir Ltd, Landsnet, Landsvirkjun, report no. LV-2007/075

Snaebjornsson J, Sigbjornsson R (2008) Earthquake hazard and seismic action for for proposed power plants in the South Iceland Lowland (No. Report LV-2008/056, Landsvirkjun, Reykjavik)

Sólnes J (2016) Assessment of earthquake hazard and seismic risk in Iceland. North Iceland 77

Solnes J, Halldorsson B (1995) Generation of synthetic earthqauke catalogs: applications in earthquake hazard and seismic risk assessment. Presented at the Proceedings of the fifth international conference on seismic zonation, ouest editions, presses academiques

Solnes J, Sigbjornsson R, Halldorsson B (1994) Assessment of seismic risk based on synthetic and upgraded earthquake catalogs of Iceland

Solnes J, Sigbjörnsson R, Eliasson J (2000) Earthquake hazard mapping and zoning of Reykjavik. In: 12ECEE

Solnes J, Sigbjörnsson R, Eliasson J (2004) Probabilistic seismic hazard mapping of Iceland. In: Proceedings of the 13th world conference on earthquake engineering, Vancouver, BC, Canada

Stefansson R, Gudmundsson GB, Halldorsson P (2008) Tjörnes fracture zone. New and old seismic evidences for the link between the North Iceland rift zone and the mid-Atlantic ridge. Tectonophysics 447:117–126

Stewart JP, Douglas J, Javanbarg MD, Di Alessandro C, Bozorgnia Y, Abrahamson NA, Boore DM, Campbell KW, Delavaud E, Erdik M et al (2012) Selection of a global set of ground motion prediction equations: work undertaken as part of task 3 of the GEMPEER global GMPEs project. PEER Report

Stewart JP, Douglas J, Javanbarg M, Bozorgnia Y, Abrahamson NA, Boore DM, Campbell KW, Delavaud E, Erdik M, Stafford PJ (2015) Selection of ground motion prediction equations for the global earthquake model. Earthquake Spectra 31:19–45

Strasser FO, Abrahamson NA, Bommer JJ (2009) Sigma: issues, insights, and challenges. Seismol Res Lett 80:40–56. https://doi.org/10.1785/gssrl.80.1.40

Tryggvason E (1973) Seismicity, earthquake swarms, and plate boundaries in the Iceland region. Bull Seismol Soc Am 63:1327–1348

Zhao JX, Zhang J, Asano A, Ohno Y, Oouchi T, Takahashi T, Ogawa H, Irikura K, Thio HK, Somerville PG (2006) Attenuation relations of strong ground motion in Japan using site classification based on predominant period. Bull Seismol Soc Am 96:898–913

Chapter 3
Bayesian Hierarchical Model of Peak Ground Acceleration for the Icelandic Strong-Motion Arrays

Sahar Rahpeyma, Benedikt Halldorsson, and Birgir Hrafnkelsson

Abstract A reliable estimation of regional ground motion plays a critical role in probabilistic seismic hazard analysis (PSHA). The earthquake resistant design of structures within a region of a small spatial scale is often based on the assumption of relatively uniform form-factors which leads to the assumption of same station condition. However, for some small-scale regions this may not be the case. In this study, we propose a new Bayesian Hierarchical Model (BHM) for peak ground acceleration (PGA) records from two small-aperture Icelandic strong-motion arrays. The proposed BHM characterizes source effect, local station effect, source-station effect, and an error term that represents the measurement error and other unaccounted factors, separately. Posterior inference is based on a Markov chain Monte Carlo algorithm that uses the Metropolis algorithm. Uncertainty in unknown parameters is assessed through their joint posterior density. Analysis of PGA records based on the proposed BHM will improve the comprehensive understanding of the source effects, localized station conditions, and wave propagation.

Keywords PGA · BHM · Strong-motion array · Spatial variability · Uncertainties

---

S. Rahpeyma
Earthquake Engineering Research Centre, and Faculty of Civil and Environmental Engineering, School of Engineering and Natural Sciences, University of Iceland, Selfoss, Iceland
e-mail: sahar@hi.is

B. Halldorsson (✉)
Earthquake Engineering Research Centre, and Faculty of Civil and Environmental Engineering, School of Engineering and Natural Sciences, University of Iceland, Selfoss, Iceland

Division of Processing and Research, Icelandic Meteorological Office, University of Iceland, Reykjavík, Iceland
e-mail: skykkur@hi.is

B. Hrafnkelsson
Department of Mathematics, Faculty of Physical Sciences, School of Engineering and Natural Sciences, University of Iceland, Reykjavik, Iceland
e-mail: birgirhr@hi.is

3.1 Introduction

Iceland is the largest subaerial part of the Mid-Atlantic Ridge where the plate boundary between North American and Eurasian crustal plates are drifting apart with an approximate rate of 2 cm/year (Sæmundsson 1979; Einarsson 1991; DeMets et al. 1994; Einarsson 2008; Metzger and Jónsson 2014). Two major fracture zones are known across the active seismic belt from southwest to the north of Iceland. The South Icelandic Seismic Zone, SISZ, lies across the densely populated South Iceland Lowland, SIL, and Tjörnes Fracture Zone (TFZ) in the north (see Fig. 3.1). Historical records indicate that the largest earthquakes in Iceland have been recorded within these transform zones the size of which have exceeded magnitude of 7.0 (Stefánsson et al. 1993, 2008; Sigbjörnsson et al. 2006). During recent decades, the seismicity of the SISZ has been well documented and characterized by arrays of N-S right lateral strike-slip faults with high potential of destructive earthquakes either as strong single or short to long sequence of seismic activities (Einarsson et al. 1981; Stefánsson and Halldórsson 1988; Sigbjornsson et al. 2009). On the other hand, the TFZ is known as a tectonically complex region which primarily subdivided into west-northwest striking seismic lineaments (Saemundsson 1974; Einarsson 1991).

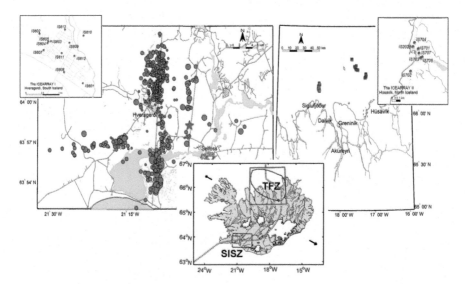

Fig. 3.1 The map in the middle shows Iceland and the approximate location of the Mid-Atlantic Ridge (dotted gray line), the South Iceland Seismic Zone (SISZ) and the Tj-rnes Fracture Zone (TFZ) outlined by the black dashed lines. Red rectangles indicate the areas shown in detail. The map on the left shows the aftershock distribution (gray circles) from 29 May 2008 Ølfus earthquake in southwest Iceland. The ICEARRAY I is located within the town of HveragerÞi (red dashed rectangle). The inset figure at the top left shows the twelve ICEARRAY I strong-motion stations (represented by red circle, along with station ID-codes). Map on the right shows the seismicity distribution for the duration of September 19, 2012 (red symbols) to April 4, 2013 (blue symbols) in the TFZ, in north Iceland. The inset figure at the top right shows the seven ICEARRAY II strong motion stations (red circle symbols, along with station IDcodes) across H°savÚk

Since 1984, the Icelandic strong-motion network (ISMN), owned and operated by the Earthquake Engineering Research Centre of the University of Iceland, has provided strong-motion data required for various earthquake engineering applications such as earthquake resistant design of structures and seismic hazard assessment. Today, the network consists of about 40 free-field stations spreading over the populated regions and infrastructural centers of the SISZ and TFZ with interstation distance range of 5–15 km (Sigbjörnsson et al. 2004; Sigbjörnsson and Ólafsson 2004). However, the general lack of dense local recordings of ground motion confines the quantitative evaluation of strong-motion variability which is a key element for a reliable Seismic Hazard Analysis (SHA) and its application in building codes. Therefore in 2007, with the aim of monitoring and quantifying spatial variability of strong-motion over short distances, the first small-aperture Icelandic strong-motion array, ICEARRAY I, was deployed in the town of Hveragerði in SISZ (Halldorsson and Sigbjörnsson 2009; Halldorsson et al. 2009). The ICEARRAY I, consists of 13 strong-motion stations, is installed in an area of approximately 1.23 km^2 with interstation distance ranging from 50 to 1900 m (Fig. 3.1). Then in 2012, the second strong-motion array, ICEARRAY II, was deployed in the TFZ in North Iceland, specifically in the town of Húsavík which is located effectively on top of the Húsavík Flatey Fault (HFF), the largest transform faults in the country. The ICEARRAY II consists of 6 free-field stations and one structural monitoring system in the regional hospital building. The free-field stations are three-component CUSP-3C instruments of Canterbury Seismic Instruments and the structural system is a CUSP-3D3 unit (Halldorsson et al. 2012). The arrays are monitored by a dedicated CUSP-HUB on which the Common-Triggering Scheme operates which optimizes the efficiency of the array's event detection (Halldorsson and Avery 2009).

Since deployment of the Icelandic arrays, they have significantly boosted the quantity of the Icelandic strong-motion data set. The ICEARRAY I was in the extreme near-fault region of the $M_w6.3$ Ölfus earthquake on 29 May 2008, the ground motions of which were characterized by intense ground accelerations (PGA of 38–89%g) of relatively short durations (~5–6 s) and large amplitude near-fault velocity pulses due to simultaneous rupture directivity and permanent tectonic displacements. Moreover, more than 1700 aftershocks were recorded during a one-year period after the main event (Halldorsson and Sigbjörnsson 2009; Douglas and Halldorsson 2010). On the contrary however, the ICEARRAY II has recorded much less data, the far-field ground motions of total of 26 small-to-moderate size earthquakes, during the largest seismic sequence over the last 30 years in North Iceland during 2012–2013 (Halldorsson et al. 2012; Olivera et al. 2014; Rahpeyma et al. 2017).

The surface geology conditions across ICEARRAY I are uniform and characterized as "rock" (see Rahpeyma et al. 2016, and references therein). However, the geological features across ICEARRAY II are notably complex and can be clustered into three main groups according to different origins of the subsoil across the town of Húsavík (Olivera et al. 2014; Waltl et al. 2014). While insignificant variability in ground motion amplitudes was expected due to uniform station condition across ICEARRAY I, considerable variations in PGA and peak-ground velocity (PGV) were observed during the recorded ground motions of the main-shock and aftershocks (Halldorsson and Sigbjörnsson 2009; Douglas and Halldorsson 2010; Rahpeyma et al. 2016). It is noteworthy that no azimuthal dependency was captured for ICEARRAY I recordings (Rahpeyma et al. 2016). Rationally, this variation is more highlighted across ICEARRAY II as a result of the variability of subsoil structure. It is well acknowledged that localized geological structure can significantly vary within short distances and influence site responses (Bessason and Kaynia 2002; Di Giacomo et al. 2005; Rahpeyma et al. 2016). Consequently, quantifying the origins of strong-motion variability over the relatively small area can have important effects on structural applications. Moreover, there is a vital requisite to enhance our understanding of the spatial relative differences in strong-motion amplitudes and the associated uncertainties to mitigate the seismic risk of potentially destructive earthquakes, specifically for urban areas.

The fundamental goal of this article is to apply a new Bayesian Hierarchical Model (BHM) for mapping the differences in PGA across ICEARRAY I and ICEARRAY II. The proposed BHM for PGA lies in the incorporating of different factors on the observed ground-motion intensities (Kuehn and Scherbaum 2015; Sigurdarson and Hrafnkelsson 2016; Kuehn and Scherbaum 2016). The BHM can be used for spatial prediction of ground motion intensities and it will improve our understanding of spatial variability as well as interpretation of the quantitative information about the source effects, localized station effects, and incoherence effect. The results indicate the importance of station effects and regionalization of the predictive seismic intensities to decrease the associated aleatory variability, which consequently affect the seismic hazard analysis.

3.2 Array Strong-Motion Data

ICEARRAY I recorded the main-shock of the 29 May 2008 $M_w 6.3$ Ölfus earthquake and more than 1700 of its aftershocks over the following year. As can be seen in Fig.3.1, the aftershock recordings occurred on two parallel, near vertical north-south striking right-lateral strike slip faults that are approximately 4.5 km apart (Sigbjornsson et al. 2009; Halldorsson et al. 2010). In the current study, we used the aftershock recordings from ten nominated stations which are directly located on the lava-rock layer under most of the town. We excluded station IS610 which is located on a considerably older bedrock on a steep hillside, and station IS613 due to relatively few recorded and unreliable data. Furthermore, two stations (IS608 and IS688) are collocated and provide essentially identical results (Halldorsson and Sigbjörnsson 2009).

In October 2012 and April 2013, the strongest earthquake sequence in North Iceland in over 30 years took place. In occurred on the northwestern part of the TFZ, the first sequence (October 2012) took place in the Eyjafjarðarál rift on the HFF, while the second (April 2013) took place on the Grimsey Oblique Rift (GOR) (see Fig. 3.1). Due to the large event-station distances, the Icelandic strong-motion array

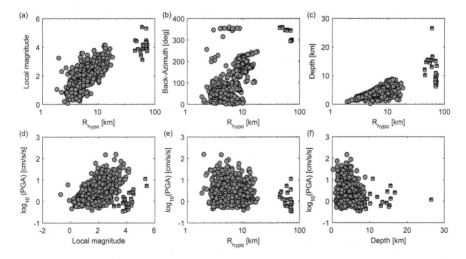

Fig. 3.2 Distribution of (**a**) local magnitude, (**b**) back-azimuth, (**c**) depth plotted vs. hypocentral distance and (**d**) local magnitude, (**e**) hypocentral distance, (f) depth vs. $\log_{10}$PGA recorded by ICEARRAY I (grey symbols) and ICEARRAY II (red and blue symbols represents swarm 2012 and swarm 2013, respectively)

in North Iceland (ICEARRAY II) only recorded the largest 26 earthquakes of the sequence. It should also be noted that stations IS707 and IS202B did not record swarm 2012 and 2013, respectively. The northernmost part of the town sits on relatively hard Tillite, while the geology along the shoreline towards South is characterized by horizontally layered fluvial sediments. On top of the horizontal sediments lies a delta formation of glacial deposits formed at the end of the last glacial period, some 10,000 years ago. The oldest sediments are glacial deposits which have over time been altered to solid Tillite rock which underlay parts of Húsavík (Sæmundsson and Karson 2006). Figure 3.2 presents the characteristics from parametric earthquake data for ICEARRAY I and ICEARRAY II.

3.3 Bayesian Multilevel Model

In order to model the spatial distribution of PGA over the strong-motion arrays, we implement a Bayesian hierarchical (multilevel) model (BHM). The Bayesian hierarchical modelling is a statistical tool written in multiple levels which estimates the posterior distribution for the model parameters in the context of the Bayesian theorem. In this article, data are structured hierarchically in three levels: data, latent, and hyper levels. The first level, data level, describes the distributional model for the observed data conditioned on the model parameters. The second level, latent level, describes the distribution of the model variables which are conditioned on the hyper level, so-called hyperparameters. The third level, hyper level, contains prior distributions for the model parameters at the data and latent levels.

The observed $\log_{10}PGA_{es}$ is modelled as having a normal distribution at the data level of the BHM. Let Y_{es} represent the base 10 logarithm of PGA for event e at station s. The proposed model, consists of independent terms, can be presented as Eq. (3.1):

$$Y_{es} = \gamma_{es} + \nu_e + \alpha_s + \varepsilon_{es} + \varphi_{es}, \quad e = 1, \ldots, N \quad s = 1, \ldots, Q \quad (3.1)$$

where γ_{es} is a linear function that depends on event e and station s, ν_e is the over-all effect of event e, α_s is the station effect for station s, ε_{es} is a spatially correlated event-station term, and φ_{es} is an independent error term representing both the measurement error and other similar uncontrollable factors, N is the number of events, and Q is the number of stations.

The term γ_{es} is modelled with a general ground motion model (GMM) as a function of the local magnitude of the e-th event, M_e, and the hypocentral distance from e-th event to s-th station, R_{es}

$$\gamma_{es} = \beta_1 + \beta_2 M_e + \beta_3 \log_{10}(R_{es}) \quad (3.2)$$

The values of parameters $\vec{\beta} = (\beta_1, \beta_2, \beta_3)$ correspond to the region, reflecting peculiarities of seismic regime and geological structure. We assumed a non-informative prior distribution for $\vec{\beta}$ coefficients, a normal distribution with the mean and the standard deviations, $\vec{\mu}_\beta$ and Σ_β.

The random effect variability, ν_e, is normally distributed with mean zero and standard deviation of σ_ν. The parameter $\vec{\nu} = (\nu_1, \ldots, \nu_N)$ quantifies the variability between-events, that is, the event dependent deviation from the model given by γ_{es}.

Let $\vec{\alpha} = (\alpha_1, \ldots, \alpha_Q)$ be a vector containing the station terms at the observed stations. These station effects are assumed to follow a mean zero Gaussian distribution with covariance matrix Σ_α based on a Matérn covariance function with marginal variance σ_α^2 which describes the station-to-station variability, smoothness parameter $\nu_\alpha = 1.5$, and range parameter ϕ_α. In this study, we used a constant range parameter since a dynamic one can highly influence the simulation process and lead to bad mixing in Markov chains. Improved estimates were not obtained even after multiple inversions where the range parameter was changed gradually. Moreover, many geo-statistical studies have reported that fixing the range parameter has little impact on the overall predictive performance of the model (Sahu et al. 2007; Zhang and Wang 2010; Gneiting et al. 2010). We therefore set the decay rate of the station term $\phi_\alpha = 0.03$ (units in km) for modeling ICEARRAY I and $\phi_\alpha = 1.5$ for ICEARRAY II.

The spatially correlated event-station terms, ε_{es}, are modelled as a mean zero Gaussian field for each event e governed by a Matérn covariance function with marginal variance σ_ε^2 which describes within-event variability, smoothness parameter $\nu_\varepsilon = 1.5$, and decay parameter ϕ_ε.

Finally, the measurement and model error term, φ_{es}, is mean zero Gaussian with standard deviation of σ_φ and the φ_{es} are independent of each other.

The total standard deviation of Y_{es} can be defined as the square root of sum of the inter-event variance, σ_ν^2, the inter-station variance, σ_α^2, the intra-event variance, σ_ε^2, and the measurement and model error variance, σ_φ^2

$$\sigma_T = \sqrt{\sigma_\nu^2 + \sigma_\alpha^2 + \sigma_\varepsilon^2 + \sigma_\varphi^2} \qquad (3.3)$$

In the proposed BHM, the vectors of $\vec{\eta} = (\vec{\beta}, \vec{\nu}, \vec{\alpha})$ and $\vec{\theta} = (\sigma_\varphi, \sigma_\nu, \sigma_\varepsilon, \phi_\varepsilon, \sigma_\alpha)$ were defined as latent parameters in the second level and hyperparameters in the third level, respectively. In order to infer η and θ we used a Gibbs sampler (Geman and Geman 1984; Casella and George 1992), an iterative Markov Chain Monte Carlo (MCMC) sampling algorithm, with Metropolis step in each iteration (Metropolis et al. 1953; Hastings 1970). In total, we analyzed 610 earthquake events recorded in 10 ICEARRAY I strong-motion stations, a total of 4620 data point measurements with considering the missing data points. Likewise, for ICEARRAY II we had 83 (from 14 events) and 66 (from 11 events) data points from the October 2012 and April 2013 sequences, respectively, recorded on 6 strong-motion stations. In order to eliminate the path effects from ICEARRAY II recording, we split the data set into two groups (i.e. October 2012 and April 2013) and set up the BHM separately for each data set.

3.4 Results and Discussion

We used 4 parallel Markov chains of total length of 8000 samples and considering the first 25% of iterations (2000) as burn-in period when analyzing the ICEARRAY I data. In the case of ICEARRAY II however, more samples were required to get the convergence due to lack of data points compared with the ICEARRAY I dataset. Thus, all four chains used for the BHM proposed for ICEARRAY II were sampled with a total length of 20,000 iterations where the first 25% iterations (5000) were used as burn-in period. The accuracy of the obtained model was diagnosed both visually in addition to the Gelman-Rubin Statistics and autocorrelation function (Gelman and Rubin 1992).

3.4.1 Hyperparameters Inference

The posterior mean, standard deviation and 95 percentiles of the hyperparameters for ICEARRAY I and ICEARRAY II are presented in Table 3.1. A key point to evaluate how well the model fits to the data is the variation of the parameters in addition to its posterior distribution. A close scrutiny on the posterior distribution of the

Table 3.1 Posterior mean, standard deviation and 95 percentiles of θ for ICEARRAY I and ICEARRAY II

θ	ICEARRAY I			ICEARRAY II (swarm 2012)			ICEARRAY II (swarm 2013)		
	mean	SD	(2.5, 97.5) %	mean	SD	(2.5, 97.5) %	mean	SD	(2.5, 97.5) %
σ_φ	0.06	0.00	(0.05, 0.06)	0.05	0.01	(0.01, 0.07)	0.05	0.02	(0.00, 0.09)
σ_ν	0.24	0.01	(0.22, 0.25)	0.21	0.05	(0.14, 0.33)	0.13	0.05	(0.04, 0.24)
σ_e	0.11	0.00	(0.11, 0.12)	0.03	0.02	(0.00, 0.08)	0.09	0.03	(0.01, 0.15)
ϕ_ε	0.27	0.02	(0.24, 0.31)	0.80	0.86	(0.02, 3.17)	0.52	0.48	(0.08, 1.75)
σ_α	0.08	0.02	(0.05, 0.15)	0.20	0.07	(0.11, 0.39)	0.26	0.09	(0.14, 0.49)

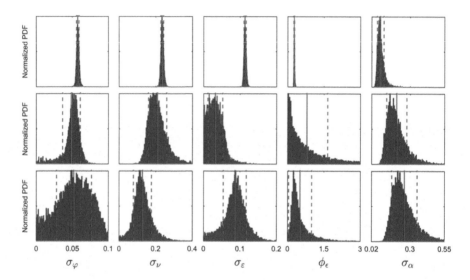

Fig. 3.3 Posterior histograms of hyperparameters for ICEARRAY I (1st row) and ICEARRAY II (2nd and 3rd rows for swarms 2012 and 2013, respectively)

hyperparameters in Fig. 3.3 indicates that the ICEARRAY I results are well obtained with low variation and narrow with distribution in comparison to the ICEARRAY II results. Undoubtedly the extensive dataset recorded on the ICEARRAY I allow a robust and more reliable estimation of the hyperparameters of the BHM. Nevertheless, we cannot exclude also contributing effects of greater variability of local geology and topography of the ICEARRAY II.

3.4.2 Latent Parameters Inference

As can be seen in Table 3.2 the inference of βs, i.e., the constant coefficients of nominated GMM (Eq. (3.2)) in the BHM formulation, resulted in well-simulated

Table 3.2 Posterior mean, standard deviation and 95 percentiles of β for ICEARRAY I and ICEARRAY II

β	ICEARRAY I			ICEARRAY II (swarm 2012)			ICEARRAY II (swarm 2013)		
	mean	SD	(2.5, 97.5) %	mean	SD	(2.5, 97.5) %	mean	SD	(2.5, 97.5) %
β_1	0.87	0.05	(0.78, 0.97)	−11.84	8.41	(−28.21, 5.09)	−1.42	1.46	(−4.31, 1.45)
β_2	0.65	0.02	(0.62, 0.68)	0.99	0.16	(0.66, 1.30)	0.47	0.08	(0.31, 0.63)
β_3	−2.25	0.07	(−2.38, −2.11)	4.15	4.45	(−4.81, 12.80)	−0.21	0.80	(−1.81, 1.36)

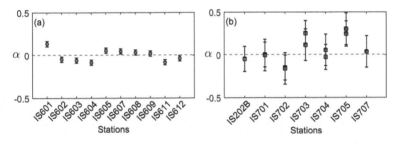

Fig. 3.4 Posterior mean and standard deviation of the station term for (**a**) ICEARRAY I (grey symbols), and (**b**) ICEARRAY II (red symbols for swarm 2012, and blue symbols for swarm 2013)

posterior samples for ICEARRAY I. However, the obtained results for ICEARRAY II datasets, indicate extremely large standard deviations, indicating unreliable results, most likely due to the lack of data.

Figure 3.4 shows the comparison of the obtained station terms from ICEARRAY I and ICEARRAY II. It is noteworthy that posterior distributions of station terms specify to what relative extent PGA can be expected to be either higher or lower than the mean PGA over the array. Contrary to the ICEARRAY I, the ICEARRAY II station terms are not well constrained as they have considerably large standard deviations (see Fig. 3.4). It is largely due to the complexity of station conditions and limited data. Even though large station term uncertainties for ICEARRAY II, mapping the spatial distribution of mean posterior of the station terms and mean $\log_{10}$PGA confirms good agreement between spatial distribution patterns across both arrays (see Figs. 3.5 and 3.6). As can be seen in Fig. 3.4b, the station terms follow almost the same trend for both data sets across ICEARRAY II; however, the obtained mean values (except IS703) and standard deviations are slightly larger for the 2013 event. It is noteworthy that although stations IS703 and IS705 with interstation distance of 210 m in the western part of the array are the closest stations, their obtained station terms are noticeably different. The only convincing explana-

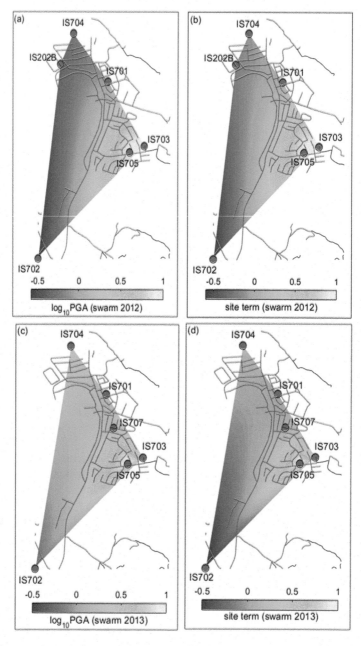

Fig. 3.5 Mapping the spatial distribution of normalized (**a**) and (**c**) mean of $\log_{10}$PGA, (**b**) and (**d**) posterior mean of station terms for ICEARRAY II stations obtained from swarm 2012 and swarm 2013

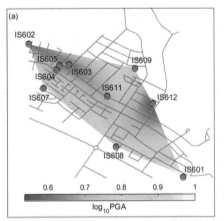 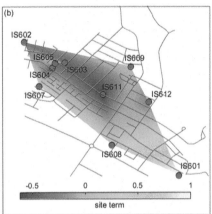

Fig. 3.6 Mapping the spatial distribution of normalized (**a**) mean of $\log_{10}$PGA and (**b**) posterior mean of station terms for ICEARRAY I stations obtained from aftershocks of 2008 Ölfus earthquake

tion for such difference can be related to the station locations in which station IS703 sits on the edge of the southern terraces of the town and different wave direction can largely affect the station term while station IS705 is placed in the middle of the terraces.

3.4.3 Uncertainties

The inference of the hyperparameters for ICEARRAY I shows that inter-event standard deviation is larger than intra-event standard deviation which consists of both inter-station and inter-record variabilities (see Table 3.1). In general, the inter-event variability in empirical GMMs is reported to be smaller than intra-event variability (Strasser et al. 2009; Kuehn and Scherbaum 2015); however, the inter-event variability obtained for ICEARRAY I is larger than the intra-event variability. This is mainly due to the relatively uniform site conditions and also similar wave propagation across the region. In contrast, from the ICEARRAY II results intra-event variability appears higher than the inter-event variability (Table 3.1). Contrary to the ICEARRAY I, the strong-motion recorded by ICEARRAY II are characterized by small spatial dimensions of source regions compared to large hypocentral distances of ~50–80 km (see Fig. 3.1). Especially for the 2012 event which is located in the same small source region on the HFF, while the 2013 event locations are distributed over the GOR lineament as a result have a wider azimuthal range. But, an important factor is the relative size differences of the datasets between ICEARRAY II compared to ICEARRAY I.

The total uncertainties using Eq. (3.3) vary in the range of $\simeq 0.28$–0.29 for ICEARRAY I and ICEARRAY II which are sensibly in the same range with the standard deviation of commonly used GMMs (0.20–0.30) in seismic hazard studies for Iceland mostly obtained from strong-motion data from South Iceland (Olafsson and Sigbjörnsson 2000; Ambraseys et al. 2005; Douglas and Halldorsson 2010; Kowsari et al. 2017). The inter-station standard deviation makes it possible to estimate single-station sigma which improves PSHA by removing the risk of counting twice some parts of the variability (Atkinson 2006). The inter-station variability, mainly reflects geological characteristics, for ICEARRAY I is lower than ICEARRAY II mainly due to the difference in geology and topography.

3.5 Summary and Conclusion

In this article, we used the earthquake strong-motion aftershock recordings from $M_w 6.3$ Ölfus earthquake recorded by ICEARRAY I stations and also the earthquake sequence in North Iceland in October 2012 and April 2013 recorded by ICEARRAY II stations to set up a Bayesian hierarchical spatial model for PGA. The proposed model offers a probabilistic framework for multi-level modelling of PGA that accounts for the effects of earthquake source as well as local station effects while at the same time quantifying their uncertainties over multiple hierarchy levels.

The proposed BHM quantifies the observed uncertainties and the relative contributions of the earthquake and path effects, localized station effects, and their interdependent effects, respectively, to those uncertainties of array peak ground accelerations. The inference of the posterior distribution of the latent parameters as well as the hyperparameters obtained for ICEARRAY I and ICEARRAY II indicates significant station-effects with respect to local geology and topography. The large difference in confidence limit in the results depending on the size of the dataset emphasizes the need for more detailed geological measurements and analysis of the ground motion characteristics across relatively small areas where relative differences of ground motion intensities are of interest and importance, such as in urban areas near active faults.

The results have direct practical implications for improving earthquake hazard assessment and site-specific design. In other words, a new hazard estimate will enable a more optimal earthquake resistant design of structures as well as serve to improve long-term urban planning. The general conclusion is that the proposed model here is detailed and accurate enough in its spatial predictions and incorporates important assumptions needed in the analysis of PGA, and shows clearly if the results are reliable or not due to the uncertainty parameters. We note that the proposed BHM is a general model and could be applied to any earthquake ground motion or structural response dataset (e.g. PGV, and spectral acceleration, SA).

Acknowledgements This work was supported by the Icelandic Centre for Research (Grant of Excellence no. 141261-051/52/53), the Icelandic Catastrophe Insurance (Grant no. S112-2013) the

University of Iceland Research Fund, and the Doctoral grants from Eimskip Fund of the University of Iceland.

References

Ambraseys NN, Douglas J, Sarma SK, Smit PM (2005) Equations for the estimation of strong ground motions from shallow crustal earthquakes using data from Europe and the Middle East: horizontal peak ground acceleration and spectral acceleration. Bull Earthq Eng 3:1–53

Atkinson GM (2006) Single-Station sigma. Bull Seismol Soc Am 96:446–455

Bessason B, Kaynia AM (2002) Site amplification in lava rock on soft sediments. Soil Dyn Earthq Eng 22:525–540

Casella G, George EI (1992) Explaining the Gibbs sampler. Am Stat 46:167–174

DeMets C, Gordon RG, Angus DF, Stein S (1994) Effect of recent revisions to the geomagnetic reversal time scale on estimates of current plate motions. Geophys Res Lett 21:2191–2194

Di Giacomo D, Maria Rosaria G, Mucciarelli M et al (2005) Analysis and modeling of HVSR in the presence of a velocity inversion: the case of Venosa, Italy. Bull Seismol Soc Am 95:2364–2372

Douglas J, Halldorsson B (2010) On the use of aftershocks when deriving ground-motion prediction equations. In: 9th US National and 10th Canadian conference on earthquake engineering (9USN/10CCEE). Paper no. 220, Toronto, Canada

Einarsson P (1991) Earthquakes and present-day tectonism in Iceland. Tectonophysics 189:261–279

Einarsson P (2008) Plate boundaries, rifts and transforms in Iceland. Jökull 58:35–58

Einarsson P, Björnsson S, Foulger G et al (1981) Seismicity pattern in the South Iceland seismic zone. Earthq Predict 4:141–151

Gelman A, Rubin DB (1992) Inference from iterative simulation using multiple sequences. Stat Sci 7(4):457–472

Geman S, Geman D (1984) Stochastic relaxation, Gibbs distributions, and the Bayesian restoration of images. IEEE Trans Pattern Anal Mach Intell PAMI-6:721–741

Gneiting T, Kleiber W, Schlather M (2010) Matérn cross-covariance functions for multivariate random fields. J Am Stat Assoc 105:1167–1177

Halldorsson B, Avery H (2009) Converting strong-motion networks to arrays via common triggering. Seismol Res Lett 80:572–578

Halldorsson B, Sigbjörnsson R (2009) The Mw6. 3 Ölfus earthquake at 15: 45 UTC on 29 may 2008 in South Iceland: ICEARRAY strong-motion recordings. Soil Dyn Earthq Eng 29:1073–1083

Halldorsson B, Sigbjörnsson R, Schweitzer J (2009) ICEARRAY: the first small-aperture, strong-motion array in Iceland. J Seismol 13:173–178

Halldorsson B, Sigbjörnsson R, Rupakhety R, Chanerley AA (2010) Extreme near-fault strong-motion of the M6.3 Ölfus earthquake of 29 May 2008 in South Iceland. In: 14th European Conference on Earthquake Engineering (14ECEE). Paper no. 1640, Ohrid, Macedonia

Halldorsson B, Jónsson S, Papageorgiou AS, et al (2012) ICEARRAY II: a new multidisciplinary strong-motion array in North Iceland. In: 15th World Conference on Earthquake Engineering (15WCEE). Paper no. 2567, Lisbon, Portugal

Hastings WK (1970) Monte Carlo sampling methods using Markov chains and their applications. Biometrika 57:97–109

Kowsari M, Halldorsson B, Hrafnkelsson B, et al (2017) On the selection of ground-motion prediction equations for Seismic Hazard assessment in the South Iceland Seismic Zone. In: 16th World Conference on Earthquake Engineering (16WCEE). Paper no. 2809, Santiago, Chile

Kuehn NM, Scherbaum F (2015) Ground-motion prediction model building: a multilevel approach. Bull Earthq Eng 13:2481–2491

Kuehn NM, Scherbaum F (2016) A partially non-ergodic ground-motion prediction equation for Europe and the Middle East. Bull Earthq Eng:1–14

Metropolis N, Rosenbluth AW, Rosenbluth MN et al (1953) Equation of state calculations by fast computing machines. J Chem Phys 21:1087–1092

Metzger S, Jónsson S (2014) Plate boundary deformation in North Iceland during 1992–2009 revealed by InSAR time-series analysis and GPS. Tectonophysics 634:127–138

Olafsson S, Sigbjörnsson R (2000) Attenuation of strong ground acceleration: a study of the South Iceland earthquakes

Olivera CI, Halldorsson B, Green RA, Sigbjörnsson R (2014) Site effects estimation using ambient noise and earthquake data in Húsavík, North Iceland. In: Proceedings of the 2nd European Conference on Earthquake and Engineering Seismology (2ECEES). Istanbul, Turkey

Rahpeyma S, Halldorsson B, Olivera C et al (2016) Detailed site effect estimation in the presence of strong velocity reversals within a small-aperture strong-motion array in Iceland. Soil Dyn Earthq Eng 89:136–151

Rahpeyma S, Halldorsson B, Green RA (2017) On the distribution of earthquake strong-motion amplitudes and site effects across the Icelandic strong-motion arrays. In: 16th World Conference on Earthquake Engineering (16WCEE). Paper no. 2762, Santiago, Chile

Saemundsson K (1974) Evolution of the axial rifting zone in northern Iceland and the Tjörnes fracture zone. Geol Soc Am Bull 85:495–504

Sæmundsson K (1979) Outline of the geology of Iceland

Sæmundsson K, Karson JA (2006) Stratigraphy and tectonics of the Húsavík-Western Tjörnes Area. Iceland Geosurvey

Sahu SK, Gelfand AE, Holland DM (2007) High-resolution space–time ozone modeling for assessing trends. J Am Stat Assoc 102:1221–1234

Sigbjörnsson R, Ólafsson S (2004) On the South Iceland earthquakes in June 2000: strong-motion effects and damage. Boll Geofis Teor Ed Appl 45:131–152

Sigbjörnsson R, Ólafsson S, Thórarinsson Ó (2004) Strong-motion recordings in Iceland. In: Proceedings of the 13th world conference on earthquake engineering. Paper no. 2370, Mira, Vancouver, BC, Canada

Sigbjörnsson R, Sigurdsson T, Snæbjörnsson JT, Valsson G (2006) Mapping of crustal strain rate tensor for Iceland with applications to seismic hazard assessment. In: First European conference on Earthquake Engineering and Seismology (1ECEES). Paper no. 1211, Geneva, Switzerland

Sigbjornsson R, Snæbjörnsson JT, Higgins S et al (2009) A note on the Mw 6.3 earthquake in Iceland on 29 may 2008 at 15:45 UTC. Bull Earthq Eng 7:113–126

Sigurdarson AN, Hrafnkelsson B (2016) Bayesian prediction of monthly precipitation on a fine grid using covariates based on a regional meteorological model. Environmetrics 27:27–41

Stefánsson R, Halldórsson P (1988) Strain release and strain build-up in the South Iceland seismic zone. Tectonophysics 152:267–276

Stefánsson R, Böðvarsson R, Slunga R et al (1993) Earthquake prediction research in the South Iceland seismic zone and the SIL project. Bull Seismol Soc Am 83:696–716

Stefánsson R, Guðmundsson GB, Halldórsson P (2008) Tjörnes fracture zone. New and old seismic evidences for the link between the North Iceland rift zone and the mid-Atlantic ridge. Tectonophysics 447:117–126

Strasser FO, Abrahamson NA, Bommer JJ (2009) Sigma: issues, insights, and challenges. Seismol Res Lett 80:40–56

Waltl P, Halldorsson B, Pétursson HG, et al (2014) The geological and urban setting of Húsavík, North Iceland, in the context of Earthquake Hazard and Risk Analysis. In: Proceedings of the 2nd European Conference on Earthquake and Engineering Seismology (2ECEES). Istanbul, Turkey

Zhang H, Wang Y (2010) Kriging and cross-validation for massive spatial data. Environmetrics 21:290–304

Chapter 4
Towards an Automated Kappa Measurement Procedure

Tim Sonnemann and Benedikt Halldorsson

Abstract We present an automatic algorithm for measuring the high-frequency acceleration spectral decay parameter (κ) on seismic records of strong ground motion. This task is often done manually due to the strongly varying quality and characteristics of earthquake recordings, such that an adaptive algorithm is required to mostly reproduce manual measurements but in a less subjective way. We prepended a P- and S-phase picking algorithm to achieve complete automation, and we discuss the common pitfalls for such attempts regarding both algorithms. To test the measurement schemes, a dataset of accelerograms from South Icelandic earthquakes is used on which all parameters have been determined manually as well. The proposed program facilitates automatic creation of κ-value databases from input record databases, which might be used in regional hazard analyses and to better constrain earthquake source parameter inversions.

Keywords Near-surface attenuation · Strong ground motion · High frequency amplitude decay · Kappa

---

T. Sonnemann
Earthquake Engineering Research Centre, and Faculty of Civil and Environmental Engineering, School of Engineering and Natural Sciences, University of Iceland, Selfoss, Iceland
e-mail: tsonne@hi.is

B. Halldorsson (✉)
Earthquake Engineering Research Centre, and Faculty of Civil and Environmental Engineering, School of Engineering and Natural Sciences, University of Iceland, Selfoss, Iceland

Division of Processing and Research, Icelandic Meteorological Office, University of Iceland, Reykjavík, Iceland
e-mail: skykkur@hi.is

4.1 Introduction

At least since the 1980s, a distinct decay of the high-frequency acceleration spectra of strong ground motions has been noted and discussed by various researchers (Hanks 1982; Papageorgiou and Aki 1983; Anderson and Hough 1984). The breakdown at higher frequencies was observed to start at what was termed f_{max}, a corner frequency attributed mainly to site conditions (Hanks 1982), while others argued for it to be a source effect (Papageorgiou and Aki 1983) resulting from non-elastic behavior at the fault with the influence on high frequencies coming from an assumed cohesive zone at the crack tip (Achenbach and Harris 1978). Then, a different parameterization of this effect was proposed by directly measuring the slope of the high frequency logarithmic acceleration spectral amplitude versus a linear frequency axis (Anderson and Hough 1984). The decay was characterized by a slope of $-\pi\kappa$, where κ has become the parameter most used today (Ktenidou et al. 2014 and references therein). This high-frequency spectral decay has also been attributed to a combination of site and source effects and path attenuation (Anderson and Hough 1984; Hough et al. 1988; Purvance and Anderson 2003; Ktenidou et al. 2014). The view that it is mostly influenced by site conditions and the upper few hundreds or thousands of meters is predominant in the literature (Ktenidou et al. 2014), so that it is also known as near-surface attenuation effect, but the most commonly used term is Kappa (κ) filter or Kappa effect.

The relevance of κ is noted in multiple areas of research and engineering. It is required in source studies to constrain the source spectrum and investigate its parameters, which is done in various ways, for example by inverting for the attenuation (Q) structure or by deriving empirical Green's functions, but also by using κ_r to correct for the high-frequency spectral decay (Irikura 1986; Lancieri et al. 2012; Ktenidou et al. 2014). It is also important for the purpose of simulating ground motion by stochastic or hybrid broadband methods to apply κ_0 for spectral high-frequency reduction as it is observed in recorded data (Boore 2003; Halldorsson and Papageorgiou 2005; Graves and Pitarka 2010; Foster et al. 2012). The use of κ extends also to engineering seismology where it is used to adapt ground-motion prediction equations (GMPEs) to target regions which are typically less active (Campbell 2003; Cotton et al. 2006). Furthermore, it is a parameter considered in adjusting GMPEs to specific site conditions for critical facilities in probabilistic seismic hazard assessment (PSHA), where such corrections using V_{S30} and κ_0 can result in differences up to a factor of 3 in the high-frequency response spectra (Biro and Renault 2012; Ktenidou et al. 2014).

A number of κ measurement approaches exist, such as broadband inversion of the acceleration spectrum for the seismic moment M_0, source corner frequency f_c and κ_r. However, this paper focuses on the classical approach of measuring kappa by a least-squares fit to the high-frequency acceleration spectra (Anderson and Hough 1984), and it is still followed by many authors (Hough et al. 1988; Tsai and Chen 2000; Purvance and Anderson 2003; Douglas et al. 2010; Van Houtte et al. 2011; Ktenidou et al. 2013; Edwards et al. 2015). The details of that method are more variable in literature, but several general assumptions are made by most authors: The main part of the S wave should be included in the time window, the frequency band used should be above the source corner frequency, conspicuous site resonance peaks (amplification bulges in spectra) should be avoided or a broad bandwidth should be used to even out this effect (Parolai and Bindi 2004).

Regarding the long-standing scientific discussion about how exactly to treat the earthquake source stress drop and the high-frequency spectral decay parameter κ in engineering seismology, it has been well noted that these two variables trade off with each other in seismological models (Scherbaum et al. 2006; Ktenidou et al. 2014). Therefore, the classical measurement approach (Anderson and Hough 1984) could help independently constraining an important seismological parameter to avoid influencing the estimates of other physical model parameters. This measurement requires time and trained personnel, as a fully automatic procedure has not been developed so far, due to non-trivial issues that often require manual determinations on many records. This problem is addressed by this work through developing a reliable algorithm to be used as a standard method for κ_r measurement on strong ground motion records or seismograms. The goal is to automatically produce databases of consistent estimates of κ_r and κ_0 for all seismically active regions using typical ground-motion records. Here, we present an algorithm for that purpose and demonstrate its efficiency using the acceleration records of earthquakes from the South Iceland Seismic Zone (SISZ, Fig. 4.1).

4.2 Data

The data used are ground acceleration measurements generated by medium-sized earthquakes from 1987 to 2008. There are 27 earthquakes with a magnitude range of $M_W 3.7$–6.5 which are all located in the South Iceland Seismic Zone (SISZ, see Fig. 4.1). These events were recorded by up to 17 accelerometric stations (Fig. 4.2), and for each record the P and S wave arrival times were determined manually. The database contains records of more events and stations, but the selection was limited in this study such that each station will have at least five recorded events, while all events were recorded by at least two stations. This selection already excludes records which did not pass the quality criteria for time series and spectral analyses. The stations belong to the Icelandic strong-motion network (ISMN) of the Earthquake Engineering Research Centre of the University of Iceland (Sigbjörnsson et al. 2004). While most stations are set up to directly measure ground response (i.e., free-field), some stations are part of structural monitoring systems located within power stations, office buildings and in bridges. All instruments record acceleration on three orthogonal components, but only the horizontal ones were used for the analysis. Most stations' sampling frequencies are 200 Hz, but some record at 100 Hz. As the instruments measure ground movement, the immediate subsurface characteristics are

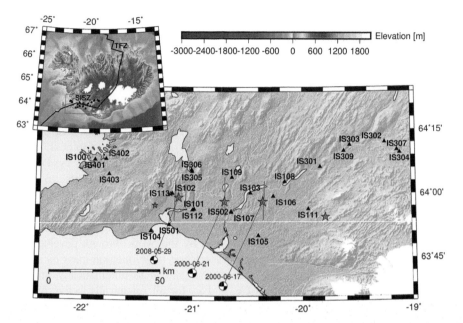

Fig. 4.1 The left-hand inset shows the plate boundary across Iceland, the locations of the South Iceland Seismic Zone (SISZ) and the Tjörnes Fracture Zone (TFZ) as well as the study area indicated by a rectangle. All earthquakes and stations used in this study are located in or close to the SISZ. The stars indicate epicenters of the studied earthquakes, which are scaled to magnitude. For the three largest events, moment tensor solutions from the USGS and fault lines are shown as well. The labeled triangles represent accelerometer stations, of which 17 have been used in this study

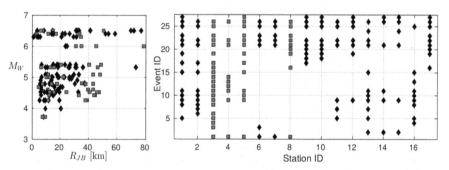

Fig. 4.2 Dataset used in this study. *Left*: Magnitude-distance distribution. *Right*: Distribution of records indicating which station has accepted data for a given event. Black diamonds and grey squares indicate rock and stiff soil sites, respectively

relevant to the measurements. Most stations are located on what is generally referred to as rock, but some stations are located on stiff soil (Þórarinsson et al. 2002; Sigbjörnsson and Ólafsson 2004).

4.3 Analysis

4.3.1 Automatic Phase Detection

As the slope of the high-frequency part of the direct shear wave spectrum is to be measured, the S-wave onset time needs to be known. As the spectra are relatively insensitive to small changes of time window start and length, various authors use fixed lengths which can start shortly before the S-phase onset and contain part of the S-wave coda. Therefore, automatic phase picking could be used to facilitate the process as well, since the presumable small differences to manual time picks would be acceptable. One of the most common algorithms for phase picking is the short-term average to long-term average (STA-LTA) filter with set trigger levels. This scheme does not perform well to detect S-phase onsets and would also fail completely on late-triggered records. Instead, a different approach is used, which mainly relies on the cumulative energy of the traces. If the overall record time window is not overly long, but contains the earthquake signal, then even without much pre- or post-event noise the cumulative energy would have to rise significantly at both the P- and S-wave onsets. Instead of setting an arbitrary threshold value for its slope to detect the phases, the cumulative energy function is detrended (called DCE from here), such that the part after the first significant energy rise would become the lowest part of the DCE function at time t_{min}. This minimum is a rough indicator for the signal position within the time window and mostly falls between the P and S onsets. The maximum of the DCE at t_{max} indicates the end of the main signal energy, after which the increase in cumulative energy becomes very small for the rest of the time window. We calculate the slope between the values of the DCE at t_0 and t_{max}, then subtract a linear function with that slope and intercept of DCE(0) from the DCE until reaching t_{max}. The thus lowered DCE has its new minimum more towards the second significant rise in cumulative energy, which we take as the S-phase onset time t_S. In our algorithm, the arithmetic mean of the two horizontal components' t_S is finally saved as the S wave arrival time of that record. We also need the P-phase onset to know where to retrieve the pre-event noise window. For that, the cumulative energy is recalculated from t_0 until somewhat past the previous t_{min}, then detrended again. This adjusts the new minimum of the short DCE towards the first significant rise in cumulative energy, which is assumed to be the P-wave energy. The time of the new minimum is accepted as t_P. As some of our records' vertical components have low resolution or low signal-to-noise ratio (SNR), the t_P estimates can be falsely triggered by irregular noise, while the two horizontal components may yield better estimates. Therefore, all three components' t_P estimates are sorted by time and the earliest value is only accepted as t_P if the time difference between the first two t_P estimates is less than twice the difference between the two last t_P estimates, otherwise the second estimate is accepted as t_P.

4.3.2 Automatic κ Measurement

The model proposed by Anderson and Hough (1984) assumes that above a certain frequency f_E the shape of the log-acceleration spectrum versus linear frequency would be fairly linear until it eventually falls below the noise level. Deviations from this model are usually considered to stem from local site effects which often are presumed sediment layer resonance phenomena. Additionally, the source spectral shape can bias κ measurements as the acceleration spectral energy increases until a certain frequency f_c above which it is theorized to remain flat. Only the physically most simplified assumption of a circular crack with full stress drop equally distributed over the source area leads to the simple shape with one corner frequency, while more complex ruptures can have more complicated spectral shapes especially at intermediate frequencies (Papageorgiou and Aki 1983). For these reasons, most authors choose only higher magnitude events, check the displacement spectra for f_c, consider the pre-event noise spectra or rather the spectral SNR, and either completely avoid site resonance peaks in the spectra or attempt to average them out by choosing wider ranges when selecting the measurement bandwidth from f_E to f_X. There is some subjectivity in this, especially for spectra that do not fit the assumed model very well. Removing the site response function has been attempted by other researchers, but yielded unfavorable results especially at higher frequencies (Van Houtte et al. 2011). Therefore, the presented algorithm is kept as simple as possible but as complex as needed to account for the non-triviality of the measurement process.

First, the S-wave signal window needs to be determined (Fig. 4.3, red waveforms). The time window is set to start 0.3 s before t_S and lasts until 95% trace

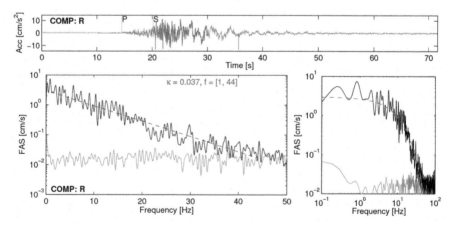

Fig. 4.3 Waveforms and Fourier amplitude spectra of horizontal component of station IS402 (rock, R_{JB} 37 km, M_W 6.3). Green and red on the time series indicate noise and signal windows, respectively. The spectra (black for signal, gray for noise) are displayed in log-linear (left) and log-log units (right). The measured spectral slope is shown as red dashed line, labeled with κ value and frequency limits

energy, but at least 3 to maximum 10 s after t_S. The noise window is taken from before t_P with a length in the range of 3.5–10 s, but only up until 0.5 s before t_P (Fig. 4.3, green waveforms). If this cannot be achieved due to the record's brevity before t_P from a late trigger, the noise window is taken instead at the end of the available record. This solution may be debatable due to coda contamination, but in the interest of data extraction under unfavorable conditions it can be attempted (Van Houtte et al. 2014). The criteria set for that case are that the noise window shall start earliest after 95% trace energy plus 5 s and still meet at least the minimum length of 3.5 s, measured from the record end. If the minimum criteria for any of the two time windows cannot be met, the record is rejected. Then, both windows are tapered with a 5% cosine taper, transformed to the Fourier domain and their spectral amplitudes are smoothed. Here, a Hamming-window is designed to have its Fourier-transformed corner at 2 s (covariance function lag time) and used as zero-phase lowpass FIR-filter applied to the log-spectrum. It is applied symmetrically to avoid phase-shift. A simple Konno-Ohmachi filter (Konno and Ohmachi 1998) with bandwidth b = 40 could be used instead, as done by a few other researchers (Douglas et al. 2010; Ktenidou et al. 2013). Hough et al. (1988) used a cosine window with half-bandwidth 5 Hz to smoothen the spectra, and they argued that the spectral fine structure is not relevant to the κ measurement while calculating a standard deviance of the obtained spectral slope would depend significantly on the chosen smoothing filter parameters. In the proposed algorithm, a range of frequencies for both f_E (1 to 15 Hz, Δf_E: 1 Hz) and f_X (10 to 50 Hz, Δf_X: 2 Hz) is used to obtain the SNR, the root-mean-squared (RMS) error of the fitted line to the spectrum and the frequency bandwidth Δf. The SNR is computed as the harmonic mean of signal to interpolated noise spectral ratios at all frequencies. This is a softer criterion than limiting the bandwidth to the first frequency where the SNR is too low, as this can happen at intermediate frequencies already if the spectra contain narrow spikes or troughs, while on average the SNR would still be acceptable. The arithmetic average is not used in this case, as it is too insensitive to low values while the harmonic mean decreases strongly in the presence of a few low values and results in a more conservative average. Human analysts would also select broader bandwidths as long as the spectral decay follows the assumed model, even if a small part is already close to the noise level, because the spectra often have dents which can bias the measurement if not enough signal is included. The SNR limit is set to 5 in this study, rejecting frequency bands which fall below this value for further analysis. The RMS residual E_{ij} between the fitted line and the smoothed spectrum is calculated for each frequency band from $f_{E,\ i}$ to $f_{X,\ j}$, and is then divided by the square-root of the corresponding bandwidth,

$$P_{ij} = \frac{E_{ij}}{\sqrt{\Delta f_{ij}}}$$

The value P_{ij} represents how well the spectral window fits a straight line adjusted by the range of values used to obtain this measure. The desire to pick a relatively straight line (low RMS residual) over a broad spectral range (high Δf) is accounted for by selecting the lowest P_{ij} value and its corresponding $f_{E,i}$ and $f_{X,j}$ for the actual κ_r measurement. A not very straight section can be selected if it spans a wide frequency range, as recommend by various researchers (Parolai and Bindi 2004). At the same time, significant curves or bulges in the spectra are interpreted as less acceptable by this algorithm, as the E_{ij} values would be significantly higher than those from a more linear section. Therefore, low-frequency site resonance peaks are avoided and high-frequency flattening due to noise or other irregularities is mostly excluded, if a reasonably linear section exists in the main part of the spectral window which fulfills the SNR criterion.

4.4 Results

4.4.1 Phase Onset Time Picker Performance

The comparison between manual and automatic time picks shows that the bulk of the phase identifications is accurate enough for this study. The sets of P phase and S phase picks which have less than 1 (or 0.5) seconds absolute error is 82% (74%) and 88% (77%), respectively. An overview for all events regarding the mean and range of timing error is given in Fig. 4.4. It can be seen that the automatic phase picks are quite accurate for most earthquakes, except for those in June 2000 which also have the most records available. 95% of the time differences (automatic − manual) are within ±3 and ± 2 s for P and S arrivals, respectively. Compared to typical accuracy values of more advanced phase time picking algorithms, our results may seem modest, and the plots in Fig. 4.5 illustrate the most typical problems encountered. Especially the events in June 2000 had multiple closely spaced ruptures, examples of which are displayed as the top and middle records in Fig. 4.5. While the top record

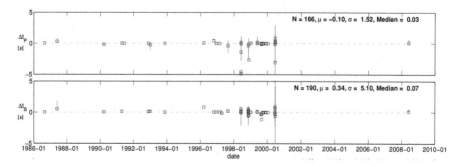

Fig. 4.4 Time differences (automatic − manual) of t_P and t_S versus event origin time. Black squares are the mean values for each event and the red, solid lines indicate the full range of differences. The displayed time ranges are limited to avoid showing a few outliers significantly beyond 2σ of the mean

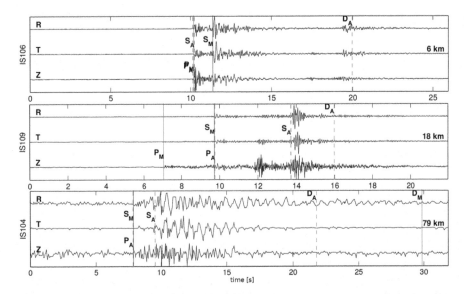

Fig. 4.5 Several examples of not well fitted phase arrivals at different stations from various events, with label subscripts "M" for manually picked and "A" for automatically detected. *Top:* Long time windows for short signals can result in P and S phases being assumed both at the signal start. *Middle:* Closely spaced multiple earthquakes are not separated and understood by the algorithm. *Bottom:* Instruments triggered late and not retaining enough pre-trigger recordings yield signal input which does not fit the assumed model of P and S waveforms

(station IS106) contains two relatively well separated earthquake signals of which the first is much stronger than the second, the middle record (station IS109) contains at least two events whose signals overlap much more with a medium-sized signal preceding a stronger one followed by smaller signals. The picker algorithm does not distinguish in such cases and misidentified the first S wave as P phase while accepting the strongest (second) S wave as S phase. The larger residuals in about 10 records can be attributed to confusion from signals of multiple earthquakes per record. Another issue is the record window length in that for signals whose S-P time difference is at least about 20 times shorter than the record length the P and S phases get less well distinguished (Fig. 4.5, top). Misidentification on short signals in long record windows also gets compounded in cases where the P phase is strong compared to the S wave due to the radiation pattern as in the top example of Fig. 4.5, and at least 3 records seem to be in this condition. Lastly, stations that triggered late and lack the P phase onset on record cannot be used to check P wave arrival pick accuracy, an example of which is shown as the bottom record in Fig. 4.5. Mostly for that reason only 166 P phase picks are compared as opposed to 190 S phase picks. Still, for κ_r measurements the S phases are often detected well enough even on late triggered records, although in such cases the automatic noise window retrieval often results in usage of strong P or S phase coda waves which then hampers the automated f_E and f_X determinations for the κ analysis.

4.4.2 Automatic κ_r Measurement Performance

The manually obtained κ_r values exhibit a significant scatter with an overall standard deviation of 0.022 s when combining data of all sites. The scatter of the automatically determined κ_r values is about 25% larger than that, but the mean and median values of manual and automated measurements are almost the same. An overall increase in κ_r with distance can be noted due to the main cluster of data falling below 0.05 s when closer than 30 km, while for greater distances the κ_r distribution becomes broader with a higher average. The only two measurements for $R_{epi} > 100$ km is from a site in North Iceland, which clearly indicates stronger attenuation at these distances. The limiting frequencies f_E and f_X determine the value of κ_r on each spectrum, and the algorithm's choices for f_E resulted in a highly skewed distribution with its mode at 1 Hz and median at 2 Hz, while the manual picks constitute a less skewed distribution with mean and median both between 6 and 7 Hz. For the upper limiting frequency f_X, the automatic values have a high peak at 50 Hz with a median of 42 Hz and an approximately uniformly distributed part from 10 to 49 Hz, while the manual values have their mode at about 44 Hz as broader peak between 40 and 50 Hz and a more flat distribution to lower frequencies tapering off towards 10 Hz. Despite these differences, the algorithm has performed relatively well as the overall mean and median values of κ_r and its distribution shape are very similar to the manually obtained values.

4.4.3 Mean Statistics and Distance Dependence

To increase the reliability of general observations with respect to the dependence of κ_0 to site class, some stations' statistics were combined to obtain their median values (Table 4.1). The criteria set for this selection are: (1) Only stations with more than 5 measurements, (2) only stations whose zero-distance intercept value standard deviation σ_{κ_0} was below 7.5 ms, (3) only if the slope with distance (κ_e) was positive, and (4) only if the distance range was above 25 km. These data seem to be the most reliable, as they cover a wider range and are less scattered. The distance dependence derived from these data corresponds to a slope κ_e between 0.13 and 0.16 ms/km with average correlation coefficients $\rho_{\kappa, R}$ of about 0.38 to 0.46 for the two groups of rock and stiff soil stations, respectively. The κ_r values versus R_{epi} for the selected stiff soil

Table 4.1 Median parameter values for the two soil types from selected stations

Type	$\overline{\kappa_r}$ [s]	κ_0 [s]	κ_e [s/km]	$\rho_{\kappa, R}$	σ_{κ_0} [s]	$N_{stations}$
Rock	0.033	0.031	0.00013	0.38	0.0048	6
Stiff soil	0.050	0.045	0.00016	0.46	0.0063	2

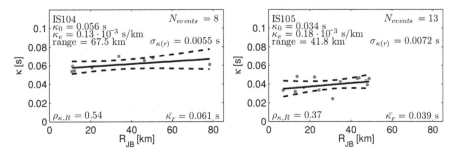

Fig. 4.6 κ_r versus distance on stiff soil sites which met the selection criteria, with linear regression lines and 95% regression intervals as dashed lines

sites are shown for each station in Fig. 4.6. For almost all of the individual κ_e estimates by station, the 95% confidence intervals for the slopes (dashed lines) indicate the possibility that the slopes may be zero, although on average they are all positive values.

4.5 Discussion

The automated phase time picks are mostly acceptable on typical local earthquake records, but the detection of problematic cases needs improvement. The algorithm cannot judge whether the signal is cut off at the start to exclude the P wave onset (late trigger) or the occurrence of multiple ruptures. If such pattern could be recognized, the automatic procedure would be adequate. A simple first update should also adjust the analysis window length to avoid misidentification on longer records with short signals from close events.

For the unbiased κ_r measurement, the source corner frequency f_c should be estimated as well. The algorithm does not incorporate any f_c analysis at this point, and trials with smaller events ($M_W < 3$) have shown significant bias towards lower κ_r values due to setting the lower limiting frequency f_E below f_c. Next, the higher scatter of the automatically determined κ_r due to inadequate choice of f_E and f_X needs to be addressed. This could be done either by adjusting the existing criteria or by modifying the trade-off parameter's functional form such that it becomes more sensitive to what a human analyst would choose. Adjusting the criteria and a possible different target function could be done through non-linear optimization using the manually picked data as reference in a simple residual function for κ_r or maybe a combination of the limiting frequencies. Furthermore, various uncertainty measures should be incorporated into the analysis of κ_r to obtain more realistic estimates of the range of values. So far, quantitative quality criteria applied by researchers have included standard deviations from each slope (Lai et al. 2016), from combining the two horizontal components' κ_r values (Purvance and Anderson 2003), using several spectral windows to obtain a mean and standard deviation (Ktenidou et al. 2016), using the fitted slope's standard error or even the variability from using many different component orientations (Van Houtte et al. 2014). Another approach is to have several analysts manually pick the limiting frequencies to derive the standard deviance (Douglas et al. 2010). Such sensitivity analyses can uncover very unreliable measurements and might help to explain some sites' highly scattered data.

Trials where all spectra were accepted for analysis regardless of SNR were done as well, such that if all bandwidths had SNR < 5, then the range with the highest SNR was used. Most of such cases turned out to be bad choices, as they included site resonance peaks and yielded nonsensical results. Often enough, this issue was caused by having an inappropriate noise window close after the earthquake signal in cases where the record would be quite short and often triggered late. The input data quality is important, as the universal principle of 'garbage in, garbage out' applies even more the simpler the algorithm. A human analyst would identify such situations, disregard the noise spectrum and manually choose an appropriate frequency range, if the signal itself seems to fit the assumed model. The algorithm could include such a decision, but then the obtained κ_r should be marked to indicate low-quality data, which should affect the above-mentioned uncertainty estimates.

4.6 Conclusion

Simple first versions of a phase onset time picker algorithm and a κ_r measurement algorithm using the acceleration spectrum have been developed and tested. The onset times of both P and S waves as well as κ_r values for each record have been manually determined to obtain a reference database. The algorithms were used to identify seismic phases and subsequently measure κ_r automatically, with the resulting values being found similar to the manually obtained ones. There are still a few problematic cases for which the time picker needs to be improved, while the κ_r analysis algorithm would also require adjustment and possibly more discerning measurement criteria, because the automatically gathered κ_r data have a 25% higher scatter than the manual values. Using a subset of 8 stations, a significant but weak κ_r increase of $\kappa_e = 0.16$ ms/km with increasing source-site distance has been discovered by regression analysis. Further investigation is indicated as improvements of the measurement algorithms is realistic and of interest to the scientific community.

Acknowledgements This study was supported by the Icelandic Centre for Research (Grant of Excellence No. 141261-051/052/053), and partially by the Icelandic Catastrophe Insurance, and the Research Fund of the University of Iceland.

References

Achenbach JD, Harris JG (1978) Ray method for elastodynamic radiation from a slip zone of arbitrary shape. J Geophys Res Solid Earth 83:2283–2291

Anderson JG, Hough SE (1984) A model for the shape of the fourier amplitude spectrum of acceleration at high frequencies. Bull Seismol Soc Am 74:1969–1993

Biro Y, Renault P (2012) Importance and impact of host-to-target conversions for ground motion prediction equations in PSHA. In: Proceedings of the 15th World conference on earthquake engineering, pp 24–28

Boore DM (2003) Simulation of ground motion using the stochastic method. Pure Appl Geophys 160:635–676

Campbell KW (2003) Prediction of strong ground motion using the hybrid empirical method and its use in the development of ground-motion (attenuation) relations in eastern North America. Bull Seismol Soc Am 93:1012–1033

Cotton F, Scherbaum F, Bommer JJ, Bungum H (2006) Criteria for selecting and adjusting ground-motion models for specific target regions: application to Central Europe and rock sites. J Seismol 10:137–156

Douglas J, Gehl P, Bonilla LF, Gélis C (2010) A κ model for mainland France. Pure Appl Geophys 167:1303–1315

Edwards B, Ktenidou O-J, Cotton F et al (2015) Epistemic uncertainty and limitations of the κ0 model for near-surface attenuation at hard rock sites. Geophys J Int 202:1627–1645

Foster KM, Halldorsson B, Green RA, Chapman MC (2012) Calibration of the specific barrier model to the NGA dataset. Seismol Res Lett 83:566–574

Graves RW, Pitarka A (2010) Broadband ground-motion simulation using a hybrid approach. Bull Seismol Soc Am 100:2095–2123

Halldorsson B, Papageorgiou AS (2005) Calibration of the specific barrier model to earthquakes of different tectonic regions. Bull Seismol Soc Am 95:1276–1300

Hanks TC (1982) f_{max}. Bull Seismol Soc Am 72:1867–1879

Hough SE, Anderson JG, Brune J et al (1988) Attenuation near Anza, California. Bull Seismol Soc Am 78:672–691

Irikura K (1986) Prediction of strong acceleration motion using empirical Green's function. In: Proceedings of the 7th Japan earthquake engineering symposium, p 156

Konno K, Ohmachi T (1998) Ground-motion characteristics estimated from spectral ratio between horizontal and vertical components of microtremor. Bull Seismol Soc Am 88:228–241

Ktenidou O-J, Gélis C, Bonilla L-F (2013) A study on the variability of kappa (κ) in a borehole: implications of the computation process. Bull Seismol Soc Am 103:1048–1068

Ktenidou O-J, Cotton F, Abrahamson NA, Anderson JG (2014) Taxonomy of κ: a review of definitions and estimation approaches targeted to applications. Seismol Res Lett 85:135–146

Ktenidou OJ, Abrahamson N, Darragh R, Silva W (2016) A methodology for the estimation of kappa (κ) from large datasets, example application to rock sites in the NGA-East database, and implications on design motions

Lai T-S, Mittal H, Chao W-A, Wu Y-M (2016) A study on kappa value in Taiwan using borehole and surface seismic Array. Bull Seismol Soc Am 106:1509–1517

Lancieri M, Madariaga R, Bonilla F (2012) Spectral scaling of the aftershocks of the Tocopilla 2007 earthquake in northern Chile. Geophys J Int 189:469–480

Papageorgiou AS, Aki K (1983) A specific barrier model for the quantitative description of inhomogeneous faulting and the prediction of strong ground motion. I. Description of the model. Bull Seismol Soc Am 73:693–722

Parolai S, Bindi D (2004) Influence of soil-layer properties on k evaluation. Bull Seismol Soc Am 94:349–356

Purvance MD, Anderson JG (2003) A comprehensive study of the observed spectral decay in strong-motion accelerations recorded in Guerrero, Mexico. Bull Seismol Soc Am 93:600–611

Scherbaum F, Cotton F, Staedtke H (2006) The estimation of minimum-misfit stochastic models from empirical ground-motion prediction equations. Bull Seismol Soc Am 96:427–445

Sigbjörnsson R, Ólafsson S (2004) On the South Iceland earthquakes in June 2000: strong-motion effects and damage. Boll Geofis Teor Ed Appl 45:131–152

Sigbjörnsson R, Ólafsson S, Thórarinsson Ó (2004) Strong-motion recordings in Iceland. In: Proceedings of the 13th World Conference on Earthquake Engineering. Paper no. 2370, Mira, Vancouver, BC, Canada

Þórarinsson Ó, Bessason B, Snæbjörnsson JT, et al (2002) The South Iceland earthquakes in 2000: Strong motion measurements. In: 12th European conference on earthquake engineering. Paper no. 321, Elsevier Science Ltd, London, UK

Tsai C-CP, Chen K-C (2000) A model for the high-cut process of strong-motion accelerations in terms of distance, magnitude, and site condition: an example from the SMART 1 Array, Lotung, Taiwan. Bull Seismol Soc Am 90:1535–1542

Van Houtte C, Drouet S, Cotton F (2011) Analysis of the origins of κ (kappa) to compute hard rock to rock adjustment factors for GMPEs. Bull Seismol Soc Am 101:2926–2941

Van Houtte C, Ktenidou O-J, Larkin T, Holden C (2014) Hard-site κ0 (Kappa) calculations for Christchurch, New Zealand, and comparison with local ground-motion prediction models

Chapter 5
Application of MASW in the South Iceland Seismic Zone

Elín Ásta Ólafsdóttir, Bjarni Bessason, and Sigurður Erlingsson

Abstract Multichannel Analysis of Surface Waves (MASW) is a seismic exploration method to evaluate shear wave velocity profiles of near-surface materials. MASW was applied at seven locations in or close to the South Iceland Seismic Zone, providing shear wave velocity profiles for the top-most 15–25 m. The profiles were utilized for seismic soil classification according to Eurocode 8. The results indicated that the sites that are characterized by sandy glaciofluvial, littoral or alluvial sediments fall into category C and the sites where the deposits are cemented to some degree belong to category B. Furthermore, the MASW measurements were used to evaluate the liquefaction potential at a site where liquefaction sand boils were observed during an M_w6.3 earthquake occurring in May 2008. The simplified procedure of assessing cyclic stress ratio to normalized shear wave velocity revealed that liquefaction had occurred down to 3–4 m depth, which is consistent with observations on site.

Keywords MASW · Shear wave velocity · Seismic site classification · Liquefaction potential

5.1 Introduction

The shear wave velocity (V_S) of near-surface materials is an important parameter in civil engineering work for characterization of natural soil sites and man-made fillings. The small-strain shear modulus (G_0) of individual soil layers is directly proportional to the square of their shear wave velocity. Furthermore, in areas where earthquakes are of concern, the shear wave velocity is fundamental in assessing both liquefaction potential and soil amplification and when defining site-specific earthquake design loading (Kramer 1996). For planning and design of structures, information about the shear wave velocity of the top few tens of meters of the soil stratum is, in general, of most importance.

Several in-situ methods can be used for determination of shear wave velocity profiles of subsoil sites (Kramer 1996). This includes methods that require a drilled borehole such as down-hole and cross-hole seismic surveys, methods where the resistance of soil to penetration is measured like the Standard Penetration Test (SPT) and the Cone Penetration Test (CPT), and surface wave analysis methods. The two-receiver seismic analysis method, Spectral Analysis of Surface Waves (SASW), has been applied at a number of sites in Iceland in the last two decades (Bessason et al. 1998; Bessason and Erlingsson 2011). Multichannel Analysis of Surface Waves (MASW) is a newer and more advanced method, based on simultaneous analysis of multiple surface wave traces acquired by an equally spaced line of receivers (Gabriels et al. 1987; Park et al. 1999). The MASW method has attracted increasingly more attention in recent years and has become one of the main surface wave analysis methods for evaluation of near-surface shear wave velocity profiles for applications in civil engineering (Xia 2014). The MASW method was first applied in Iceland in 2013 (Olafsdottir et al. 2014).

The main objective of the study presented in this paper was to map the shear wave velocity profiles of several test sites in South Iceland, in or close to the South Iceland Seismic Zone, by the MASW method and compare profiles obtained at different sites. The results of the MASW measurements were exploited to classify the soil at the test sites into ground types according to Eurocode 8. Furthermore, the results were used to evaluate the liquefaction hazard at one of the sites.

5.2 Seismic Activity in South Iceland

The seismicity in Iceland is related to the Mid-Atlantic plate boundary that crosses the island. Within the country, the boundary is shifted towards the east through two complex fracture zones (Einarsson 1991). One is located in the South Iceland Lowland, where it crosses the biggest agricultural region in the country, while the other is mostly located off the northern coast. The first is usually termed the South Iceland Seismic Zone (SISZ) and the second the Tjörnes Fracture Zone (TFZ). The largest earthquakes in Iceland have occurred within these zones, mostly associated with a strike-slip motion at shallow depth (5–10 km).

Earthquakes in the SISZ tend to occur in sequences, typically every 100 years. One such started in 1896, when five quakes of magnitude greater than six struck the area within two weeks. Another event occurred in 1912 and is by some scientists considered to have finished the sequence. A new sequence started in 2000 when two earthquakes of magnitude $M_w6.5$ struck on the 17th and 21st of June. They were followed by the third one, an $M_w6.3$ earthquake on the 29th of May, 2008, referred to as the Ölfus earthquake. The epicentral distances between these three strike-slip quakes were less than 35 km, and they were all shallow with a hypocentral depth of less than 7 km. A great deal of structural damage occurred in these earthquakes, but no residential buildings collapsed and there was no loss of life (Bessason and Bjarnason 2016). In all these events rock fall and landslides occurred, as well as lateral spreading and liquefaction.

5.3 Multichannel Analysis of Surface Waves

Multichannel Analysis of Surface Waves is a relatively new technique for determining shear wave velocity profiles of near-surface soil materials based on analysis of Rayleigh waves. Rayleigh waves are surface waves that have the strongest wave motion closest to the surface and decay rapidly with depth (Fig. 5.1). Short wavelength (high frequency) Rayleigh wave components mainly propagate through the top-most soil layers, while longer wavelength (lower frequency) components travel deeper.

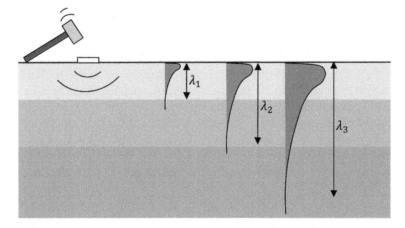

Fig. 5.1 Rayleigh wave components with different wavelengths propagating through a layered medium. Wave components with different frequencies reflect soil properties at different depths

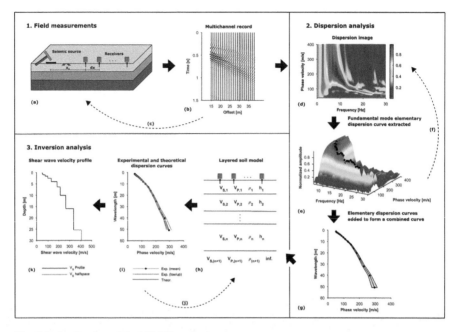

Fig. 5.2 Application of the MASW method

The phase velocity of a Rayleigh wave component propagating through a layered medium is determined by the average properties of the soil layers that it travels through. Hence, wave components with different frequencies propagate at different velocities and reflect material properties of soil layers at different depths. Waves that have a propagation velocity that depends on frequency are called dispersive waves. A dispersion curve describes the relation between frequency (wavelength) and phase velocity.

The investigation depth of a MASW survey is determined by the longest Rayleigh wave wavelength that is retrieved. A commonly used rule-of-thumb for interpretation of fundamental mode Rayleigh wave dispersion curves is that the maximum depth of the shear wave velocity profile is approximately one third to half the longest wavelength (Garofalo et al. 2016). The investigation depth that can be achieved by MASW is, in general, few tens of meters, assuming that the surface waves are generated by a reasonably heavy impact load, e.g. a sledgehammer. The observed difference between results obtained by MASW and direct borehole measurements has been evaluated as approximately 15% or less and random (Xia et al. 2002).

In its basic form, the MASW method consists of three steps as illustrated in Fig. 5.2 (Park et al. 1999).

1. Data acquisition in the field.
2. Data processing to extract an experimental Rayleigh wave dispersion curve from the acquired data (dispersion analysis).

3. Evaluation of the shear wave velocity profile by inversion of the experimental dispersion curve (inversion analysis).

For data acquisition, multiple (12 or more) low frequency geophones are lined up on the surface of the test site with equal receiver spacing dx (Fig. 5.2a). A wave is generated by an impact load that is applied at a distance x_1 from one end of the geophone line and the wave propagation is recorded (Fig. 5.2b).

It is commonly recognized that the length of the receiver spread and the source offset can have a substantial effect on the acquired records (e.g. Dikmen et al. 2010; Ivanov et al. 2008; Olafsdottir et al. 2016; Park and Carnevale 2010; Zhang et al. 2004). Applications of the MASW method at Icelandic sites have indicated that an increased range in investigation depth can be obtained by combining data recorded by measurement profiles of different lengths (Olafsdottir et al. 2016). Furthermore, combined or repeated analysis of surface wave records acquired by using several different measurement profile configurations can help confidently identifying the fundamental mode dispersion curve in the subsequent dispersion analysis. Hence, in this work, the field measurements are repeated several times using measurement profiles with different dx and/or x_1 while keeping the midpoint of the receiver spread fixed (Fig. 5.2c). Computations were carried out by using the open source Matlab software MASWaves (Olafsdottir et al. 2018) [see also masw.hi.is].

A dispersion image (Fig. 5.2d) is obtained based on each multichannel surface wave record by using the phase shift method (Park et al. 1998). The dispersion image visualizes the dispersion properties of all types of waves contained in the recorded data. Subsequently, the fundamental mode of the Rayleigh wave propagation is identified based on the spectral high-amplitude (peak energy) bands observed in the dispersion image and the fundamental mode Rayleigh wave dispersion curve is extracted (Fig. 5.2e).

The previous analysis steps (Fig. 5.2d, e) are repeated for each surface wave record (Fig. 5.2f) resulting in multiple elementary dispersion curves that are subsequently combined into a single experimental curve (Fig. 5.2g). Here, the combined (average) curve is obtained by grouping the data points included in the elementary dispersion curves together within 1/3 octave wavelength intervals. The phase velocity values within each interval are added up and their mean used as a point estimate of the phase velocity of the Rayleigh wave components belonging to the given wavelength range. The accuracy of the estimated mean phase velocity is evaluated in terms of the $p\%$ classical confidence interval for the sample mean within each wavelength interval.

Inversion problems involving the dispersion of Rayleigh waves in a layered elastic medium are by nature both non-linear and non-unique, and must be solved by iterative methods (Foti et al. 2015). Computations are based on Rayleigh wave propagation theory assuming a plane-layered earth model. The parameters required to describe the properties of each layer are shear wave velocity (V_S), compressional wave velocity (V_P) (or Poisson's ratio (ν)), mass density (ρ) and layer thickness (h). The last layer is assumed to be a half-space.

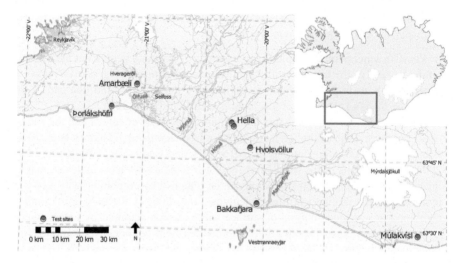

Fig. 5.3 Location of MASW test sites in South Iceland. The map is based on data from the National Land Survey of Iceland

The first step of the inversion analysis is to make an initial estimate of the layered soil model for the test site (Fig. 5.2h). A theoretical dispersion curve is obtained based on the assumed model and compared to the experimental (combined) curve (Fig. 5.2i). Here, the stiffness matrix method (Kausel and Roësset 1981) is used for computations of theoretical dispersion curves. Subsequently, the soil model is updated and the theoretical dispersion curve recomputed until the misfit between the theoretical and experimental curves has reached an acceptably small value (Fig. 5.2j). The shear wave velocity profile and the layer structure that result in an acceptable fit, and are believed to realistically represent the characteristics of the test site, are finally taken as the results of the survey (Fig. 5.2k).

5.4 MASW Measurements

MASW measurements have been conducted at several locations in Iceland from 2013 to 2016. The seven test sites reported in this paper are all located in South Iceland (Fig. 5.3). Some of the sites were in the near-fault area of the South Iceland earthquakes of June 2000 and May 2008. They are all at sites which were created by some mixture of ordinary fluvial and Aeolian processes, glacial outwash, sub-glacial outburst floods, coastal sediment transport and ash fall from volcanic eruptions. The exact origin is not well known. At some of the sites, the deposits are cemented to some degree whilst at others they are looser. The main characteristics of the seven test sites are summarized in Table 5.1.

At the seven MASW test sites, multichannel surface wave records were collected using 24 vertical geophones with a natural frequency of 4.5 Hz and a critical

Table 5.1 Overview of site characteristics and test configuration at the MASW test sites in South Iceland

Site		Arnarbæli	Bakkafjara	Hella – Gaddstaðaflatir	Hella – sandnáma	Hvolsvöllur	Múlakvísl	Þorlákshöfn
Site characteristics								
Soil type		Holocene glaciofluvial sand	Modern littoral sand	Late-glacial cemented Aeolian silty sand	Late-glacial (slightly) cemented Aeolian silty sand	[Unknown]	Modern littoral sand	Alluvial sand
USCS classification		SW-SM[a]	SW[c]			[Unknown]	SW[c]	SP[c]
Estimated location of groundwater table		At surface	At 4 m depth	At great depth	At great depth	At great depth	At 3 m depth	At surface
Mass density	ρ [kg/m³]		1850	2200	2200	2000	1850	
Saturated mass density	ρ_{sat} [kg/m³]	1800	2000				2000	1880
Poisson's ratio[b]	ν [–]		0.35	0.35	0.35	0.35	0.35	
Field measurements								
No. geophones	N	24	24	24	24	24	24	24
No. profiles		3	3	3	2	2	3	1
Receiver spacing (source offsets)	dx (x_1) [m]	0.5 (3.0–10.0) 1.0 (5.0–30.0) 2.0 (5.0–30.0)	0.5 (3.0–10.0) 1.0 (5.0–30.0) 2.0 (5.0–50.0)	0.5 (3.0–10.0) 1.0 (5.0–30.0) 2.0 (5.0–30.0)	0.5 (3.0–10.0) 1.0 (5.0–30.0)	0.5 (3.0–5.0) 1.0 (5.0–30.0)	0.5 (5.0–7.5) 1.0 (5.0–30.0) 2.0 (5.0–30.0)	2.0 (10.0–40.0)
No. elementary dispersion curves		95	71	113	58	43	85	26
Recording time	T [s]	2.4	1.2	2.4	2.4	2.4	2.4	1.2

[a]Green et al. (2012)
[b]The compressional wave velocity of the saturated soil layers is specified as $V_P = 1440$ m/s
[c]Assumed

damping ratio of 0.5. Up to three measurement profiles with the same midpoint but different receiver spacing (receiver spread length) were used at each site. For each receiver spacing, several different source offsets (x_1) were used. A summary of the main parameters related to the field measurements at each site is provided in Table 5.1. The impact load was in all cases created by a 6.3 kg sledgehammer, except at the Bakkafjara site where a single jump at the end of the measurement profile was also used. No systematic difference was observed between surface wave records where the impact load was created by a sledgehammer blow and where it was the result of a jump.

5.4.1 Results of MASW Measurements

Figure 5.4a shows the combined (average) experimental dispersion curves for the seven MASW test sites in South Iceland. The shaded areas in Fig. 5.4a correspond to the 95% confidence intervals for the mean phase velocity within each wavelength interval. In general, the longer wavelength (lower frequency) parts of the dispersion curves are characterized by higher uncertainty than the shorter wavelength (higher frequency) regions.

The results of the inversion analysis are illustrated in Fig. 5.4b. The half-space shear wave velocity in each case is indicated by a broken line. The misfit between the experimental dispersion curves and the optimum theoretical curves was in all cases less than 2.5%, except at the Hvolsvöllur site where it was approximately 4.5%.

The average shear wave velocity $V_{S,d}$ of the uppermost $d = [5, 10, 20, 30]$ m at each test site is further provided in Table 5.2. The $V_{S,d}$ velocity was obtained as (CEN 2004)

$$V_{S,d} = \frac{d}{\sum_{j=1}^{N} \frac{h_j}{V_{S,j}}} \tag{5.1}$$

where $V_{S,j}$ and h_j denote the shear wave velocity and thickness of the j-th layer, respectively, for a total of N layers down to depth d. When an estimated shear wave velocity profile went down to a depth less than d, the profile was extrapolated using the half-space velocity (Fig. 5.4b) down to depth d.

The shear wave velocity profiles and the average values presented in Fig. 5.4b and Table 5.2 indicate that the test sites can be divided into three main groups. First, the loose sand sites at Arnarbæli, Bakkafjara, Múlakvísl and Þorlákshöfn showed similar characteristics with V_S increasing from around 60–150 m/s close to ground level to 300–400 m/s at 20–25 m depth. Second, the two test sites close to Hella,

5 Application of MASW in the South Iceland Seismic Zone 61

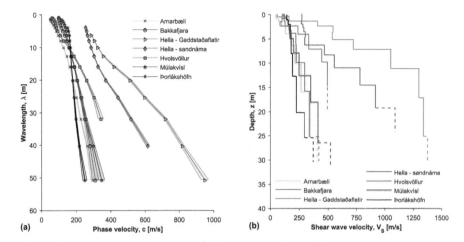

Fig. 5.4 (a) Combined (average) dispersion curves with 95% confidence intervals and (b) estimated shear wave velocity profiles for the MASW test sites in South Iceland

Table 5.2 Average shear wave velocity of the uppermost 5m, 10m, 20m and 30m at the MASW test sites

Site	$V_{S,5}$ [m/s]	$V_{S,10}$ [m/s]	$V_{S,20}$ [m/s]	$V_{S,30}$ [m/s]
Arnarbæli	93.9	129.0	178.4	215.0
Bakkafjara	142.1	174.0	223.5	258.0
Hella – Gaddstaðaflatir	284.2	425.5	637.5	774.2
Hella – sandnáma	283.6	338.8	478.3	588.2
Hvolsvöllur	127.9	192.4	277.0	(324.5)
Múlakvísl	153.5	180.0	231.2	275.3
Þorlákshöfn	163.5	173.3	192.0	222.2

characterized by cemented silty sand, showed much higher velocities, reaching values above 750 m/s and 1000 m/s, respectively, at around 10 m depth. Third, the Hvolsvöllur site was characterized by low values of V_S at shallow depths which increased rapidly with increasing depth. At the Hvolsvöllur site, however, the maximum retrievable Rayleigh wavelength was close to 30 m (Fig. 5.4a), limiting the maximum depth of the shear wave velocity profile to approximately 15 m. The increase in shear wave velocity with depth might therefore be substantially more than indicated by the estimated half-space velocity (Fig. 5.4b), leading to a higher estimate of $V_{S,30}$. Hence, for characterization of the properties of the soil at the Hvolsvöllur site based on $V_{S,30}$, further measurements are considered necessary (see also Sect. 5.5.1).

5.5 Seismic Hazard Application

5.5.1 Seismic Site Classification

In Eurocode 8, the European standard for design of structures in seismic zones, (CEN 2004) construction sites are classified into five categories, referred to as ground types A, B, C, D and E, based on the average shear wave velocity of the uppermost 30 m at the site (Eq. 5.1 with $d = 30$ m). The ground type is further used to account for the effects of the local ground conditions on the seismic action and, hence, fundamental for determination of site-specific design spectrums. The definition of ground types A, B, C, D and E, as presented in Eurocode 8, is provided in Table 5.3. The soil classification groups of the seven MASW test sites in South Iceland (Fig. 5.3) are provided in Table 5.4.

5.5.2 Liquefaction Hazard Analysis at the Arnarbæli Site

Liquefaction is attributed to the tendency of soils to compact during shaking. Furthermore, the resistance to shearing strains is highly linked to the material shear modulus. During the May 2008 M_w6.3 Ölfus earthquake, liquefaction was observed at a few places in South Iceland, among them at Arnarbæli where sand boils were found on the riverbank of the Ölfus River (Fig. 5.5a).

The Arnarbæli site is located less than 1 km from the active fault that ruptured in the Ölfus earthquake and the estimated PGA for the site (a_{max}) was in the range 0.6–0.7 g. The soil at the Arnarbæli site consists of glaciofluvial volcanic sand deposited on the western bank of the estuary of the Ölfus River. Based on a soil sieve analysis it has been concluded that the material has a 6–7% fine content (FC) and lies within the boundaries identified as potentially liquefiable soils (Green et al. 2012; Tsuchida 1970).

The simplified procedure to assess liquefaction resistance of soils based on measurements of shear wave velocity (Andrus and Stokoe 2000) was used to evaluate the liquefaction hazard at the Arnarbæli site. The method involves assessment of three parameters: (1) The cyclic stress ratio (CSR), describing the cyclic loading imposed on the soil by the earthquake

$$CSR = 0.65 \left(\frac{a_{max}}{g}\right) \left(\frac{\sigma_{v0}}{\sigma'_{v0}}\right) r_d \tag{5.2}$$

where σ_{v0} is total overburden stress, σ'_{v0} is effective overburden stress and r_d is a shear stress reduction coefficient, here estimated based on the work of Idriss (1999). (2) The overburden stress corrected shear wave velocity (V_{s1})

5 Application of MASW in the South Iceland Seismic Zone

Table 5.3 Identification of ground types based on $V_{S,30}$, after CEN (2004)

Ground type	Description of stratigraphic profile	$V_{S,30}$ [m/s]
A	Rock or other rock-like geological formation, including at most 5 m of weaker material at the surface.	> 800
B	Deposits of very dense sand, gravel, or very stiff clay, at least several tens of metres in thickness, characterized by a gradual increase in mechanical properties with depth.	360–800
C	Deep deposits of dense or medium-dense sand, gravel or stiff clay with thickness from several tens to many hundreds of metres.	180–360
D	Deposits of loose-to-medium cohesionless soil (with or without some soft cohesive layers), or of predominantly soft-to-firm cohesive soil.	<180
E	A soil profile consisting of a surface alluvium layer with V_S values of type C or D and thickness varying between about 5 and 20 m, underlain by stiffer material with V_S > 800 m/s.	

Table 5.4 Soil classification groups for the MASW test sites in South Iceland

Site	Ground type
Arnarbæli	C
Bakkafjara	C
Hella – Gaddstaðaflatir	B
Hella – sandnáma	B
Hvolsvöllur	C/B/E[a]
Múlakvísl	C
Þorlákshöfn	C

[a]Further measurements are needed

$$V_{S1} = \min\left(V_S \left(\frac{p_a}{\sigma'_{v0}}\right)^{0.25}; 1.4 V_S\right) \quad (5.3)$$

where $p_a = 100$ kPa and (3) the cyclic resistance ratio (*CRR*), which for a given V_{s1} is the value of *CSR* separating liquefaction and non-liquefaction occurrences

$$CRR = \left\{a\left(\frac{V_{S1}}{100}\right)^2 + b\left(\frac{1}{V_{S1}^* - V_{S1}} - \frac{1}{V_{S1}^*}\right)\right\}\left(\frac{M_w}{7.5}\right)^n \quad (5.4)$$

with $a = 0.022$, $b = 2.8$ and $n = -2.56$ (Andrus and Stokoe 2000). V_{S1}^* is the limiting upper value of V_{S1} for cyclic liquefaction occurrence estimated as $V_{S1}^* = 215 - 0.5(FC - 5)$ m/s for sands with a fine content between 5% and 35%.

A liquefaction resistance curve (V_{S1}–*CRR* curve) scaled to account for the effects of an $M_w 6.3$ earthquake is presented in Fig. 5.5b. The cyclic stress ratio and the normalized shear wave velocity were calculated for reference points at depth $z = [0.3, 0.8, 1.3, 2.1, 3.3, 5.2, 8.2, 13.0, 20.7]$ m, i.e. at the centre of each of the finite thickness layers indicated for the Arnarbæli site in Fig. 5.4b, and compared to the liquefaction resistance curve. The PGA of the site was estimated as 0.65 g. The

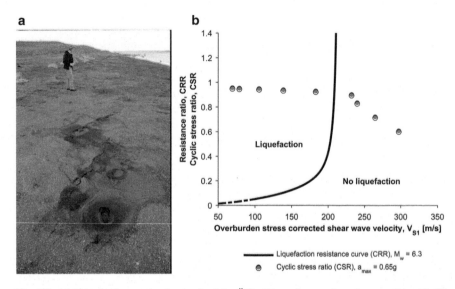

Fig. 5.5 (a) Liquefaction on the riverbank of the Ölfus River close to the epicentre of the $M_w 6.3$ 2008 Ölfus earthquake (photo: Oddur Sigurðsson). (b) Liquefaction evaluation chart for the Arnarbæli site

results, presented in Fig. 5.5b, indicate that liquefaction had occurred at the site down to around 3–4 m depth.

5.6 Concluding Remarks

MASW is a relatively new seismic exploration method to evaluate shear wave velocity profiles of near-surface materials. In this paper, the results of MASW measurements carried out at seven test sites in South Iceland are presented. Twenty-four 4.5 Hz geophones were used for recording and a 6.3 kg sledgehammer or a single jump at the end of the geophone line was used as an impact source. Several different source offset and/or receiver spacing configurations were used at each site in order to maximize the investigation depth range of the survey. This resulted in 26–113 multichannel surface wave records for each of the seven sites. Dispersion analysis and inversion were then applied to evaluate the combined (average) dispersion curve, along with upper and lower boundaries, and the shear wave velocity profile for each site. Using this set-up, shear wave velocity profiles down to 15–25 m could be obtained.

The results of the MASW measurements were utilized for seismic classification of the seven test sites according to Eurocode 8. The results indicated that the loose sand sites at Arnarbæli, Bakkafjara, Múlakvísl and Þorlákshöfn fall into category C and that the two sites close to Hella, where the deposits are cemented to some degree, belong to category B. At the Hvolsvöllur site, further measurements are considered necessary in order to carry out seismic classification. Furthermore, the results obtained by MASW were used to evaluate the liquefaction potential of the soils at the Arnarbæli site where liquefaction sand boils were observed during the M_w6.3 Ölfus earthquake in 2008. The simplified procedure of assessing cyclic stress ratio (CSR) to normalized shear wave velocity reveals that liquefaction has occurred at the site down to 3–4 m depth, which is consistent with observations on site.

Future work includes continuing development of the MASW method, e.g. further analysis of the effects of the measurement profile configuration and a more detailed study of the uncertainty associated with the combined dispersion curves and the subsequent shear wave velocity profiles. The ongoing build-up of a database of MASW shear wave velocity profiles for the South Iceland Seismic Zone is also of importance, along with seismic soil classification and evaluation of liquefaction potentials.

Acknowledgements The project is financially supported by grants from the University of Iceland Research Fund, the Icelandic Road and Coastal Administration and the Energy Research Fund of the National Power Company of Iceland.

References

Andrus RD, Stokoe KH II (2000) Liquefaction resistance of soils from shear-wave velocity. J Geotech Geoenviron Eng 126(11):1015–1025. https://doi.org/10.1061/(ASCE)1090-0241(2000)126:11(1015)

Bessason B, Bjarnason JÖ (2016) Seismic vulnerability of low-rise residential buildings based on damage data from three earthquakes (M_w6.5, 6.5 and 6.3). Eng Struct 111:64–79. https://doi.org/10.1016/j.engstruct.2015.12.008

Bessason B, Erlingsson S (2011) Shear wave velocity in surface sediments. Jökull 61:51–64

Bessason B, Baldvinsson GI, Thórarinsson Ó (1998) SASW for evaluation of site-specific earthquake excitation. In: Proceedings of the 11th European conference on earthquake engineering, Paris, CD-ROM

CEN (2004) EN 1998-1 Eurocode 8: design of structures for earthquake resistance, part 1: general rules, seismic actions and rules for buildings. European Committee for Standardization

Dikmen Ü, Arisoy MÖ, Akkaya I (2010) Offset and linear spread geometry in the MASW method. J Geophys Eng 7(2):211–222. https://doi.org/10.1088/1742-2132/7/2/S07

Einarsson P (1991) Earthquakes and present-day tectonics in Iceland. Tectonophysics 189 (1–4):261–279. https://doi.org/10.1016/0040-1951(91)90501-I

Foti S, Lai CG, Rix GJ, Strobbia C (2015) Surface wave methods for near-surface site characterization. CRC Press, Taylor & Francis Group, Boca Raton, FL

Gabriels P, Snieder R, Nolet G (1987) In situ measurements of shear-wave velocity in sediments with higher-mode Rayleigh waves. Geophys Prospect 35(2):187–196. https://doi.org/10.1111/j.1365-2478.1987.tb00812.x

Garofalo F, Foti S, Hollender F, Bard PY, Cornou C, Cox BR et al (2016) InterPACIFIC project: comparison of invasive and non-invasive methods for seismic site characterization. Part I: intra-comparison of surface wave methods. Soil Dyn Earthq Eng 82:222–240. https://doi.org/10.1016/j.soildyn.2015.12.010

Green RA, Halldorsson B, Kurtulus A, Steinarsson H, Erlendsson OA (2012) Unique liquefaction case study from the 29 May 2008, M_w6.3 Olfus earthquake, Southwest Iceland. In: 15th World conference on earthquake engineering, Lisbon, Portugal

Idriss IM (1999) An update of the Seed-Idriss simplified procedure for evaluating liquefaction potential. In: Transportation Research Board '99 workshop on new approaches to liquefaction analysis, Publ. No. FHWARD-99-165. Federal Highway Administration, Washington, DC

Ivanov J, Miller RD, Tsoflias G (2008) Some practical aspects of MASW analysis and processing. In: Proceedings of the 21st EEGC symposium on the application of geophysics to engineering and environmental problems, Philadelphia, PA, pp 1186–1198. https://doi.org/10.4133/1.2963228

Kausel E, Roësset JM (1981) Stiffness matrices for layered soils. Bull Seismol Soc Am 71(6):1743–1761

Kramer SL (1996) Geotechnical earthquake engineering. Prentice-Hall, Upper Saddle River

Olafsdottir EA, Bessason B, Erlingsson S (2014) Multichannel analysis of surface waves for estimation of soils stiffness profiles. In: Proceedings of the 23rd European young geotechnical engineers conference, Barcelona, Spain, pp 45–48

Olafsdottir EA, Erlingsson S, Bessason B (2016) Effects of measurement profile configuration on estimation of stiffness profiles of loose post glacial sites using MASW. In: Proceedings of the 17th Nordic geotechnical meeting, Reykjavik, Iceland, pp 327–336

Olafsdottir EA, Erlingsson S, Bessason B (2018) Tool for analysis of multichannel analysis of surface waves (MASW) field data and evaluation of shear wave velocity profiles of soils. Can Geotech J 55(2):217–233. https://doi.org/10.1139/cgj-2016-0302

Park CB, Carnevale M (2010) Optimum MASW survey–revisit after a decade of use. In: Proceedings of GeoFlorida 2010, West Palm Beach, FL, pp 1303–1312. https://doi.org/10.1061/41095(365)130

Park CB, Miller RD, Xia J (1998) Imaging dispersion curves of surface waves on multi-channel record. In: 68th Annual International meeting, SEG, Expanded abstracts 17, pp 1377–1380. doi: https://doi.org/10.1190/1.1820161

Park CB, Miller RD, Xia J (1999) Multichannel analysis of surface waves. Geophysics 64(3):800–808. https://doi.org/10.1190/1.1444590

Tsuchida H (1970) Prediction and Countermeans against the liquefaction in sand deposits. In: Abstract of the seminar in the Port and Harbour Research Institute, Yokohama, Japan, pp 3.1–3.33

Xia J (2014) Estimation of near-surface shear-wave velocities and quality factors using multichannel analysis of surface-wave methods. J Appl Geophys 103:140–151. https://doi.org/10.1016/j.jappgeo.2014.01.016

Xia J, Miller RD, Park CB, Hunter JA, Harris JB, Ivanov J (2002) Comparing shear-wave velocity profiles inverted from multichannel surface wave with borehole measurements. Soil Dyn Earthq Eng 22(3):181–190. https://doi.org/10.1016/S0267-7261(02)00008-8

Zhang SX, Chan LS, Xia J (2004) The selection of field acquisition parameters for dispersion images from multichannel surface wave data. Pure Appl Geophys 161(1):185–201. https://doi.org/10.1007/s00024-003-2428-7

Chapter 6
Experimental Study of All-Steel Buckling-Restrained Braces Under Cyclic Loading

Ahmad Fayeq Ghowsi and Dipti Ranjan Sahoo

Abstract Buckling-restrained braces (BRBs) are the type of braces which capable of yielding in tension and compression under cyclic loading. BRBs provide the nearly symmetrical hysteretic response under cyclic loading and higher energy dissipation. Though the all-steel BRBs are considered as cost-effective and lightweight, the main parameters influence their cyclic performance are the flexibility of steel restraining elements, friction between core the restrainers, and interlocking mechanism. In this study, the cyclic performance of all-steel BRBs (ABRB) with angle restrainers has been investigated experimentally. Two reduced-scale ABRB specimens with and without welded stiffeners are tested under cyclic displacements in accordance with AISC 341-10 (2010) provisions Both specimens are subjected to axial strain of 3.5%. The main parameters studied are hysteretic response, energy dissipation response, and displacement ductility. A finite element model has also been developed to predict the cyclic response of ABRB specimens and to compare the experimental results.

Keywords Buckling-restrained braces · Cyclic loading · Finite element analyses · Hysteretic energy · Experiment

6.1 Introduction

Conventional steel braced frames (CBFs) used as the lateral load-resisting systems have high stiffness with less lateral displacement under earthquake and wind loadings. However, the buckling failure of steel braces can limit their ductility and energy dissipation potential. Buckling-restrained brace (BRB) is becoming popular because of the high ductility, symmetrical hysteretic behaviour, and good energy dissipation. A BRB exhibit significant inelastic deformation while yielding in both tension as well as compression. The restraining mechanism prevents the global

---

A. F. Ghowsi (✉) · D. R. Sahoo
Department of Civil Engineering, Indian Institute of Technology Delhi, New Delhi, India

© Springer International Publishing AG, part of Springer Nature 2019
R. Rupakhety et al. (eds.), *Proceedings of the International Conference on Earthquake Engineering and Structural Dynamics*, Geotechnical, Geological and Earthquake Engineering 47, https://doi.org/10.1007/978-3-319-78187-7_6

buckling of inner core undergoing into the higher mode of local buckling (Wu et al. 2014; Kersting et al. 2015). This restraining mechanism results in the superior seismic performance of buckling-restrained braced frames (BRBFs) by developing the symmetrical hysteretic behaviour in BRBs under tension and compression and providing significant energy dissipation under earthquakes Clark et al. (1999).

The typical BRBs used in USA consisted of hollow square, circular, and rectangular steel sections filled with cement mortar (Kim et al. 2015). Other BRBs developed till date consisted of all-steel restraining elements with welded or bolted connections (Xie 2005; Tremblay et al. 2006; Usami et al. 2008; Palazzo et al. 2009; Chao and Chen 2009, Chou and Chen 2010; Takeuchi et al. 2010; Corte et al. 2015; Wu and Mei 2015; Deng et al. 2015; Hoveidae and Rafezy 2015; Metelli et al. 2016; Khoo et al. 2016; Shen et al. 2016; Midorikawa et al. 2016; Chen et al. 2016). All-steel BRBs can be made lighter compares to the conventional BRBs (Dusicka and Tinker, 2013). The big challenge in steel BRB is minimization of friction between the core and restrainers parts. The friction can cause an increase in the compressive overstrength of BRBs. Friction can also transfer the load to the casing parts of BRBs which may cause of global buckling in the BRBs (Judd et al. 2014). More use of welding can cause more friction between the core and frictions interfaces (Eryasar and Topkaya 2010; Eryaşar 2009). The global buckling of BRB may occur due to the weak flexural stiffness of restraint members (Hoveidae and Rafezy 2012). Further, the position of stoppers, interlocking, and end rotation may influence the hysteretic response and energy dissipation potential of all-steel BRBs. Since a steel BRB system is light in weight, minimizing the friction and end rotation without interlocking has not been explored till date. Hence, more experimental studies are required.

6.2 Scope and Objectives

In this study, an experimental study has conducted on all-steel angle BRB (ABRB). ABRB is an all steel BRB which uses four angle restrained the core to resist global flexural as well as rotation the brace and forced the core to the higher mode buckling. In this study, a cyclic test of two ABRB conducted to investigate the hysteretic behaviour under both axil and rotational demands. The sub-assemblages of ABRB specimens are subjected under quasi-static displacement control loading in accordance with AISC 341-10 (2010) provisions. A finite element model has been developed and validated using the experimental results.

6.3 Design of Test Specimens

Two ABRB specimens of same dimensions with two different stoppers detailing are considered in this study. Figure 6.1 shows the component of ABRB in which the BRB core of rectangular core cross-section and cruciform shape of ends and transition zones are used. The core cross-section of ABRB specimens is 40 mm × 8 mm and the length of core is 1000 mm. Figure 6.2 shows the combined assembly of ABRB specimens with cross-section details. Eight millimetre diameter bolts at a spacing of 30 mm on centres are used along the length of specimens. The

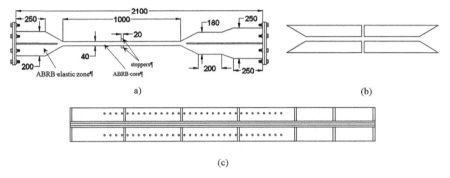

Fig. 6.1 Various components of ABRB specimen. (**a**) ABRB core dimension (mm), (**b**) Orientation of gap control plates, (**c**) Angle restrainers

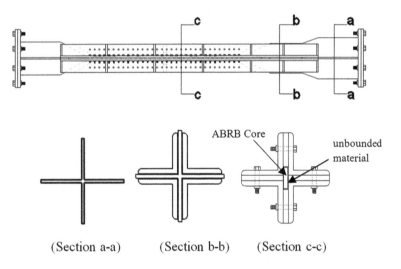

Fig. 6.2 Details of ABRB assembly

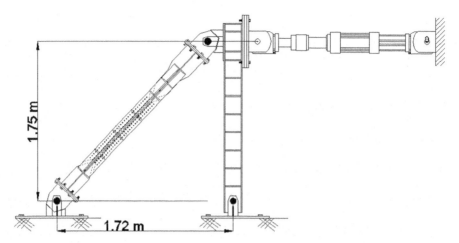

Fig. 6.3 ABRB test set-up

restrainer angles are attached to each other using bolted connections. A spacer (gap control) plate is placed between every two angles to limit the gap between the core and restrained parts. The cruciform shaped end regions are designed to remain elastic even at the ultimate load of the core. These angles are extended to cover almost ends portions to prevent the possible end rotations. Additional stiffening plates are welded to the angle restrainer. To minimize the friction, the core plate is wrapped with a 1 mm thick Polytetrafluoroethylene (PTFE) sheets.

Figure 6.3 shows the test set-up and overall ABRB assembly subjected under cyclic loading. The total length between the work point to work point of braces is 2.45 m, with a column high 1.75 m. The brace is fixed at both ends and not allowed to rotate. The bottom column and bottom of braced end is connected to a plate which is fastened to the rigid floor. A servo-controlled hydraulic actuator of 500 kN capacity is used to apply the cyclic loading to the ABRB specimens.

6.3.1 Material Testing

A material of grade Fe410 with specified yield stress of 250 MPa is used as the steel core of the specimens. Coupon tests are carried out to determine the tensile stress-strain characteristics of the plates. Figure 6.4 shows the coupon test results. The measured yield and ultimate stress values are 269 MPa and 397 MPa, respectively.

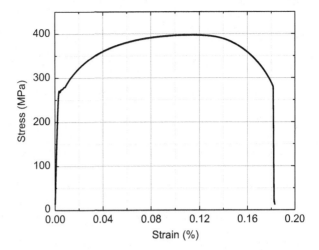

Fig. 6.4 Tensile stress-strain characteristics of core plate

Table 6.1 Assumed design parameters

Description	Value
Plastic Poisson's ratio (γ)	0.5
Expected maximum core strain (ε_c)	0.04
Width of ABRB core (w_c)	40 mm
Core thickness (t_c)	8 mm
Gap between core and restrainers on strong side (S_s)	2 mm
Gap between core and restrainers on weak side (S_w)	2 mm
Center-to-center distance between two bolts (L_c)	100 mm
Center-to-center distance between two continues bolts (L_w)	100 mm
Material overstrength (R_y)	1.076
Strain-hardening adjustment factor (ω)	1.54
Compression strength adjustment factor (β)	1.04

6.3.2 Design of Strength of Specimens

Normal and shear design load on bolts are computed based on recommendation of Wu et al. (2014). Bolts used are of high strength of grade 8.8 with 8 mm diameter. The maximum compressive axial strength in steel BRB for the purpose of transferring load trough the higher mode buckling can defined as follows:

$$P_{max} = R_y \omega \beta P_y \tag{6.1}$$

Where, R_y is the overstrength factor, ω is the strain-hardening adjustment factor, β is the compression strength adjustment factor, and P_y is the axial yield strength of steel. Table 6.1 summarizes the assumed design parameters used in this study. The core area of ABRB is 320 mm^2 and the over-strength factor is computed as 1.076.

Maximum load can be carrying by the ABRB at the ultimate level, $P_{max} = 137.9\ kN$.

6.3.2.1 Design of Bolts

The wave length and the wave shape within BRB core produce a lateral force over the restrainers which should be considered in the design on the weak and strong axis. The maximum tensile load demand over the bolts through the higher mode buckling on the strong axis surface of the core is given by (Wu et al. 2014).

$$N_S = \frac{4S_S + 2\gamma\varepsilon_c w_c}{L_c} P_{max} = 13.2\ kN \tag{6.2}$$

Maximum tensile strength (F_t) of 8 mm bolt with using bolt stress P_t as 560 MPa is computed as follows:

$$F_t = P_t A_{bolt} = 28.14\ kN > 13.2\ kN$$

The load from the higher mode of buckling of BRB acting on the weak surface axis over the bolts, which can cause shear failure, is computed as follows:

$$N_w = \frac{4S_w + 2\gamma\varepsilon_c t_c}{L_w} P_{max} = 11.5\ kN \tag{6.3}$$

Maximum shear strength (V) of the bolt using bolt strength P_v as 375 MPa is given by

$$V = P_v A_{bolt} = 18.85\ kN > 11.5\ kN$$

6.3.2.2 ABRB Core Design Strength

The design axial strength of ABRB is given by.

$$\phi P = \phi R_y \omega \beta P_y \tag{6.4}$$

Where, ϕ is taken 0.9 as design strength factor.
Accordingly, the design axial strength is computed as 124 kN.

6.4 Experimental Results

Figure 6.5a shows the test set-up used in the cyclic testing of ABRB specimens. LVDTs are used for measuring the axial deformation as well as lateral displacement of specimens. The horizontal load from the actuator has been inclined in the axial direction of brace and assumed that the angle as constant angle. As shown in Fig. 6.5b, the selected loading protocol consisted of two cycles at each deformation of axial strain of 0.5, 1, 1.5, 2, 2.5, 3, and 3.5% with an increment of 0.5% after every two cycles. The strain rate of loading is 0.25 mm/sec for the first four cycles and 0.5 mm/sec for the last 5 cycles.

ABRB with welded stoppers performed well for the loading cycles up to 3% of axial strain. The stable hysteretic response is noted in the first cycles of 3.5% axial strain as shown in Fig. 6.6a. Tensile fracture on the second cycle of 3.5% as shown in

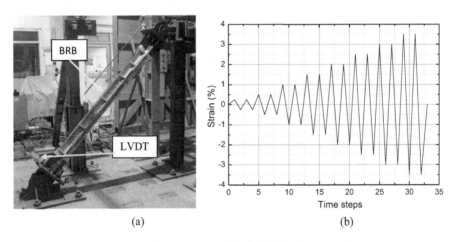

Fig. 6.5 (a) ABRB assembly with instrumentations and (b) loading protocol

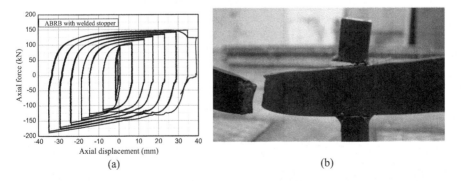

Fig. 6.6 Test results of ABRB without stopper (a) Hysteretic response, (b) core fracture

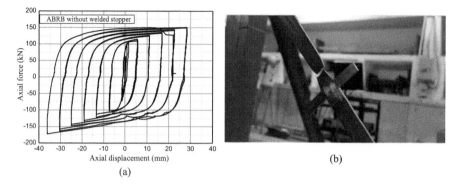

Fig. 6.7 Test results of ABRB without stopper (**a**) Hysteretic response, (**b**) core fracture

Fig. 6.6b. Load control test has been done for the purpose of effective stiffness of the ABRB. The effective stiffness of ABRB is calculated as follows: (FEMA-356 2000)

$$k_{eff} = \frac{|F^-| + |F^+|}{\Delta^- + \Delta^+} \tag{6.5}$$

Where, K_{eff} is the effective stiffness, F^+ is the maximum in positive load, F^- is the minimum in negative of compressive load, Δ^+ is the maximum deflection in elastic range, and Δ^- is the minimum deflection of brace in elastic range. For ABRB with wilding stopper, the effective stiffness is 45.1 kN/mm.

Figure 6.7 shows the hysteretic response and fracture of brace core. ABRB without welded stopper performed well up to 3% of axial strain. Though stable hysteretic is noted in compression cycle of 3.5% axial strain, the brace core fracture is noted in the tension cycle. The effective stiffness of ABRB without welded stoppers is noted as 43.6 kN/mm.

The theoretical elastic stiffness of ABRB can be calculated as follows:

$$k_1^c = \frac{1}{\frac{1}{k_{yi}} + \sum \frac{1}{k_{tr}} + \sum \frac{1}{k_{con}}} \tag{6.6}$$

Where, $k_{yi} = \frac{EA_{yi}}{L_{yi}}$ is elastic stiffness of ABRB core portion, $k_{con} = \frac{EA_{con}}{L_{con}}$ is stiffness of end portions of ABRB, and $k_{tr} = \frac{EA_{tr}}{L_{tr}}$ is the stiffness of transition portions. The overall theoretical stiffness is 56.32 kN/mm. A comparison of theoretically computed and experimentally observed values of elastic stiffness of ABRB specimens has been summarized in Table 6.2. Experimental values are found to be about 20% smaller than the theoretical ones. The difference between the theoretical and experimental stiffness values may be due to the non-inclusion of the influence of gussets plates, bolts, and others components of ABRB specimens.

Table 6.2 Comparison of elastic stiffness

Specimen name	Theoretical stiffness K_1^c (kN/mm)	Computed stiffness, K_1^e (kN/mm)	Difference (%)
ABRB with stopper	55.6	45.1	−19.1
ABRB without stoppers		43.6	−22.2

Table 6.3 Comparison of strength adjustment parameters

Specimen	ϕP_n(kN)	T_{max}(kN)	C_{max}(kN)	β	ω	$\beta\omega_{max}$
ABRB with stopper	124	148	188	1.27	1.49	1.89
ABRB without stoppers	124	151	172	1.14	1.38	1.57

Post-yield stiffness of ABRB can be expressed as follows:

$$K_2^c = \alpha K_1^c \tag{6.7}$$

Where, K_2^c is the post-yield (secondary) stiffness, K_1^c is the elastic stiffness, and α is the ratio of post-yield stiffness ratio.

The value of α is computed as 2.79% for ABRB specimen with welded stoppers and 2.98% for ABRB specimens without stoppers. These values are found to be 10–15% higher than the conventionally assumed values of 2% for BRBs. The maximum tensile force, T_{max} and the maximum compressive force, C_{max} are also determined from the experimental results. These axial strengths are computed corresponding to the peak deformed position of specimens. The compressive adjustment factor, β can expressed as follows:

$$\beta = \frac{C_{peak}}{T_{peak}} \tag{6.8}$$

Where, C_{peak} is the peak compressive and T_{peak} is the peak tension force after first significant yield. The strength hardening adjustment factor of ω is given by.

$$\omega = \frac{T_{peak}}{F_{ysc}A_{ysc}} \tag{6.9}$$

Where F_{ysc} is the yield stress of core plate and calculated based on coupon test result, and A_{ysc} is the brace area of ABRB core. The hardening adjustment factor and compressive adjustment factor has been compared in Table 6.3.

6.5 Finite Element Modeling

Finite element ABRB models are developed in ABAQUS CAE (2010) platform to predict their cyclic response under two different stopper consideration. Steel cores of BRB are modelled using eight-node solid (C3D8R) elements having reduced-integration technique. Combined isotropic and kinematic strain hardening properties is considered for the non-linear material modelling elastic and plastic material to account for their cyclic hardening behaviour. The stoppers are modelled as the same as the core segments on both sides of weak surfaces of ABRB models, which are provided at centre of the core to keep safe from creeping of the core in the casing. One millimetre gaps are provided on each side. The restrainers are also modelled as eight-node solid (C3D8R) elements with reduced-integration technique and with elastic properties. Surface-to-surface contact with tangential behaviour having friction coefficient of 0.1, normal behaviour as hard contact and the tangential behaviour used with penalty friction formulation. In normal behaviour, the maximum stiffness value is used as default with stiffness scale factor of one, Initial/Final stiffness ratio as 0.01, upper quadratic limit scale factor as 0.03, lower quadratic limit ratio as 0.3333, and the zero clearance at which contact pressure is zero. The restrainer parts are connected by tie constraints with each other and considered to remain elastically during analysis.

The material yield stress at zero plastic strain is taken as 269 MPa as found from excremental results for ABRB with weld stoppers and without stoppers. The combined hardening parameters used in the ABAQUS modelling are: $C1 = 10$ GPa, $\gamma 1 = 48$, $Q_\infty = 45$ MPa and $b = 4$ (Korzekwa and Tremblay 2009). An initial imperfection of total BRB length/1000 is assigned based on the first mode of buckling scaling. The same loading protocol as used in experiment has also been applied to the ABRB models. Ductile damage has been considered for the prediction of fracture of brace cores in the analytical model. Figure 6.8 shows finite element assembly of the ABRB specimens and their final deformed states.

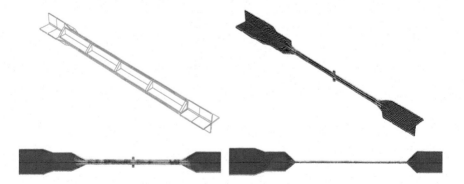

Fig. 6.8 FEM assembly of ABRB

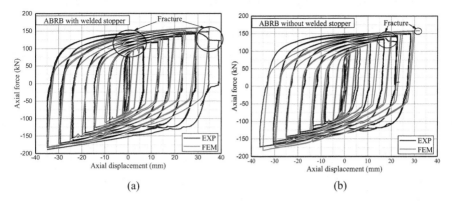

Fig. 6.9 Comparison of hysteretic response of ABRB model (**a**) with stopper and (**b**) without stopper

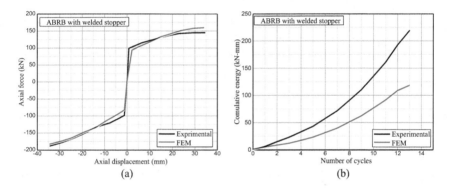

Fig. 6.10 Comparison of numerical with experimental ABRB with welding (**a**) backbone curve (**b**) energy

Figure 6.9 shows the comparison of predicted hysteretic response with the test results. The predicted hysteresis loops matched very well with the test results. The peak strengths at each axial strain levels also matched well with test results. The finite element models predicted the brace core fracture at the same cycle of 3% axial strain in tension.

Figure 6.10 shows the comparison of backbone curves of hysteretic response and energy dissipation response of ABRB with weld stoppers. The predicted backbone curve matched reasonable well with the experimental results though some minor deviation in noted in the tension part. This resulted some difference in the energy dissipation response. The similar comparison for ABRB specimen without stopper has been shown in Fig. 6.11. A better match is obtained between the predicted and experimental results in terms of backbone curves and energy dissipation. ABRB models with stoppers showed better axial resistance and energy dissipation response.

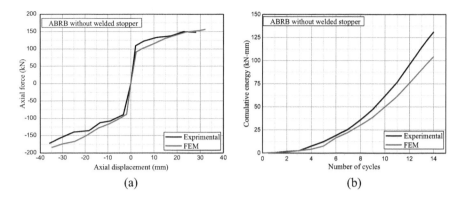

Fig. 6.11 compression of numerical with experimental ABRB without stopper (a) backbone curve (b) energy

Table 6.4 Compression of displacement ductility

Brace type	Maximum ductility	Cumulative ductility
ABRB with weld stoppers	38.28	779
ABRB without weld stoppers	34.99	647

In addition, ABRB specimen with stopper also exhibited the maximum ductility and cumulative displacement ductility (Table 6.4).

6.6 Conclusions

Based on experimental and analytical studies, the following conclusions can be drawn.

- The proposed all-steel BRB exhibited the excellent axial strength, hysteretic response, and energy dissipation. As compared to conventional BRB, ABRBs are light in weight.
- The use of welded stoppers in the brace core significantly influence the overall cyclic response of ABRB specimens. The welded stoppers help in enhancing the deformability prior to their fracture.
- ABARB specimens with and without stoppers exhibited stable hysteretic energy till 3% core strain. The core fracture is noted at 3.5% axial strain level for the ABRB specimens.
- The post-yield stiffness ratio of ABRB specimen is found about 2.8%. The maximum displacement ductility of about 40 can be obtained in the ABRB specimens.
- The proposed finite element model successfully captured the hysteretic response and fracture behaviour of ABRB specimens.

References

ABAQUS 2010. ABAQUS/CAE user's manual. Version 6.10, Dassault Systèmes
AISC (2010) Seismic provisions for structural steel buildings (ANSI/AISC 341-10). American Institute for Steel Construction, Chicago, Illinois
Chao CC, Chen SY (2009) Subassemblage tests and finite element analyses of sandwiched buckling-restrained braces with a replaceable core. In: 6th International Conference on behaviour of steel structures in seismic areas (STESSA 2009), Philadelphia, The Netherlands
Chen Q, Wang CL, Meng SP, Zeng B (2016) Effect of the unbonding materials on the mechanic behavior of all-steel buckling restrained braces. Eng Struct 111(1):478–493
Chou CC, Chen SY (2010) Subassemblage tests and finite element analyses of sandwiched buckling-restrained braces. Eng Struct 32:2108–2121
Clark P, Aiken I, Kasai K, Ko E, Kimura I (1999) Design procedures for buildings incorporating hysteretic damping devices. In: 68th Annual convention of structural engineers association of California (SEAOC), Santa Barbara, CA, pp 229–254
Corte GD, D'Aniello M, Landolfo R (2015) Field testing of all-steel buckling-restrained braces applied to a damaged reinforced concrete building. ASCE J Struct Eng 141(1):D4014004 1–11
Deng K, Pan P, Nie X, Xu X, Feng P, Ye L (2015) Study of GFRP steel buckling-restraint braces. ASCE J Comp Constr 19(6):04015009(8)
Dusicka P, Tinker J (2013) Global restraint in ultra-lightweight buckling-restrained braces. J Constr Steel Res 17(1):139–150
Eryaşar M (2009) Experimental and numerical investigation of buckling-restrained braces. Master's thesis, Middle East Technical University, Ankara, Turkey
Eryasar M, Topkaya C (2010) An experimental study on steel-encased buckling restrained brace hysteretic damper. Earthq Eng Struct Dyn 39:561–581
Hoveidae N, Rafezy B (2012) Overall buckling behavior of all-steel buckling restrained braces. J Constr Steel Res 79:151–158
Hoveidae N, Rafezy B (2015) Local buckling behavior of core plate in all-steel buckling restrained braces. Int J Steel Struct 15(2):249–260
Judd JP, Eatherton MR, Charney FA, Phillips AR (2014) Cyclic testing of all-steel web-restrained buckling-restrained brace (WRB) subassemblages-Part II. Report No. CE/ VPI-ST-14/05. Department of Civil and Environmental Engineering, Virginia Tech, Blacksburg, Virginia
Kersting RA, Fahnestock LA, López WA (2015) Seismic design of steel buckling-restrained braced frames: a guide for practicing engineers. NIST GCR 15-917-34, NEHRP Seismic Design Technical Brief No. 11, National Institute of Standards and Technology (NIST), Gaithersburg, Maryland
Khoo H, Tsai K, Tsai C, Tsai CY, Wang K (2016) Bidirectional substructure pseudo-dynamic tests and analysis of a full-scale two-story buckling-restrained braced frame. Earthq Eng Struct Dyn 45:1085–1107
Kim DH, Lee CH, Ju YK, Kim SD (2015) Subassemblage test of buckling-restrained braces with H-shaped steel core. Struct Des Tall Spec Build 24(4):243–256
Korzekwa A, Tremblay R (2009) Numerical simulation of the cyclic inelastic behavior of buckling restrained braces. In: 6th International Conference on behaviour of steel structures in seismic areas (STESSA 2009), Philadelphia, The Netherlands
Metelli G, Bregoli G, Genna F (2016) Experimental study on the lateral thrust generated by core buckling in bolted-BRBs. J Constr Steel Res 122:409–420
Midorikawa M, Hishida S, Iwata M, Okazaki T, Asari T (2016) Bending deformation of the steel core of buckling-restrained braces. In: ASCE Geotechnical and structural engineering congress, Phoenix, Arizona, pp 613–623
Palazzo G, López-Almansa F, Cahís X, Crisafulli F (2009) A low-tech dissipative buckling restrained brace: design, analysis, production and testing. Eng Struct 31:2152–2161
Shen J, Seker O, Sutchiewcharn N, Akbas B (2016) Cyclic behavior of buckling-controlled braces. J Constr Steel Res 121:110–125

Takeuchi T, Hajjar JF, Matsui R, Nishimoto K, Aiken ID (2010) Local buckling restraint condition for core plates in buckling-restrained races. J Constr Steel Res 66:139–149

Tremblay R, Bolduc P, Neville R, DeVall R (2006) Seismic testing and performance of buckling-restrained bracing systems. Can J Civ Eng 33(2):183–198

Usami T, Ge H, Luo X (2008) Overall buckling prevention condition of buckling restrained braces as a structural control damper. In: 14th world conference on earthquake engineering, Beijing, China

Wu B, Mei Y (2015) Buckling mechanism of steel core of buckling-restrained braces. J Constr Steel Res 107:61–69

Wu AC, Lin PC, Tsai KC (2014) High-mode buckling responses of buckling-restrained brace core plates. Earthq Eng Struct Dyn 43:375–393

Xie Q (2005) State of the art of buckling-restrained braces in Asia. J Constr Steel Res 61(6):727–748

Chapter 7
Seismic Design Procedure for Staggered Steel Plate Shear Wall

Abhishek Verma and Dipti Ranjan Sahoo

Abstract Steel plate shear wall (SPSW) is used as a lateral force-resisting system capable of dissipating the input seismic energy through metallic hysteresis. High axial force demand on the vertical boundary elements (VBEs) results in relatively heavier sections at the lower story levels of SPSWs. This lead to a non-uniform interstory drift distribution over the height, which may exceed the acceptable drift limits. Staggering of web plates can reduce the axial force demands in VBEs, ensuring better drift distribution and improved energy dissipation. This study aims at developing a design methodology for the staggered SPSW systems having the similar over-strength as the conventional counterpart to ensure an acceptable yield mechanism. A linear static analysis procedure is used to predict the axial and flexural demand in VBEs. The effectiveness of the proposed procedure is validated through the non-linear analysis for a six-story SPSW.

Keywords Earthquake · Nonlinear static analysis · Seismic design · Steel plate shear wall · Steel structures

7.1 Introduction

Steel Plate Shear Wall (SPSW) can be used as an effective lateral force-resisting system, which has excellent energy dissipating capabilities and high initial lateral stiffness. It consists of steel plates (webs) surrounded by boundary elements. Input seismic energy is dissipated by the yielding of webs in tension and the formation of flexural plastic hinges at the ends of horizontal boundary elements (HBEs). ANSI/AISC 341-10 (AISC 2010) recommends the design of SPSWs by assuming the entire lateral load to be resisted by the web plates only. The boundary elements of this system are designed as per the capacity design approach. HBEs are designed to

---

A. Verma (✉) · D. R. Sahoo
Department of Civil Engineering, Indian Institute of Technology Delhi, New Delhi, India
e-mail: drsahoo@civil.iitd.ac.in

restrict the formation of flexural plastic hinges at both ends, whereas the vertical boundary elements (VBEs) are considered to be essentially elastic. The stiffer and elastic VBEs can result in the better drift distribution and reduce the possibility of soft story mechanisms in SPSWs (Qu et al. 2013; Verma and Sahoo 2016a).

VBEs of a high-rise SPSW experience high overturning forces. The magnitude of VBE axial forces is significantly higher in case of narrow (lower aspect ratio) web plates. This may result in the drift concentration or even the formation of soft story mechanisms in the stories with relatively lower inter-story stiffness. Horizontal restrainers in VBEs are found to be effective in reducing the shear and bending forces (Lin et al. 2010). Similarly, the axial force demand in VBEs can be reduced by using coupled SPSWs (Borello and Fahnestock 2012). Coupling increases the lever arm resisting the overturning moment. This increase in the lever arm of the coupled forces can also be achieved by increasing the width of SPSWs (Verma and Sahoo 2016b). ANSI/AISC 341-10 (AISC 2010) recommends the aspect ratio of web plates of SPSWs in the range of 0.8–2.5. However, there are no guidelines available for a designer to choose a particular value of aspect ratio in order to achieve the better seismic performance and economy.

Figure 7.2b shows SPSW with staggered arrangement of web plates (henceforth, referred as S-SPSW). Such an arrangement can reduce the axial force demand in VBEs (Verma and Sahoo 2016b). Figure 7.1 shows the comparison of mean inter-story drift response of six-, nine- and twenty-story SPSWs under the design basis earthquake (DBE) level ground motions. The influence of variation in aspect ratios, i.e., narrow (N), medium (M), wide (W) and extra wide (WW) has been considered in these studies. The suffix "-S" is used for the corresponding staggered systems. Staggering of web plates considerably reduces the interstory drift response of medium to high-rise SPSW systems. S-SPSWs exhibit a better drift distribution and a more uniform yielding of components over the height as compared to the conventional SPSWs (Conv-SPSW). Though the S-SPSW systems have shown a promising seismic response, the main challenge is how to design the various components of such systems in order to achieve the desired drift and yielding mechanism. This study proposes a design methodology for an S-SPSW system to achieve these objectives.

7.2 Design Approach

ANSI/AISC 341-10 (AISC 2010) recommends the use of moment-resisting connections between the boundary elements of a conventional SPSW. Number of such connections is doubled if the web plates are staggered in S-SPSW. In a Conv-SPSW system, forces from the web plates connected to the top and the bottom flanges of HBE act in the opposite directions. In an S-SPSW system, HBEs are connected to the web plates either at the top or at the bottom flanges. These HBEs need to resist high flexural and shear demand to avoid the formation of in-span flexural hinges. Thus, an S-SPSW is expected to have heavier HBEs and more number of moment-resisting

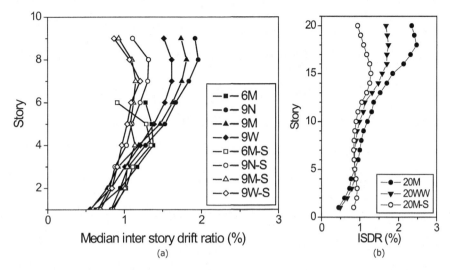

Fig. 7.1 Comparison of interstory drift response of (**a**) six-story, nine-story, and (**b**) twenty-story conventional and staggered SPSW under DBE level ground motions (Verma and Sahoo 2016b)

connections than the corresponding Conv-SPSW. This leads to high overstrength and an uneconomical system.

7.2.1 Overstrength

As per the current design practice, the entire load is assumed to be resisted by the plates only. The boundary elements are designed as per the capacity-based approach. While this approach is appropriate for a Conv-SPSW, it results in a considerably high over-strength for an S-SPSW system. The overstrength is majorly due to two reasons. Firstly, the number of moment-resisting connections between HBE-to-VBEs is twice when compared to a Conv-SPSW. Secondly, as the HBEs in such a system have plates connected to them either at the top or at the bottom flange, they must be designed to resist higher force demand, resulting in further over-strengthening of the system. In a Conv-SPSW, forces exerted by the plate, on any intermediate HBE, balance out each other, whereas it is not the case in S-SPSW. Figure 7.2 shows the forces acting on the boundary elements due to tension in the plates. Such a system would not only be over designed but also be uneconomical and inefficient as few components might not even yield during a design based earthquake (DBE) hazard level. A design approach is proposed in this study which aims to counter the overstrength due to staggering of web plates.

The resistance to lateral force provided by the boundary frame is not considered during the current design. This results in a certain amount of overstrength. For a case where the web plate thicknesses of conventional and staggered SPSW is same, plates

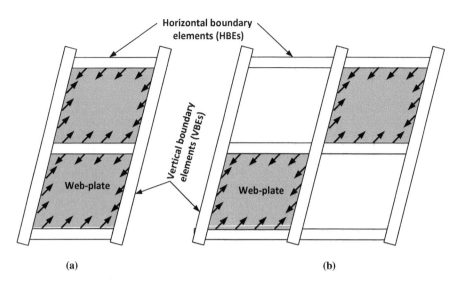

Fig. 7.2 Force exerted by web plate on boundary elements in (**a**) conventional SPSW (**b**) staggered SPSW

would impart the same strength to both the systems. The capacity-based design would lead to frames of different capacities. Thus, total capacity of both the systems would be different. The ratio of the total capacity of S-SPSW to the total capacity of its conventional counterpart would give an approximation of the overstrength due to staggering of web plates. An analytical method was proposed by Verma and Sahoo (2016b) to predict the overstrength due to staggering. The reciprocal of this overstrength, also known as the base shear reduction factor (γ) is calculated using Eq. (7.1). γ is applied to the design base shear. SPSW is then designed for this reduced design base shear following the conventional design procedure.

$$\gamma = \frac{\sum_{i=1}^{n} C_{vi} H_i + \frac{L}{\varphi_b}}{\sum_{i=1}^{n} C_{vi} H_i + \frac{L}{\varphi_b} \left[0.5 + \frac{\sum_{i}^{n} V_i}{V_d} + \frac{F_n}{2V_d} \right]} \quad (7.1)$$

where C_{vi}, H_i and V_i are the vertical distribution factor for seismic force (ASCE 2010), the story height and the design base shear for any *ith* story, respectively. L is the distance between the VBE centre lines, φ_b is the resistance factor for moment capacity of HBEs used in the design, V_d is the design base shear, F_n is the design shear of the top story and n is the number of stories.

7.2.2 Design of Boundary Elements

In Conv-SPSWs, boundary elements are designed to resist the axial, shear and flexural forces to ensure a proper yield mechanism. It is desired that the formation of flexural plastic hinge is restricted to HBE ends only, whereas the columns should remain elastic. The formation of in-span plastic hinges may prevent or restrict the yielding of web plates. This may result in a weak story mechanism. Similarly, elastic columns prevent concentration of story drift and soft story mechanism.

A similar design philosophy is suggested for boundary elements of S-SPSW also. HBEs may be designed to prevent in-span plastic hinges. In Conv-SPSW, web plates are connected to both top and bottom flanges of intermediate HBEs, which pull the HBEs, through tension field in opposite directions. Whereas, HBEs in S-SPSW have plate connected to either the top or the bottom flange, resulting in more flexural and shear demand. Design of HBEs in S-SPSW is thus quite similar to the design of top and bottom HBEs in Conv-SPSW. In Conv-SPSW, due to symmetry, lateral load share is equally distributed between the two columns. S-SPSW have three VBEs against two VBEs in the Conv-SPSW. Thus, load distribution at each level of S-SPSW is not exactly similar as the Conv-SPSW.

A computer-aided linear analysis has been performed to estimate the VBE forces. All the three VBEs of S-SPSW are modelled as elastic line elements and forces are applied as shown in Fig. 7.3. Three nodes at each story level are constrained to have equal displacements in the horizontal direction. Horizontal (ω_{xc}) and vertical (ω_{yc}) components of the distributed load due to tension field action of the web plates are applied to the VBEs. Bending moments equal to the plastic moment capacity of HBEs are applied at each node. Shear forces in the HBEs, resulting from the frame sway action and the tension field action of the web plates, are also applied as axial forces on VBEs. Like Conv-SPSW, the base shear capacity of the S-SPSW can be represented by the following equation.

$$V_{BS} = \frac{\sum_{i=1}^{n} F_{Wi} H_i + \sum_{i=0}^{n} (2M_{ai} + 2M_{bi})}{\sum_{i=1}^{n} C_{vi} H_i} \quad (7.2)$$

where, $F_{Wi} = l\omega_{xb(i+1)} - l\omega_{xbi}$, M_{ai} and M_{bi} are the flexural yield capacity of the left and the right HBEs at any floor level i, respectively and l is the width of the web plate. The base shear force is distributed along the height using the factor C_v for each story. This force at each story is resisted by web plate tension field action and frame sway action. The capacity of web plate at each story is known. Thus, the force to be resisted by the frame (F_{Fi}), at any ith story, is computed using the following equation.

$$F_{Fi} = F_i - F_{Wi} \quad (7.3)$$

where, F_{Wi} is the lateral force capacity of the webs only and F_i is the total lateral load capacity of the S-SPSW system. The axial load in the HBEs of an S-SPSW is considerably less as compared to a Conv-SPSW. Also, the flexural demand of

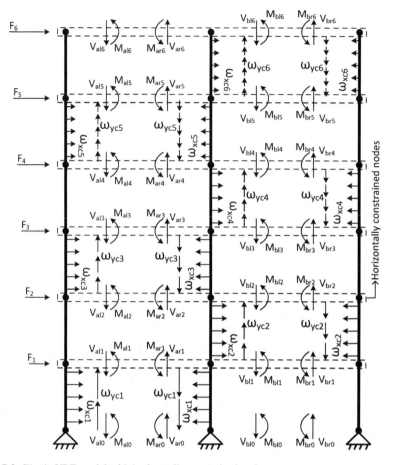

Fig. 7.3 Elastic VBE model with horizontally constrained nodes

HBEs in S-SPSW is high. Thus, the reduction of flexural capacity of HBEs due to axial force is significantly less and may be considered negligible.

7.3 Example Estimation of VBE Design Loads

A six-story S-SPSW system has been considered as the study frame. Nonlinear pushover analysis is performed using the computer software OpenSEES (Mazzoni et al. 2006). The proposed procedure to estimate VBE forces is validated with the results obtained. An example of the proposed procedure to estimate the VBE forces in an S-SPSW is discussed in this section. Firstly, the considered building is described. The base shear for the building is calculated and then reduced by applying the factor γ to calculate the reduced design base shear for S-SPSW. It is then

7 Seismic Design Procedure for Staggered Steel Plate Shear Wall

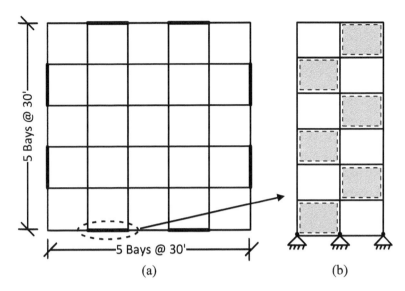

Fig. 7.4 (a) Plan of the 6-story example building (b) S-SPSW for the building

designed to obtain the web sizes and sections for boundary elements. The forces in the VBEs are estimated using the proposed computer aided linear static analysis. For this purpose, computer software SAP2000 is used. These results are then compared with the nonlinear static analysis using OpenSEES (Mazzoni et al. 2006).

7.3.1 Structure Description and S-SPSW Design

Figure 7.4 shows the plan and elevation of the six-story building considered in this study. Height of all the stories is taken as 3.96 m. Site specific parameters, S_s and S_1 are taken as 1.61 g and 0.79 g, respectively. The response modification factor is considered as 7 (ASCE 2010). Seismic weights at the roof and floor levels are assumed as 5730 kN and 5400 kN, respectively. Design base shear for the building is calculated to be 9495 kN. S-SPSW is provided in four bays in each loading direction. The aspect ratio of web plate of S-SPSW is considered as 1.15. Base shear reduction factor (γ) is calculated to be 0.52. Hence, the reduced design base shear for S-SPSW is 4938kN.

The S-SPSW is designed by reducing the design base shear obtained for a Conv-SPSW. Nominal yield strengths of plates and boundary elements are assumed to be 248 MPa and 345 MPa, respectively. Expected yield strength of plates is assumed to be 1.3 times the nominal yield strength. An average value of angle of inclination of tension field (α) with respect to the vertical direction is assumed for all the stories, which is calculated for each story using Eq. (7.4) (AISC 2010). A resistance factor of 0.9 is used for both plates and boundary elements. Like anchor beams in a Conv-

SPSW, HBEs are designed to avoid any in-span plastic hinges. VBEs are designed to remain elastic to avoid soft story mechanism. The procedure used for calculating the VBE demands is discussed in the next section. Figure 7.5 shows the web plate thicknesses and section sizes obtained for the S-SPSW.

$$\alpha = \tan^{-1} \sqrt[4]{\frac{1 + \frac{tL}{2A_C}}{1 + th\left(\frac{1}{A_b} + \frac{h^3}{360I_C L}\right)}} \tag{7.4}$$

where t is the web plate thickness, L is the distance between the VBE center lines, h is the distance between the HBE center lines, A_b is the cross-sectional area of an HBE, A_C is the cross-sectional area of a VBE and I_C is the moment of inertia of a VBE taken perpendicular to the direction of the web plate line.

7.3.2 Calculating VBE Design Loads

In case of Conv-SPSW system, both column resist about the same amount of lateral force. However, it is not the case for S-SPSW system. In the latter case, three VBEs are asymmetrically loaded, leading to an uneven distribution of lateral force among them. The columns experience load due to yielding of web plates, HBE end forces and inertia of mass at each story. An average value of α obtained for all the stories, using Eq. (7.4) is computed as 42.7°. Horizontal and vertical uniformly distributed forces exerted by web plate on the VBEs and HBEs are tabulated in Table 7.1. The beam end reactions are tabulated in Table 7.2. As discussed in the previous section, reduction of moment capacity of the HBEs due to axial forces is not considered because the axial forces are relatively low in case of S-SPSW. Table 7.3 shows the calculations for the lateral forces resisted by the frame at each story level.

7.3.3 Comparison of Results

For validation of the proposed procedure, the results obtained from the linear analysis are compared with the nonlinear analysis results obtained using OpenSEES (Mazzoni et al. 2006). A nonlinear strip model is developed in OpenSEES using center line approach. HBEs and VBEs are modelled using force-based beam-column elements with fiber sections to account for axial force (P) and moment (M) interaction. A fiber section ensures distributed plasticity in the boundary elements. Seventy-two fibers are used to model each section. Diagonal strips are modelled as tension only truss elements using *hysteretic* material of OpenSEES library module. *Steel02* is assigned as the material for the boundary elements. Both the materials are assumed

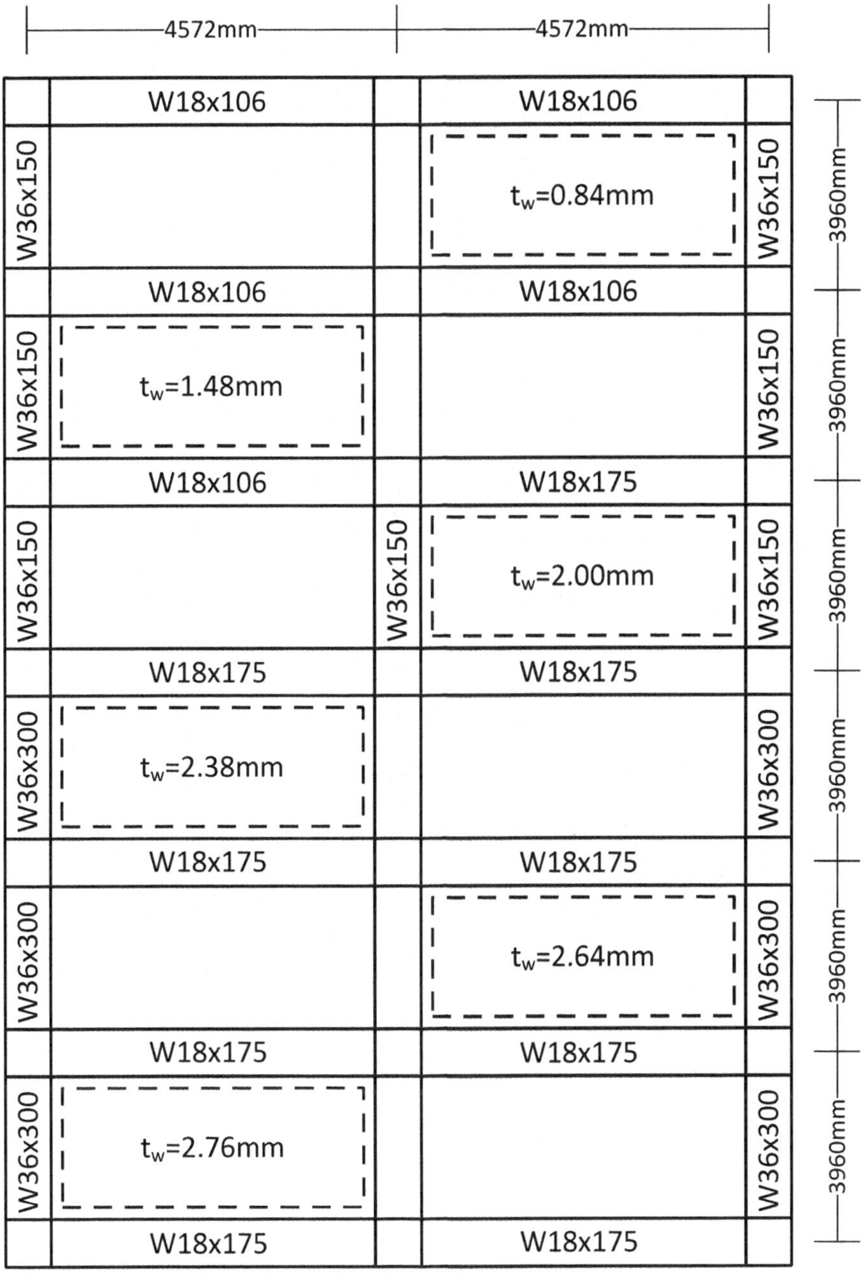

	4572mm		4572mm		
	W18×106		W18×106		
W36×150			$t_w=0.84$mm	W36×150	3960mm
	W18×106		W18×106		
W36×150	$t_w=1.48$mm			W36×150	3960mm
	W18×106		W18×175		
W36×150		W36×150	$t_w=2.00$mm	W36×150	3960mm
	W18×175		W18×175		
W36×300	$t_w=2.38$mm			W36×300	3960mm
	W18×175		W18×175		
W36×300			$t_w=2.64$mm	W36×300	3960mm
	W18×175		W18×175		
W36×300	$t_w=2.76$mm			W36×300	3960mm
	W18×175		W18×175		

Fig. 7.5 Example S-SPSW

Table 7.1 Distributed forces on the boundary elements due to plate yielding

Story level	W_{xc} (kN/m)	W_{yc} (kN/m)	W_{xb} (kN/m)	W_{yb} (kN/m)
6	124.5	135.1	135.1	146.5
5	219.4	238.0	238.0	258.1
4	296.5	321.6	321.6	348.8
3	352.9	382.7	382.7	415.1
2	391.4	424.5	424.5	460.4
1	409.2	443.8	443.8	481.3

Table 7.2 Forces applied by HBEs on VBEs

Story level	M_{al} (kN-m)	M_{ar} (kN-m)	M_{bl} (kN-m)	M_{br} (kN-m)	V_{al} (kN-m)	V_{ar} (kN-m)	V_{bl} (kN-m)	V_{br} (kN-m)
6	13,985	13,985	13,985	13,985	568	568	232	902
5	13,985	13,985	13,985	13,985	−26	1154	902	232
4	13,985	13,985	24,200	24,200	1154	−26	181	1776
3	24,200	24,200	24,200	24,200	28	1926	1776	181
2	24,200	24,200	24,200	24,200	1926	28	−77	2028
1	24,200	24,200	24,200	24,200	−126	2075	2028	−77
0	24,200	24,200	24,200	24,200	2075	−126	984	984

to have no strain hardening beyond yield point. An average value of $\alpha = 42°$ is assumed and strips are inclined at the same for all the stories. Yield strengths of 345 MPa and 323 MPa are assigned to the boundary elements and the web plates, respectively.

Figure 7.6 show the comparison of the bending moment and shear force of the three columns obtained from the linear and nonlinear analyses. Figure 7.7 shows the axial demands for the same. Bending moment diagram for exterior columns is linear at alternate stories, where there is no web plate attached to them. Similarly, constant shear and axial forces are also observed in the exterior columns at these stories. The axial force demand of the interior column is considerably less than the exterior columns. The maximum value of axial force in the exterior columns is about 8500kN, whereas it is 1400kN for the middle column.

It should be noted that the graph for axial and shear forces obtained from nonlinear analysis are step like because of the strip model approach used. The results show a good match for axial and bending forces, but discrepancies can be observed in case of shear forces of exterior columns. The discrepancies are almost equal and opposite in nature. Despite the minor mismatch, the proposed procedure seems appropriate for the estimation of forces in the VBEs of S-SPSW.

7 Seismic Design Procedure for Staggered Steel Plate Shear Wall

Table 7.3 Calculation of lateral capacity of frame at each story

Story Level	Height from ground, H (m)	Force resisted by web, F_W (kN)	Moment of force resisted by web (kN-m)	Sum of beam end moments (kN-m)	$C_V \times H$ (m)	Base shear capacity (kN)	Story force, F (kN)	Force resisted by frame, F_F (kN)
6	23.77	618	14,682	5197	23.77	5114	1548	930
5	19.81	471	9322	5197	19.81		1195	724
4	15.85	382	6059	7095	15.85		953	571
3	11.89	279	3321	8993	11.89		712	433
2	7.92	191	1515	8993	7.92		472	281
1	3.96	88	350	8993	3.96		234	146
0	0.00	–	–	8993	–			

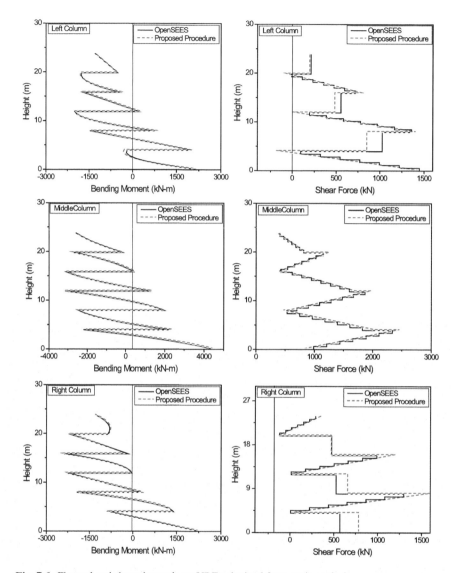

Fig. 7.6 Flexural and shear demands on VBEs obtained from static analysis

Fig. 7.7 Axial demands on VBEs obtained from static analysis

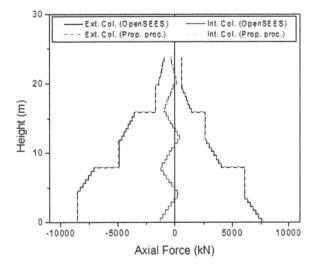

7.4 Conclusions

Staggering of web plates of a steel plate shear wall system (SPSW) reduces axial demand in the VBEs. This improves drift distribution along the height and improve system performance. The present study focusses on the design methodology of such staggered steel plate shear wall systems. The following conclusions can be made from the study.

- A design methodology is proposed in which aims to design an S-SPSW with an overstrength similar to its conventional counterpart. For this, the design base shear is reduced by a factor γ, termed as base shear reduction factor. Web plates are designed to resist the reduced base shear. HBEs and VBEs are the designed as per capacity design approach.
- The axial forces in HBEs of a staggered SPSW are low and the flexural demands are high when compared to a conventional SPSW. Reduction in flexural capacity of HBEs in presence of axial loads is less than 3.5% for the 6-story staggered SPSW considered in the study. Thus, this reduction in Flexural capacity of HBEs can be neglected for design purpose.
- For the estimation of forces in VBEs, a procedure is proposed using a computer aided linear static analysis.
- The VBE forces obtained using the proposed procedure are compared with the results of nonlinear static analysis performed using OpenSEES. The proposed procedure does a good estimate of the VBE forces.

References

AISC (2010) Seismic provisions for structural steel buildings. American Institute of Steel Construction, Chicago, ANSI/AISC 341-10, Chicago

ASCE (2010) Minimum design loads for buildings and other structures. American Society of Civil Engineers, ASCE/SEI 7–10

Borello DJ, Fahnestock LA (2012) Behavior and mechanisms of steel plate shear walls with coupling. J Construct Steel Res 74:8–16

Lin CH, Tsai KC, Qu B, Bruneau M (2010) Sub-structural pseudo-dynamic performance of two full-scale two-story steel plate shear walls. J Construct Steel Res 66(12):1467–1482

Mazzoni S, McKenna F, Scott MH, Fenves GL (2006) Open system for earthquake engineering simulation user command-language manual -Version 1.7.3. Pacific Earthquake Engineering Research Center, Berkeley, CA

Qu B, Guo X, Pollino M, Chi H (2013) Effect of column stiffness on drift concentration in steel plate shear walls. J Construct Steel Res 83:105–116

Verma A, Sahoo DR (2016a) Estimation of lateral force contribution of boundary elements in steel plate shear wall systems. Earthq Eng Struct Dyn:657–675

Verma A, Sahoo DR (2016b) Evaluation of seismic performance of steel plate shear walls arranged in staggered configuration. In: Proceedings of structural engineering convention, Chennai, India, pp 328–333

Chapter 8
Passive-Hybrid System of Base-Isolated Bridge with Tuned Mass Absorbers

Said Elias and Vasant Matsagar

Abstract Nowadays, improved versions of earthquake response modification devices are being introduced to maximise efficacy in dynamic vibration abatement in structures. Here, hybrid system has been proposed to be used for earthquake response modification of bridges by combined use of two passive devices: base isolation systems and tuned mass absorbers. The efficacy of the passive-hybrid system is verified by implementing it in a reinforced concrete (RC) bridge subjected to earthquake ground motions. The RC bridge has three continuous spans and supported on two piers in the middle and abutments at the ends. In the developed numerical model, the flexibility of the founding soil has been accounted for. The numerical model is analysed to determine the dynamic response of the bridge equipped with the passive-hybrid system and a comparison is made with the dynamic response determined without installing such systems. Primarily, it is concluded that the passive-hybrid system exhibits significantly improved performance in dynamic response abatement of the bridge. Nonetheless, the founding soil flexibility at the bottom end of the piers influences the efficacy of the tuned mass absorbers provided at the mid-span of the bridge deck because it affects the modal response quantities.

Keywords Bridge · Passive-hybrid system · Foundation flexibility

8.1 Introduction

Bridges are important connecting links in the civic life, serving in day-to-day life as well as during calamities. In fact, in rehabilitation works undertaken especially after earthquakes, bridges help in providing necessary life-saving materials and commodities to the hazard affected areas, therefore bridges are aptly treated as lifeline

---

S. Elias (✉) · V. Matsagar
Department of Civil Engineering, Indian Institute of Technology (IIT) Delhi, New Delhi, India
e-mail: matsagar@civil.iitd.ac.in

© Springer International Publishing AG, part of Springer Nature 2019
R. Rupakhety et al. (eds.), *Proceedings of the International Conference on Earthquake Engineering and Structural Dynamics*, Geotechnical, Geological and Earthquake Engineering 47, https://doi.org/10.1007/978-3-319-78187-7_8

structures. In order to ensure their continual functioning during and immediately after an earthquake, it is essential to incorporate specialised earthquake resistant features in the design of bridges.

Among the earthquake resistant techniques, the use of base isolation systems is quite successful in reducing the forces induced in the bridges (Tongaonkar and Jangid 2003; Jangid 2004). Notably, it was shown by Tongaonkar and Jangid in 2003 that the flexibility at the bridge pier, caused on account of the soil underneath, affects the dynamic behaviour of the bridges considerably. Later, several studies were reported to show effectiveness of the base isolation technology applied in bridges, which includes contributions from Matsagar and Jangid (2006), Dicleli and Buddaram (2007), and Dicleli (2007). Elias and Matsagar (2017a) have reported that matching of the frequencies contained in an earthquake ground motion signal with the modal frequencies of the bridge leads to increased dynamic response. They concluded that effectiveness of structural control using tuned mass damper (TMD) could be improved by including soil-structure interaction (SSI).

Otherwise, tuned mass absorber is also verified to serve as an effective means of abatement of detrimental vibrations in the bridges. Using such tuned mass absorbers was proved effective in the pedestrian bridges by Daniel et al. (2012). Later, for high-speed train bridges Luu et al. (2012) have concluded that the tuned mass absorbers serve as an alternative solution in controlling the vibration response. It was shown by Debnath et al. (2015) that by controlling dynamic response in different modes of a truss bridge, its overall response could be drastically reduced. The robustness of using the tuned mass absorbers in bridges was examined by Miguel et al. (2016). Installing several tuned mass friction dampers on bridges was shown to perform effectively by Pisal and Jangid (2016).

Apart from this, earlier Li (2000, 2002) showed that multiple tuned mass absorbers/ dampers (MTMDs) are effective in attenuating undesirable oscillations of structures under earthquake. For tall and flexible buildings Aly et al. (2011) and Aly (2014a, b) showed that providing the TMDs prove to be useful in mitigating their dynamic response. Dynamic response reduction in different structures achieved by using the tuned mass absorbers was studied by Lu et al. (2011, 2012). More than one tuned mass absorbers were used for achieving improved efficacy in dynamic response reduction for flexible structures (Elias and Matsagar 2014a, b, 2015, 2017a, 2018). They recommended that the dynamic response reduction can be improvised by installing more such tuned mass absorbers while tuning their frequencies with the modal frequencies of the parent structure and distributing the MTMDs spatially along the structure (Elias et al. 2015, 2016, 2017a, b, c). Their newly proposed structural control theory based on the multi-mode dynamic response control was implemented successfully for the tall buildings, chimneys, and alike long period structures. Even in terms of robustness, such multiple tuned mass absorbers prove to be preferred choice as concluded by Gill et al. (2017). Lately, Matin et al. (2014, 2017) have also reported that such tuned mass absorbers help in abating the earthquake-induced forces in the reinforced concrete (RC) bridges. More details on the tuned mass absorbers employed in various structural engineering applications could be found in Elias and Matsagar (2017b).

Both, the base isolation systems and tuned mass absorbers, are categorised under the well-known passive earthquake response reduction devices, which were previously shown to be effective when installed independently. A study on the use of two different passive devices together for abatement of the vibrations induced in the bridges due to earthquakes is therefore deemed essential. In view of this, the present investigation was undertaken with an objective of studying the efficacy of the passive-hybrid system consisting of base isolation system along with the tuned mass absorbers in bridges. Herein, an RC bridge is analysed with different structural control systems to provide protection against earthquake-induced forces, and the dynamic response obtained is compared to that with the RC bridge without any such system in place. Furthermore, the number of tuned mass absorbers employed also increased to examine if it helps in improving their efficacy when used in bridges. The bridges equipped with the different systems are designated as follows:

1. Base-isolated bridge (IB);
2. IB with a single TMD (IB + STMD);
3. IB with two TMDs (IB + 2TMDs); and
4. IB with three TMDs (IB + 3TMDs).

8.2 Theory and Numerical Model of Bridge

The efficacy of using the passive-hybrid system of base-isolated bridge with tuned mass absorbers is investigated by developing a numerical model of the bridge for conducting time domain analysis. For that purpose, an RC bridge having three spans is considered as shown in Fig. 8.1. The numerical model of the RC bridge consists of a number of small discrete beam elements interconnected at nodes. Two translation degrees of freedom are considered at each node; in the longitudinal (x) and transverse (y) directions of the bridge. The material properties assigned to the finite

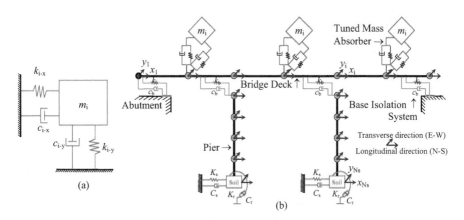

Fig. 8.1 (a) Modelling of the tuned mass absorber; and (b) Numerical model of the RC bridge with passive-hybrid control system installed

elements correspond to the RC members used in the bridges for deck and piers. As seismic response control devices are proposed to be used, which generally have mandate to limit the response of the structure within elastic range, no material and geometric nonlinearities are considered in the present numerical modelling for the bridge. Figure 8.1a shows the representation used for the tuned mass absorber with mass (m_i) connected to the bridge through stiffness, $k_{i\text{-}x}$ and damping, $c_{i\text{-}x}$ in the longitudinal (N-S) direction as well as stiffness, $k_{i\text{-}y}$ and damping, $c_{i\text{-}y}$ in the transverse (E-W) direction. Figure 8.1b shows a numerical model of the RC bridge installed with the passive-hybrid system for vibration response control. The base isolation systems are provided at four locations for the bridge deck supported on the two intermediate piers and abutments at either end. The relative displacement between their upper and lower nodes is referred here as relative bearing displacement. The tuned mass absorbers shown by the mass, m_i are attached to the bridge deck at its mid-span.

The piers are supported on soil mass at their bottom end, which has certain flexibility, appropriately taken in to consideration here in the numerical model of the bridge. Such flexibility of the founding soil has been incorporated in the numerical modelling of the bridge using spring and dashpot elements by assuming their frequency independent parameters. For circular massless foundation, the soil stiffness and damping are respectively represented by the spring and dashpot elements. The relative displacement across the soil spring-dashpot element corresponds to the degrees of freedom of the soil, x_{Ns} and y_{Ns}, respectively in the N-S and E-W directions, as shown in Fig. 8.1b. The parameters of the spring-dashpot elements are determined based on the prevalent type of soil, as recommended by Spyrakos (1990). If K_s denotes soil stiffness of the swaying spring and K_r denotes soil stiffness of the rocking spring, and corresponding damping of the dashpots in the swaying and rocking directions are denoted by C_s and C_r then

$$\text{for } \bar{H} > 1, \quad K_s = \frac{8Ga}{2-\upsilon}\left(1 + \frac{1}{2\bar{H}}\right); \tag{8.1}$$

$$\text{for } 1 < \bar{H} \leq 4, \quad K_r = \frac{8Ga^3}{3(2-\upsilon)}\left(1 + \frac{1}{6\bar{H}}\right); \tag{8.2}$$

$$C_s = \frac{4.6Ga^2}{(2-\upsilon)\,V_s}; \tag{8.3}$$

$$\text{and} \quad C_r = \frac{0.4Ga^4}{(1-\upsilon)\,V_s}. \tag{8.4}$$

Here, G denotes the soil shear modulus; a denotes the radius of circular footing; υ denotes soil Poisson's ratio; V_s denotes the soil seismic/ shear wave velocity; H denotes depth of the soil stratum overlying a rigid bedrock; and $\bar{H} = H/a$. As per Tongaonkar and Jangid (2003), Eqs. (8.1, 8.2, 8.3, and 8.4) are applicable for large soil stratum as well where $\bar{H}$ may also diminish.

Here, the base-isolated bridge equipped with tuned mass absorbers is numerically modelled by writing governing dynamic equilibrium equations when the bridge is excited under two mutually orthogonal horizontal components of a real earthquake ground motion. If $[M_s]$, $[C_s]$, and $[K_s]$ respectively denote mass, damping, and stiffness matrices of the bridge; whereas, $\{Q\} = \{X_1, X_2, \ldots X_N, x_{1b} \ldots x_{nb}, x_1 \ldots x_n, x_{Ns}, x_\theta, Y_1, Y_2, \ldots Y_N, y_{1b} \ldots y_{nb}, y_1 \ldots y_n, y_{Ns}, y_\theta\}^T$, $\{\dot{Q}\}$, and $\{\ddot{Q}\}$ respectively denote displacement, velocity, and acceleration vectors, then

$$[M_s]\{\ddot{Q}\} + [C_s]\{\dot{Q}\} + [K_s]\{Q\} = -[M_s][r]\{\ddot{Q}_g\} \tag{8.5}$$

where the matrices are of order: $(2N + 2n + 2n_b + n_s) \times (2N + 2n + 2n_b + n_s)$. Here, N, n, n_b, and n_s respectively denote the degrees of freedom of the bridge, the TMDs, isolators, and the soil. Moreover, $\{\ddot{Q}_g\}$ denotes the vector of earthquake ground acceleration components respectively in the longitudinal (N-S), $\ddot{x}_g$ and transverse (E-W), $\ddot{y}_g$ directions multiplied with an influence coefficients matrix $[r]$ to the mass matrix. The displacements of any node (*i*) of the bridge in the N-S and E-W directions are respectively denoted by $\{X_i\}$ and $\{Y_i\}$, where the subscript, *i* denotes the node under consideration. In order to solve the governing equations of motion of the RC bridge, Newmark's linear time integration scheme is adopted for all the cases considered in this study.

Even a highly nonlinear base isolation system could be idealised as an equivalent linear system with certain approximations (Matsagar and Jangid 2004). Therefore, a simplified equivalent linear model of the base isolation system is adopted in the present numerical model developed for the bridge such that the stiffness of the system in horizontal direction is k_b and corresponding damping is c_b, as shown in Fig. 8.1b. For the laminated rubber bearing (LRB) used in the bridge, the damped isolation time period (T_s) and the damping ratio of the isolator (ζ_s) are calculated based on the equivalent linearization, such as

$$T_s = 2\pi \sqrt{\frac{M_d}{\sum k_b}}; \tag{8.6}$$

$$\text{and } \zeta_s = \frac{\sum c_b}{2M_d \omega_b}. \tag{8.7}$$

Here, M_d denotes mass of the bridge deck; $\sum k_b$ denotes total stiffness of the isolation systems; $\sum c_b$ denotes the total viscous damping of the isolation systems; and $\omega_s = 2\pi/T_s$ denotes angular frequency of the base isolation systems. Note that, the base isolation system essentially modifies first/ fundamental mode of the structure. For a fixed-base (i.e. non-isolated) bridge, the relative displacement across the two nodes of the bearing is suitably modified based on the conventional bridge bearings used to accommodate temperature dependent thermal stresses, which affects T_s and ζ_s.

An Eigen value problem solved by taking mass and stiffness matrices from Eq. (8.5) leads to determination of modal frequencies of the bridge equal to its number of the degrees of freedom. Following classical normal mode superposition theory, it is possible to calculate total response under the earthquake excitation. The real earthquake excitation can also be expressed as sum of large number of sinusoids having different frequencies. When a modal frequency of the bridge coincides with one of the frequency contained within the earthquake excitation, modal response in the particular mode tends to amplify (theoretically, unbound for undamped case). Therefore, for abatement of the dynamic response of the RC bridge, it is possible and recommendable to control individual modal response, whereby the total seismic response of the bridge could eventually be minimised. Using this approach, dynamic response control, only in first few predominant modes in longitudinal and transverse directions, may lead to substantial response reduction in the RC bridge. Hence, the tuned mass absorbers are installed and designed in such a manner that their frequencies match with the modal frequency, corresponding to the i-th mode in which modal response is to be suppressed. Their tuning frequency ratios respectively in the N-S and E-W directions for $i = 1$ to n are calculated as

$$f_{i-x} = \frac{\omega_{i-x}}{\Omega_{i-x}}; \tag{8.8}$$

$$\text{and} \quad f_{i-y} = \frac{\omega_{i-y}}{\Omega_{i-y}}. \tag{8.9}$$

Note that, the frequencies of the individual tuned mass absorbers and modal frequencies of the base-isolated bridge are respectively denoted as ω_{i-x} and Ω_{i-x} in the longitudinal (N-S) direction; whereas, ω_{i-y} and Ω_{i-y} respectively denote it in the transverse (E-W) direction. The design stiffness of the tuned mass absorbers in the N-S and E-W directions are respectively calculated based on the tuning frequencies in different vibration modes of the bridge for $i = 1$ to n to control the respective modal response, such that

$$k_{i-x} = \frac{m_t}{\left(\frac{1}{\omega_{1-x}^2} + \frac{1}{\omega_{2-x}^2} + \cdots + \frac{1}{\omega_{n-x}^2}\right)}; \tag{8.10}$$

$$\text{and} \, k_{i-y} = \frac{m_t}{\left(\frac{1}{\omega_{1-y}^2} + \frac{1}{\omega_{2-y}^2} + \cdots + \frac{1}{\omega_{n-y}^2}\right)}. \tag{8.11}$$

Based on selecting the most suitable mass for the tuned mass absorbers, based on maximising the dynamic response reduction, their total mass, m_t is calculated. In order to achieve the highest possible seismic response reduction, the characteristic parameters of the tuned mass absorbers can suitably be optimised. Hence, the mass ratio, µ expressed as the relative weights of all the tuned mass absorbers as compared to the total mass of the bridge superstructure (M_t) is used in determining individual masses of the controller for $i = 1$ to n in the N-S and E-W directions as

$$m_{i-x} = \frac{k_{i-x}}{\omega_{i-x}^2}; \tag{8.12}$$

$$\text{and } m_{i\text{-}y} = \frac{k_{i\text{-}y}}{\omega_{i\text{-}y}^2}. \tag{8.13}$$

Furthermore, for all the tuned mass absorbers, $i = 1$ to n, the damping ratio, $\zeta_d = \zeta_1 = \zeta_2 = \cdots = \zeta_n$ is maintained constant; thereby, the damping coefficients are determined in the longitudinal (N-S) and transverse (E-W) directions as

$$c_{i\text{-}x} = 2\zeta_d m_i \omega_{i\text{-}x}; \tag{8.14}$$

$$\text{and } c_{i\text{-}y} = 2\zeta_d m_i \omega_{i\text{-}y}. \tag{8.15}$$

8.3 Numerical Study

A passive-hybrid system is used here for seismic response modification of a three-span continuous RC bridge by employing in-combination, base isolation systems and tuned mass absorbers. Such base-isolated bridge is applied base excitation with two horizontal components of earthquake ground motions. The Imperial Valley, 1940 earthquake ground motion used in this study contains relatively high excitation frequency components which may be detrimental to the bridge causing more seismic damage (Elias and Matsagar 2017a). This earthquake was recorded at the El Centro station in the USA and had peak ground acceleration (PGA) of 0.21 g in the longitudinal direction; and PGA of 0.34 g in the transverse direction, where g denotes the acceleration due to gravity. The digital time history records of the Imperial Valley, 1940 earthquake ground motion are applied in the longitudinal (N-S) and transverse (E-W) directions of the bridge as bi-directional base excitation. The efficacy of the passive-hybrid system in seismic response reduction of the base-isolated RC bridge resting on flexible founding soil is verified numerically.

The RC structural elements of the three-span continuous base-isolated bridge are adopted from the study reported by Tongaonkar and Jangid (2003). The 90 m long bridge has three spans of 30 m each and height of both the piers is 10 m. It consists of a square deck and two circular piers with cross-sectional areas 3.57 m^2 and 1.767 m^2, respectively. The moment of inertia of the circular piers in the N-S and E-W directions is 0.902 m^4; and, the moment of inertia of the square deck is 2.08 m^4. Elastic modulus and density of concrete for the RC members are respectively assumed as 3.6×10^7 kN/m^2 and 23.536 kN/m^3.

The numerical model of the RC bridge consists of beam elements for modelling the deck and piers connected at nodes. Each node has two degrees of freedom in the longitudinal, N-S and transverse, E-W directions of the bridge. The deck and piers are sub-divided in four elements each, connected at common nodes. Thus, the numerical model of the bridge mainly consists of 20 beam elements and 21 nodes, thus totalling 42 degrees of freedom. When the passive-hybrid control systems are introduced, the number of degrees of freedom added therein corresponds to the displacement across the base isolation systems and the tuned mass absorber.

Moreover, at the bottom ends of the piers, when the soil flexibility is accounted for, two additional degrees of freedom at each node corresponding to the soil are introduced in the numerical model of the bridge, as shown in Fig. 8.1b. By solving the Eigen value problem again, the modal properties of the RC bridge are determined in these different conditions. For 5% damping in the fixed-base (i.e. non-isolated) bridge, the time periods calculated are: 0.53, 0.49, 0.25, and 0.15 s in the longitudinal (N-S) direction; and 0.53, 0.47, 0.24, and 0.16 s in the transverse (E-W) direction, without considering founding soil flexibility.

A linearized base isolation system is employed at four locations, with isolation stiffness chosen such that the resulting isolation time period, $T_s = 2$ s with isolation damping of $\zeta_s = 0.125$ calculated using Eqs. (8.6 and 8.7), respectively. The tuned mass absorbers are installed at the mid-span locations where the displacement amplitude is typically larger. These structural control devices also have two degrees of freedom mutually perpendicular to each other so that vibration response reduction is achieved in both the longitudinal and transverse directions of the RC bridge. Hence, the passive-hybrid system comprising of the base isolation and tuned mass absorbers help in controlling the pier base shear, deck acceleration, and deck displacement under the bi-directional earthquake excitations. In this study, single tuned mass absorber (TMD) is used in the base-isolated bridge (IB), which is referred as: IB + STMD. Similarly, IB + 2TMDs and IB + 3TMDs refer the base-isolated bridge with two and three tuned mass absorbers, respectively. It is notable that the tuned mass absorbers are designed in such a manner that their frequencies in the two horizontal directions are different, and match with the corresponding modal frequencies of the bridge, wherein the response are to be controlled.

The damping ratio in the tuned mass absorbers is considered as 5% and suitable mass ratio (μ) of 1% is taken in the numerical study for comparing seismic response abatement achieved. Note that, a suitable optimisation approach may be followed to arrive at most optimum ζ_d and μ for the tuned mass absorbers, maximising dynamic response reduction. In the longitudinal and transverse directions, the seismic response quantities of the base-isolated RC bridge are evaluated, when the dynamic response is controlled using the STMD, 2TMDs, and 3TMDs placed at the three mid-spans of the bridge deck.

As shown in Fig. 8.1b, the flexibility of the founding soil has been accounted for in the numerical model developed for the base-isolated RC bridge. Based on the seismic/shear wave velocity (V_s), shear modulus (G), and similar properties of the soil, it is categorised as hard soil, medium soil, and soft soil. Tongaonkar and Jangid (2003) had recommended taking the Poisson's ratio (v) as 0.4, which is considered here. While duly considering flexibility of the founding soil, the modal time periods of the base-isolated RC bridge in its longitudinal (N-S) and transverse (E-W) directions for the rigid, hard, medium, and soft soils are calculated and shown in Fig. 8.2.

It can be observed from the trends in the plots that the tuned mass absorbers applied in the IB + 2TMDs and IB + 3TMDs cases are designed such that the shift in the modal frequencies in all the three schemes remains insignificant relatively. It is notable from Fig. 8.2 that with increasing founding soil flexibility, the modal time

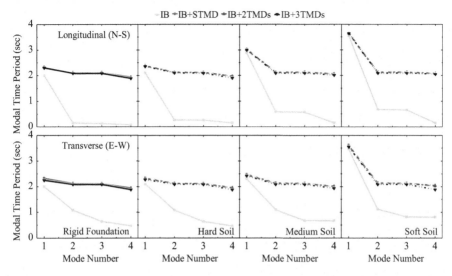

Fig. 8.2 Effect of passive-hybrid control systems and founding soil flexibility on the modal time periods of the base-isolated RC bridge

periods of the RC bridge have increased, which is as anticipated. Especially, the fundamental frequency (corresponding to the longest time period) has increased with the hard-stiffer soil conditions.

Owing to the flexibility of the underlying soil, the modal properties of the bridge are affected, which may cause detuning of the tuned mass absorbers. Hence, seismic performance of the tuned mass absorbers employed in abatement of the dynamic response of the RC bridge is investigated under varying founding soil conditions. Seismic response of the bridge is obtained in the three cases: IB, IB + STMD, IB + 2TMDs, and IB + 3TMDs and plotted in Fig. 8.3 to assess the effect of founding soil flexibility on the efficacy of the passive-hybrid control system utilised. The figure contains plots of seismic base shear induced in the piers (PBS) normalised with total weight of the superstructure, $W_d = M_d \times g$. The figure also contains plots of the deck acceleration experienced at the mid-span, and the isolator displacements at abutment and pier locations. The dynamic response is obtained in both the longitudinal (N-S) and transverse (E-W) directions of the bridge under the Imperial Valley, 1940 earthquake and shown in Fig. 8.3.

From Fig. 8.3, it can be observed that the seismic response of the base-isolated RC bridge with tuned mass absorbers has been considerably influenced by the variation in the founding soil conditions. With increasing flexibility of the bridge due to the founding soil flexibility, i.e. increasing T_s, the seismic base shear induced in the bridge pier has reduced in both longitudinal and transverse directions. The passive-hybrid control scheme has been effective in reducing the seismic base shear induced in the bridge pier, especially when the number of tuned mass absorbers are increased from one to three, i.e. IB + STMD through IB + 3TMDs, effectively controlling seismic response in the higher modes as well. Thus, earthquake response

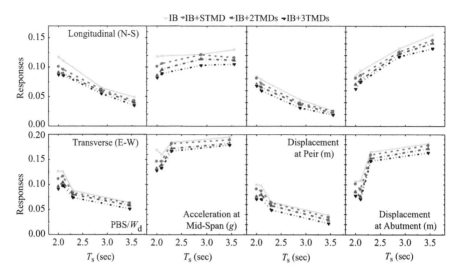

Fig. 8.3 Absolute peak seismic response of the RC bridge with different passive-hybrid control schemes with varying founding soil conditions

modification achieved in the higher modes of the structure has proved to be beneficial.

An increase in bearing displacement is noticed at the abutments with increased flexibility of the bridge; on the other hand, the bearing displacement at the piers has reduced with increasing T_s, in both the longitudinal and transverse directions of the RC bridge. It clearly shows the influence of the founding soil flexibility, i.e. increasing T_s, on the displacement response of the RC bridge at abutment and pier locations. The absolute peak bearing displacement has increased by about 30% at the abutments in the longitudinal (N-S) direction, whereas it has increased by about 20% at the abutments in the transverse (E-W) direction for the base-isolated bridge (IB) resting on flexible soil conditions. On the other hand, the absolute peak bearing displacement has reduced from 0.09 to 0.02 m at the piers in the longitudinal (N-S) direction, whereas it has reduced from 0.1 to 0.05 m at the piers in the transverse (E-W) direction for the base-isolated bridge (IB) resting on flexible soil conditions. Therefore, it is concluded that the bridge bearing displacement increases at the abutments and reduces at the piers with founding soil flexibility.

The variation in the bearing displacement with the founding soil conditions has significant influence on the design of the base isolation system and consequently on the pier caps on top of which it rests. The design displacement of the bearings at the abutments may be underestimated if the flexibility of the founding soil is ignored in the base-isolated bridges. However, the design displacement of the bearings at the piers is reduced due to the consideration of the founding soil flexibility. In order to control the displacement response in both the longitudinal and transverse directions of the base-isolated RC bridge at the abutment as well as pier locations, the tuned mass absorbers are employed. Three cases of such tuned mass absorbers are studied here: when only one device is installed, IB + STMD; when two devices are installed, IB + 2TMDs; and when three devices are installed, IB + 3TMDs.

The IB + STMD case shows reduction in the bearing displacement response as compared to the IB case in presence of the founding soil flexibility. Nevertheless, effectiveness in bearing displacement response control increases with increasing number of tuned mass absorbers, i.e. IB + 2TMDs and IB + 3TMDs cases as compared to the IB case in presence of the founding soil flexibility. Thus, the undesirable bearing displacement response in both the longitudinal and transverse directions of the base-isolated RC bridge at the abutment as well as pier locations is effectively reduced by installation of the tuned mass absorbers, controlling modal response in the higher vibration modes. Nonetheless, it is important to note that, the modal frequencies modified due to the flexibility of the founding soil must be used aptly for frequency tuning of the tuned mass absorbers to achieve increased seismic response reduction.

The improved performance of the tuned mass absorbers in the base-isolated RC bridge is evident in seismic response abatement. When the founding soil flexibility is increased, the seismic base shear induced in the bridge piers has reduced. Similarly, when the founding soil flexibility is increased, the acceleration induced in the bridge deck has reduced. Due to the installation of the tuned mass absorbers in the base-isolated RC bridge, both the seismic base shear induced in the bridge piers and the acceleration induced in the bridge deck have further reduced considerably. Most accurate calculation of the modal frequencies for tuning purpose is important in the cases when tuned mass absorbers are used in structures for dynamic response abatement. A tuned mass absorber is most effective when its tuned frequency exactly matches with the (modal) frequency of the main/ parent structure. By taking in to account the changes in the modal frequencies of the base-isolated bridge due to the founding soil properties for frequency tuning of the tuned mass absorbers, adverse effects due to the possible detuning have been evaded.

Selecting the design parameters for the bridge is important such that, the bearing displacement response reduces; whereas, at the same time the base shear induced in the piers and acceleration induced in the bridge deck are reduced as well. In order to study the effects of the flexible founding soil conditions on the base shear induced in the piers of the base-isolated RC bridge, the fast Fourier transform (FFT) of the normalised PBS/W_d are plotted in Fig. 8.4. The FFT variation of the base shear induced in the pier of the RC bridge, $T_s = 2$ s and $\zeta_s = 0.125$, with varying soil is shown in Fig. 8.4 under Imperial Valley, 1940 ground motion in longitudinal and transverse directions. The Fourier amplitude is observed to reduce for the flexible founding soil conditions. Moreover, the performance of the IB + STMD scheme of the passive-hybrid dynamic response control has been affected significantly when the flexibility of the soil underneath is accounted for. With this due consideration, the multiple tuned mass absorbers (MTMDs) controlling response in different vibration modes have indeed helped in reducing the dynamic response of the bridge. Hence, the design of the passive-hybrid control system in the base-isolated bridge with tuned mass absorbers should necessarily account for the effects of the flexible

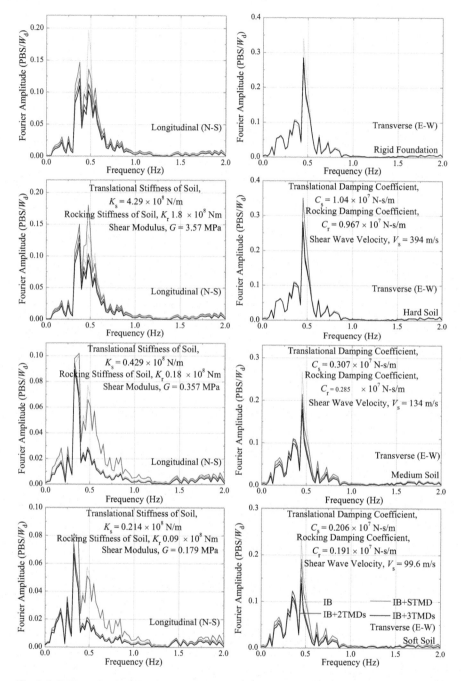

Fig. 8.4 FFT amplitudes of pier base shear in the bridge with varying soil under Imperial Valley, 1940 ground motion in longitudinal and transverse directions ($T_s = 2$ s, $\zeta_s = 0.125$)

founding soil on its modal properties, especially that of the altered frequencies in the lower vibration modes.

Robustness of the tuned mass absorbers in seismic response control in the base-isolated RC bridge is evaluated herein. If the effectiveness in seismic response abatement is maintained relatively unaffected even after changing the founding soil condition then the vibration control scheme is considered to qualify as robust. Thus, detuning may occur in the tuned mass absorbers due to the changed modal frequencies of the bridge; however, higher modal frequencies are not considerably affected. Hence, because seismic response reduction is achieved in the higher modes, the passive-hybrid control scheme with multiple tuned mass absorbers shows improved robustness comparatively. Thereby, IB + 3TMDs and IB + 2TMDs are more robust seismic control schemes as compared to the IB + STMD in reducing the base shear induced in the piers of the bridge, albeit the founding soil is flexible, causing detuning.

8.4 Conclusions

Passive-hybrid earthquake response modification using base isolation systems and tuned mass absorbers in the reinforced concrete (RC) bridge is presented. The flexibility of the founding soil underneath the RC bridge is taken in to account in the design of the structural control systems. More than one modal response is modified by tuning the tuned mass absorber or damper (TMD) with the higher modal frequencies. For achieving the intended purpose, the RC bridge is equipped with the single (STMD), two (2TMDs), and three (3TMDs) installed at the mid-spans. Real earthquake ground motion is used to conduct time domain response analysis of the bridge with the passive-hybrid system.

From the results presented through this numerical study, the following conclusions are drawn.

1. The bearing displacement response is reduced considerably in both the longitudinal and transverse directions of the base-isolated RC bridge at the abutment as well as pier locations by installation of the tuned mass absorbers, controlling modal response in the higher vibration modes.
2. Due to the installation of the tuned mass absorbers in the base-isolated RC bridge, both the seismic base shear induced in the bridge piers and the acceleration induced in the bridge deck have significantly reduced.
3. Frequency tuning of the tuned mass absorbers is maintained unaffected by taking in to account the changes in the modal frequencies of the base-isolated bridge due to the founding soil properties to evade adverse effects caused due to the possible detuning.
4. Because seismic response abatement is achieved in the higher vibration modes of the base-isolated RC bridge, the passive-hybrid control scheme with multiple tuned mass absorbers shows improved robustness relatively.

5. The effects of flexible founding soil on the modal properties of the bridge are required to be accounted for in the design of the passive-hybrid control system in the base-isolated bridge with tuned mass absorbers to achieve higher dynamic response reduction.

References

Aly AM (2014a) Proposed robust tuned mass damper for response mitigation in buildings exposed to multi-directional wind. Struct Design Tall Spec Build 23(3):664–691
Aly AM (2014b) Vibration control of high-rise buildings for wind: a robust passive and active tuned mass damper. Smart Struct Syst 13(3):473–500
Aly AM, Zasso A, Resta F (2011) Dynamics and control of high-rise buildings under multi-directional wind loads. Smart Mater Res 2011:549621. https://doi.org/10.1155/2011/549621
Daniel Y, Lavan O, Levy R (2012) Multiple tuned mass dampers for multimodal control of pedestrian bridges. J Struct Eng 138(9):1173–1178
Debnath N, Deb SK, Dutta A (2015) Multi-modal vibration control of truss bridges with tuned mass dampers and general loading. J Vib Control 22(20):4121–4140. https://doi.org/10.1177/1077546315571172
Dicleli M (2007) Supplemental elastic stiffness to reduce isolator displacements for seismic-isolated bridges in near-fault zones. Eng Struct 29(5):763–775
Dicleli M, Buddaram S (2007) Equivalent linear analysis of seismic-isolated bridges subjected to near-fault ground motions with forward rupture directivity effect. Eng Struct 29(1):21–32
Elias S, Matsagar VA (2014a) Wind response control of a 76-storey benchmark building installed with distributed multiple tuned mass dampers. J Wind Eng 11(2):37–49
Elias S, Matsagar VA (2014b) Distributed multiple tuned mass dampers for wind vibration response control of high-rise building. J Eng http://www.hindawi.com/journals/je/aip/198719/
Elias S, Matsagar VA (2015) Optimum tuned mass damper for wind and earthquake response control of high-rise building. In: Advances in structural enginerring, Dynamics 2, pp 1475–1487. ISBN: 978-8-13-222192-0 (Print), 978-8-13-222193-7 (Online). doi: 10.1007/978-81-322-2193-7_113
Elias S, Matsagar VA (2017a) Effectiveness of tuned mass dampers in seismic response control of isolated bridges including soil-structure interaction. Lat Am J Solids Struct. https://doi.org/10.1590/1679-78253893
Elias S, Matsagar VA (2017b) Research developments in vibration control of structures using passive tuned mass dampers. Annu Rev Control 44:129–156. https://doi.org/10.1016/j.arcontrol.2017.09.015
Elias S, Matsagar VA (2018) Wind response control of tall buildings with a tuned mass damper. J Build Eng 15:51–60. https://doi.org/10.1016/j.jobe.2017.11.005
Elias S, Matsagar VA, Datta TK (2015) Effectiveness of distributed multiple tuned mass dampers in along wind response control of chimney. In: 14th International Conference on Wind Engineering (ICWE14), Porto Alegre, Brazil, June 21–26
Elias S, Matsagar VA, Datta TK (2016) Effectiveness of distributed tuned mass dampers for multi-mode control of chimney under earthquakes. Eng Struct 124:1–16. https://doi.org/10.1016/j.engstruct.2016.06.006
Elias S, Matsagar VA, Datta TK (2017a) Distributed tuned mass dampers for multi-mode control of benchmark building under seismic excitations. J Earthq Eng. https://doi.org/10.1080/13632469.2017.1351407
Elias S, Matsagar VA, Datta TK (2017b) Distributed multiple tuned mass dampers for seismic response control of chimney with flexible foundation.In: 16th World Conference on Earthquake Engineering (16WCEE), Santiago, Chile, January 9–13

Elias S, Matsagar V, Datta TK (2017c) Distributed multiple tuned mass dampers for wind response control of chimney with flexible foundation. Procedia Eng 199:1641–1646. https://doi.org/10.1016/j.proeng.2017.09.087

Gill D, Elias S, Steinbrecher A, Schröder C, Matsagar VA (2017) Robustness of multi-mode control using tuned mass dampers for seismically excited structures. Bull Earthq Eng 15:1–25. https://doi.org/10.1007/s10518-017-0187-6

Jangid RS (2004) Seismic response of isolated bridges. J Bridg Eng 9(2):156–166

Li C (2000) Performance of multiple tuned mass dampers for attenuating undesirable oscillations of structures under the ground acceleration. Earthq Eng Struct Dyn 29(9):1405–1421

Li C (2002) Optimum multiple tuned mass dampers for structures under the ground acceleration based on DDMF and ADMF. Earthq Eng Struct Dyn 31(4):897–919

Lu Z, Masri SF, Lu XL (2011) Parametric studies of the performance of particle dampers under harmonic excitation. Struct Control Health Monit 18(1):79–98

Lu Z, Lu XL, Lu W, Masri SF (2012) Experimental studies of the effects of buffered particle dampers attached to a multi-degree-of-freedom system under dynamic loads. J Sound Vib 331(9):2007–2022

Luu M, Zabel V, Könke C (2012) An optimisation method of multi-resonant response of high-speed train bridges using TMDs. Finite Elem Anal Des 53:13–23

Matin A, Elias S, Matsagar VA (2014) Seismic control of continuous span concrete bridges with multiple tuned mass dampers. In: Proceeding of 2nd European Conference on Earthquake Engineering and Seismology (2ECEES), Istanbul, Turkey

Matin A, Elias S, Matsagar VA (2017) Seismic response control of reinforced concrete bridges with soil-structure interaction. Bridge Struct Eng 47(1):46–53

Matsagar VA, Jangid RS (2004) Influence of isolator characteristics on the response of base-isolated structures. Eng Struct 26(12):1735–1749. https://doi.org/10.1016/j.engstruct.2004.06.011

Matsagar VA, Jangid RS (2006) Seismic response of simply supported base-isolate bridge with different isolators. Int J Appl Sci Eng 4(1):55–71

Miguel LFF, Lopez RH, Torii AJ, Miguel LFF, Beck AT (2016) Robust design optimisation of TMDs in vehicle-bridge coupled vibration problems. Eng Struct 126:703–711

Pisal AY, Jangid RS (2016) Vibration control of bridge subjected to multi-axle vehicle using multiple tuned mass friction dampers. Int J Adv Struct Eng 8(2):1–15

Spyrakos CC (1990) Assessment of SSI on the longitudinal seismic response of short span bridges. Const Build Mater 4(4):170–175

Tongaonkar NP, Jangid RS (2003) Seismic response of isolated bridges with soil-structure interaction. Soil Dyn Earthq Eng 23:287–302

Chapter 9
Monitoring and Damage Detection of a 70-Year-Old Suspension Bridge – Ölfusá Bridge in Selfoss, Case Study

Gudmundur Valur Gudmundsson, Einar Thor Ingólfsson, Kristján Uni Óskarsson, Bjarni Bessason, Baldvin Einarsson, and Aron Bjarnason

Abstract Ölfusá Bridge in South Iceland is located in the town of Selfoss on the ring road Route 1 in Iceland, approximately 50 km from Reykjavik. It is a suspension bridge across the Ölfusá glacial river and while being a very important link in the road transport network in Iceland, it also serves as a vital link for the community of Selfoss for both vehicles and pedestrians. Inspections of the main cables of the suspension bridge have indicated an inadequate level of safety due to corrosion, as well as significantly heavier traffic loads and increased self-weight due to rehabilitation of the bridge deck. A system identification of the bridge was conducted in 2012 and a permanent monitoring system was installed in 2014 with the aim of identifying changes in the structural performance of the bridge. The measurements have provided valuable information which are used in the operation of the bridge. The measured modal properties have been used to calibrate the FE models of the bridge, which are used for structural analysis and

---

G. V. Gudmundsson (✉) · A. Bjarnason
Icelandic Road and Coastal Administration, Reykjavik, Iceland
e-mail: gudmundur.v.gudmundsson@vegagerdin.is; aron.bjarnason@vegagerdin.is

E. T. Ingólfsson
Krabbenhoft+Ingólfsson Aps, Copenhagen, Denmark
e-mail: eti@krabbenhoft.eu

K. U. Óskarsson
EFLA consulting engineers, Reykjavik, Iceland
e-mail: kristjan.uni.oskarsson@efla.is

B. Bessason
Faculty of Civil and Environmental Engineering, University of Iceland, Reykjavik, Iceland
e-mail: bb@hi.is

B. Einarsson
EFLA consulting engineers, Reykjavik, Iceland

Faculty of Civil and Environmental Engineering, University of Iceland, Reykjavik, Iceland
e-mail: baldvin.einarsson@efla.is

the structural reliability assessment of the bridge. The continuous monitoring of the cable forces gives the possibility of identifying changes in the structural response of the bridge in addition to the regular visual inspections.

Keywords Suspension bridges · Structural monitoring · Damage detection · Modal testing · System identification

9.1 Introduction

The Ölfusá Bridge in South Iceland is located in the town of Selfoss approximately 50 km east of Reykjavik on the main road Route 1. It is a 132 m long suspension bridge, with a main span of 84 m crossing the Ölfusá glacial river (Fig. 9.1). It has two lanes and a narrow walkway for pedestrians, with a total width of 8.5 m. It was opened for traffic in 1945. The average daily traffic today is more than 10,000 vehicles (3.6 million vehicles per year) and while being a very important link in the road transport network in Iceland, it also serves as a vital link for the community of Selfoss for both vehicles and pedestrians. The nearest crossings of the River Ölfusá are 16 km downstream and approximately 40 km upstream. Given the distance to alternative crossings, the bridge is very important, both for businesses and the inhabitants of the area. There are plans for a new bridge crossing about 1.5 km upstream in the near future, which will then take over most of the existing traffic.

Inspections and assessment of the condition of the main cables of the suspension bridge carried out in 2011 indicated reduced structural capacity due to corrosion. The widening of the roadway as well as heavily increased traffic since the bridge was built have led to a significantly lower factor of safety for the main cables.

The objectives of this paper are to give an overview of measurements and studies of the Ölfusá Bridge that have been carried out in the last few years as well as to present a new monitoring system that has been installed to keep track of the health of the bridge. The main focus is on the suspension cables which are considered to be most critical for the performance and durability of the bridge.

Fig. 9.1 Ölfusá suspension bridge in Selfoss

9.2 Main Cables

The main suspension cables are the primary load carrying members of the Ölfusá Bridge, although the stiffening girder also has an influence on the global behaviour of the bridge deck.

The cables are made of 6 locked coil strands which are in a parallel configuration, as can be seen in Fig. 9.2. The strands are made of 183 wires which are not galvanized. Age together with increased demands due to heavier traffic than was envisioned when the bridge was built have resulted in deformation of the strands; the outer Z-shaped wires of the locked-coil strands now have gaps in between. As a consequence, water now has access to the inner wire, making the strands susceptible to corrosion. The protective paint layer has been renewed several times during the 70 years of service.

9.2.1 Level of Safety and Reliability

The question of what should be the level of safety for the bridge is of paramount importance. Originally, the main cables were designed with a traditional factor of safety of four. The first bridge deck was made of concrete, and several asphalt layers were added later, resulting in a significantly heavier self-weight of the bridge deck. The old concrete deck and the asphalt layers were replaced in the 1990s, when the bridge was widened slightly to accommodate a new pedestrian walkway. Overall, the self-weight of the bridge deck is approximately 60% greater than originally built. The bridge also serves as a support for several water pipes, including geothermal district heating pipes, which have added significant weight.

The problem with the pass-fail criteria for either using the level of safety factor or any limit state design approach is the nonlinearity of the problem. When the cables are stretched above the linear range, their response becomes non-linear and a permanent extension of the cables is noted. Corrosion can also influence the fatigue

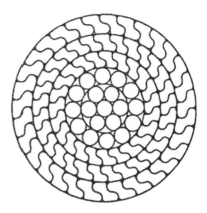

Fig. 9.2 Cross section of locked-coil strand

resistance of the individual wires (Nakamura and Suzumura 2011). The corrosion has therefore a significant influence, not only on the ultimate limit state but also the serviceability limit state.

Any means of diverting traffic or reducing the weight limits are costly. A new bridge is being planned approximately 1 km upstream and, when built, it is likely that all heavy traffic will be diverted to the new bridge.

9.2.2 Corrosion of Cables

Compared to other structural steel elements such as solid bars and rolled wide flange sections, cable bundles comprising a large number of thin wires are very susceptible to surface corrosion. Inaccessibility of internal wires makes inspection and maintenance problematic, which allows for uninterrupted accumulation of cavities between wires (Gimsing 1998).

Nakamura and Suzumura (2011) conducted extensive research on the deterioration of bridge cables and the effectiveness of different protection methods. The investigation included estimation of tensile and fatigue strength and elongation of corroded steel wires on three corrosion levels which were reproduced in laboratories. Specimens of galvanized and bare steel wires were subjected to aggressive conditions for different time frames producing levels one, two and three corrosion, defined by mass loss and appearance. Measurements revealed that both the galvanized and bare specimens experienced exponentially increasing loss of mass with elapsed time. Static testing showed that the tensile strength did not decrease in response to level of corrosion for both galvanized and bare steel specimens, while decrease in elongation was detected as the steel part of the specimen started to corrode. The resulting condition causes serious problems in terms of material ductility due to the relation between elongation and ductility properties. Cyclic loading tests showed a significant decrease in fatigue strength as the bare steel started corroding after the galvanized layer had corroded away.

Inspections of the main cables of the Ölfusá Bridge have been carried out several times in the past. An example of the condition of the locked coil strands beneath one of the hanger clamps is shown in Fig. 9.3.

9.3 Modal Identification and Cable Force Estimation

9.3.1 Modal Identification

System identification of the Ölfusá Bridge was conducted in 2012 where modal properties of the deck, cables and hanger forces were determined from measured vibrations. Natural vibrations of the bridge caused by ambient excitation transmitted through the foundations as well as being excited by wind forces were recorded by a

Fig. 9.3 Corrosion in main cables of the Ölfusá Bridge

number of accelerometers located in the bridge structure. Four uniaxial accelerometers and three triaxial accelerometers (13 channels) were used in eight different setups during the measurement programme. Some of the accelerometers had a fixed location and were used as reference points, while others were moved between setups. For each setup vibrations were recorded for 20 min with a sampling frequency of 512 Hz for each channel. The measurements were conducted in collaboration with the Technical University of Denmark (DTU BYG) and University of Iceland. Four MSc students participated in the measurements and in post-processing of the data. This resulted in four MSc theses (Óskarsson 2012; Pálsson 2012; Andersen 2012; and Bjarnason 2014).

The system identification was then conducted by the software ARTeMIS from the firm Structural Vibration Solution. Frequency domain decomposition and stochastic subspace iteration were used to compute modal shapes, natural periods and damping ratios (see for instance Brincker and Andersen 2006). Figures 9.4 and 9.5 show, as examples, the first mode and the fourth mode, respectively, based on the recorded data and the ARTeMIS analysis.

The modal properties have been used to update and calibrate the existing FE models of the bridge (Pálsson 2012). With a calibrated FE model it is possible to perform sensitivity analysis as well as to perform a classical response analysis. Figure 9.6a shows how changes in the cross-sectional area of suspension cables, for instance due to corrosion, will affect the natural frequencies of the four first modes of the bridge. The first mode is more sensitive than the other modes. In Fig 9.6b, the sensitivity of changing the masses of the cables is shown, where the main finding was that the cables are relatively insensitive to such changes, except mode no. 4, which is sensitive to increased mass.

9.3.2 Cable Forces

Estimation of the cable force based on the measured frequency of the cables is a very well-known method, as described by Caetano (2011) and Geier (2004). In addition

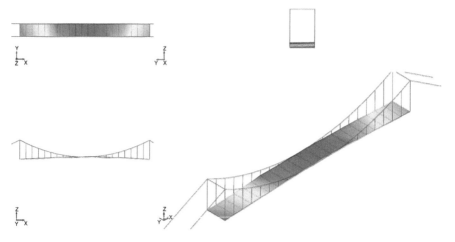

Fig. 9.4 Identification of the first mode shape (vertical mode), $T_1 = 0.982$ s and damping ratio $\xi_1 = 0.58\%$ (Bjarnason 2014)

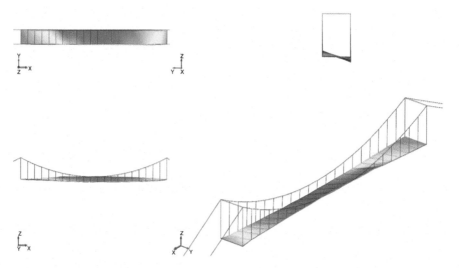

Fig. 9.5 Identification of the fourth mode (torsional mode) $T_4 = 0.479$ s and damping ratio $\xi_1 = 0.64\%$ (Bjarnason 2014)

to the traditional formulae for a vibration chord, an FE model (local model) of the backstays was used to establish a relationship between force and frequency with the aim of detecting changes in the frequency of the backstays of the cables (Fig. 9.7).

The use of direct transformation by employing commonly used chord theory (Geier 2004) is highly sensitive to support conditions and element mode shape length when correlating frequency and cable force. The global finite element

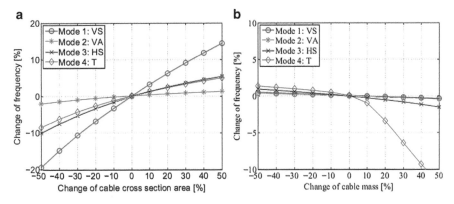

Fig. 9.6 Sensitivity analysis of parameters in the FE mode (*VS* Vertical Symmetrical mode, *VA* Vertical anti-symmetrical mode, *HS* Horizontal Symmetrical mode, *T* Torsional mode) (Pálsson 2012)

Fig. 9.7 Extruded perspective view of the local finite element model of the backstay

model has been updated after modal identification of the bridge deck and is considered to provide an adequate estimate of forces due to vehicle loading while only indirectly describing forces due to the weight of the structure itself.

The local model of the backstays is considered to reduce the sensitivity of calculated forces to the element length and length of the consistent part of the mode shape under evaluation while maintaining the direct coupling to the measured frequencies. The model itself comprises the six strings of the south end backstays, restrained at one end simulating the position at the tower saddle while at the other end distributed into the anchor block. A relatively stiff frame element is employed to simulate the splice connection, separating the strings from the main cable bundle out towards the anchor block. The elements representing the strings are frame elements of equivalent cross section to the ø60 locked coil cables. The estimated weight and elastic modulus of the steel cable section are taken as 82 kN/m^3 and 135.000 N/mm^2, respectively.

In order to obtain an estimate of the forces acting in the strings of the cable bundle, a target force of different amplitudes is defined giving a matching modal frequency. The chosen modal frequency is the first horizontal frequency (mode 1) and the first vertical frequency, (mode 3) of the local model which relates to the previously conducted measurements. Plotted values of force against frequency provide a relationship to which the measured frequencies may be coupled to estimate the cable force.

Analysis of installed cable forces in the Älvsborg Suspension Bridge conducted by Andersson et al. (2004) showed small influence of cable bending stiffness and different end restrains for long cables and suspenders. For shorter cables this effect was of considerable influence indicating the importance of proper evaluation of boundary conditions and stiffness according to the length of the cable under consideration.

9.4 Monitoring and Damage Detection

A monitoring system was installed in 2014 with the aim of detecting changes in the structural behaviour of the bridge by means of measuring the frequencies of the two backstays of the main cable on the south side (Fig. 9.8).

The cable frequencies are monitored continuously and the results stored as a 10-min average acceleration, measured in three local cable directions. That is one axial direction (Z-axis), and two perpendicular cable directions, i.e. transverse horizontal (Y-axis) and transverse inclined direction (X-axis). However, only results for the two perpendicular cable directions are stored on the server and displayed on-line. A single data file contains 10 min of data for these two directions.

An example of an acceleration time history and its corresponding single sided Auto Spectral Density (ASD) is shown in Fig. 9.9 for one direction.

The ASD shows how the energy of the vibration signal is distributed across the frequency range. This graph is used to identify the natural frequencies of the cable. The lowest natural frequency is in the range of 2.5–5.0 Hz and therefore Fig. 9.10 shows the ASD particularly in that frequency range (a zoom in). It

Fig. 9.8 Monitoring of frequencies of the backstays

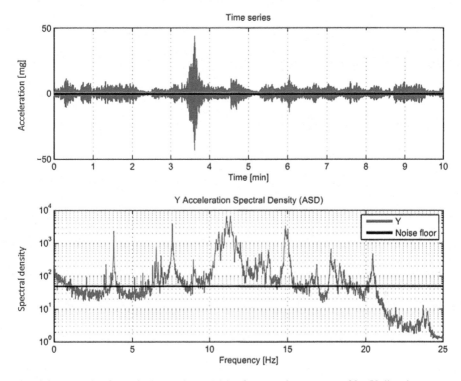

Fig. 9.9 Example of 10-min time series and ASD from accelerometer no. 20 – Y-direction

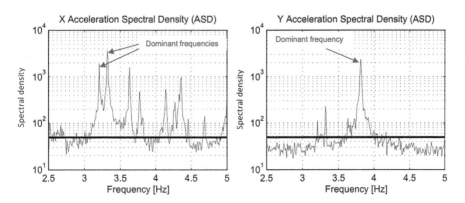

Fig. 9.10 Monitoring of frequencies of the backstays

should be noted that the horizontal black line represents the sensor noise floor and therefore data below and around this line are not reliable.

For the longitudinal vibrations, only one dominant mode appears around 3.7 Hz, whereas the horizontal vibrations feature a number of distinct peaks in the frequency range. The lowest frequencies are in the range of 3.2–3.3 Hz.

These natural frequencies will be used as key indicators related to the condition of the cables. The natural frequencies are strongly related to tension force in the back stays and thereby the temporal variations of the frequencies indicate a proportional variation in the cable force. As a follow-up the relationship between the natural frequencies and the cable forces will be established.

Changes in temperature influence the modal properties of the bridge. The effect of temperature on the frequencies of the backstays has therefore been quantified and simulated in a global finite element model of the bridge. Non-linear P-delta large displacement analysis is conducted with varying temperatures to obtain an estimation of the expected changes in the cable force due to changes in temperature. The relationship between changes in temperature and effect on the cable force can be seen in Fig. 9.11.

In Fig. 9.12 the changes in the cable force are plotted for each cable plane (upstream and downstream), and in the graph the predicted cable force (corrected for temperature) is compared to the cable force calculated from the measured frequencies. It can be seen that the predicted cable force has been relatively consistent with the measured changes in the cable force, although the magnitude of the measured force has generally been slightly smaller than the predicted force. An example of results from the monitoring can be seen in Fig. 9.13 where the ratio of the cable force in the upstream and downstream cable plane is plotted for data from 2014–2015. It can be seen that there was a difference in the cable force for the different cable planes, which seem to have been fairly constant with the upstream cable plane having a 6–7% higher frequency. There are some irregularities which are being analysed with a larger data set. Changes in the ratio of the cable forces between cable planes could give an indication of deterioration in the stiffness of the cables. Another reason could be that the bridge has an asymmetric structural response to temperature changes. Currently the results from the long-term monitoring are being analysed with the aim of defining a suitable limit of acceptable changes.

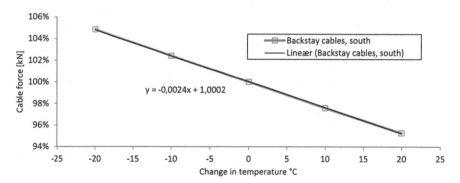

Fig. 9.11 Relationship between temperature changes and cable force

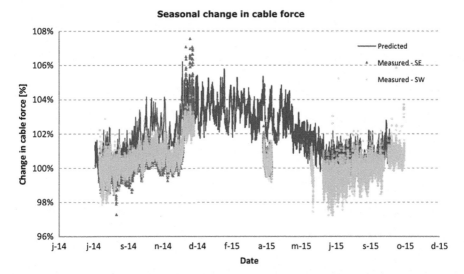

Fig. 9.12 Seasonal variations of the cable force from June 2014 (j-14) to December 2015 (d-15) (*SE* Upstream cable plane, *SW* Downstream cable plane)

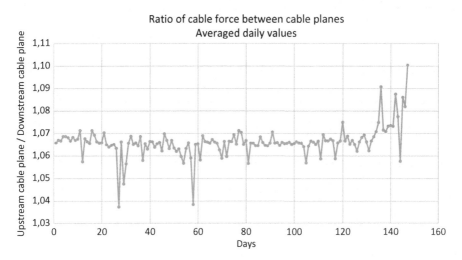

Fig. 9.13 Monitoring of the ratio of cable force between cable planes

9.5 Conclusions

The ageing Ölfusá suspension bridge in Selfoss serves as a vital link in the road network in the South of Iceland and for the local traffic in the town of Selfoss.

The measurements of modal properties and the monitoring of the backstay frequencies have provided valuable information which are used in the operation of the bridge. The measured modal properties have been used to calibrate the FE models of the bridge, which are used for structural analysis and the structural reliability assessment of the bridge.

The continuous monitoring of the cable forces gives the possibility of identifying changes in the structural response of the bridge in addition to the regular visual inspections. Currently there exist more than 2 years of data from continuous monitoring which are being analysed with the aim of improving the reliability of the predictions of the cable force, including correcting for temperature changes.

Acknowledgements This study has been supported by the Icelandic Road and Coastal Administration Research Fund.

References

Andersen JF (2012) Output-only modal identification of bridges. MSc thesis, Technical University of Denmark (DTU BYG), Copenhagen, Denmark

Andersson A, Sundquist H, Karoumi R (2004) Evaluating cable forces in cable supported bridges using the ambient vibration method. In: Proceedings of the 4th international cable supported bridge Operator's conference, June 16–19th, Copenhagen, Denmark

Bjarnason RTh (2014) System identification of Ölfusá bridge. MSc thesis, University of Iceland, Reykjavik, Iceland. (In Icelandic)

Brincker R, Andersen P (2006) Understanding Stochastic subspace identification. In: Proceedings of the 24th International Modal Analysis Conference (IMAC). St. Louis, Missouri

Caetano E (2011) On the identification of cable force from vibration measurements. In: IABSE-IASS symposium report, London 2011

Geier R (2004) Evolution of stay cable monitoring using Ambient vibration. In: IABSE symposium report, Shanghai 2004: Metropolitan Habits and Infrastructure

Gimsing NJ (1998) Cable supported bridges – concept and design, 2nd edn. Wiley, Hoboken

Nakamura S, Suzumura K (2011) Corrosion of bridge cables and the protection methods. IABSE-IASS symposium report, London 2011

Óskarsson KU (2012) Structural health monitoring of the Ölfusá suspension bridge - damage detection and monitoring aspects. MSc thesis, University of Iceland, Reykjavik, Iceland

Pálsson GP (2012) Finite element modelling and updating of medium span road bridges – case study of Ölfusá bridge in Iceland. MSc thesis, Technical University of Denmark (DTU BYG), Copenhagen, Denmark

Chapter 10
Performance of Base Isolated Bridges in Recent South Iceland Earthquakes

Bjarni Bessason, Einar Hafliðason, and Guðmundur Valur Guðmundsson

Abstract Since 1983, 15 Icelandic bridges have been base isolated for seismic protection. Lead-rubber bearings have been used in all cases. Nine of these bridges are located in the South Iceland Lowland, which is an active seismic zone where earthquakes of magnitude seven can be expected. In June 2000 two Mw6.5 earthquakes struck in the area and in May 2008 a Mw6.3 event hit the area again. Four of the base isolated bridges were located in the near-fault area of these earthquakes with a fault-to-site distance of less than 15 km, and most likely three of them were subjected to strong low-frequency near-fault velocity pulses. None of the bridges collapsed or were severely damaged and all of them were open for traffic immediately after these events. Post-earthquake response analysis has indicated that the base isolation was important for the performance of the bridges.

Keywords Base isolation · Lead-rubber bearings · Near-fault effect · Bridge design · Case history

10.1 Introduction

The seismicity in Iceland is related to the Mid-Atlantic plate boundary which bisects the island. Within Iceland, the N-S boundary shifts eastward through two complex fracture zones. One is located in the South Iceland Lowland, called the South Iceland Seismic Zone, while the other, the Tjörnes Fracture Zone, lies mostly off the northern coast of Iceland (Einarsson 1991). The largest earthquakes in Iceland have occurred within these zones. On the 17th and 21st of June 2000 two M_w6.5

B. Bessason (✉)
Faculty of Civil and Environmental Engineering, University of Iceland, Reykjavík, Iceland
e-mail: bb@hi.is

E. Hafliðason · G. V. Guðmundsson
The Icelandic Road and Costal Administration, Reykjavik, Iceland
e-mail: gudmundur.v.gudmundsson@vegagerdin.is

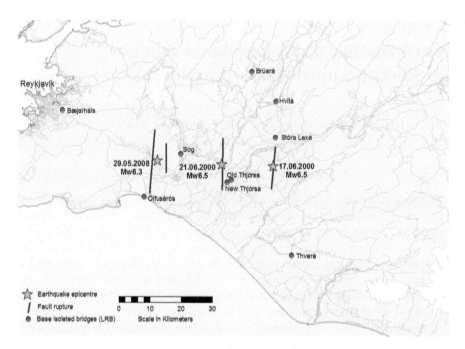

Fig. 10.1 Epicentres and fault rupture of the South Iceland Earthquakes of June 2000 and 2008 as well as location of base isolated bridges in the region (red dots)

earthquakes struck South Iceland and on the 29th of May 2008 an M_w6.3 earthquake shook the area again (Fig. 10.1). The epicentres of these three earthquakes were in the middle of the largest agricultural region in Iceland with a number of farms, small villages and the infrastructure of modern society. Lot of damage was reported, but no residential buildings collapsed and there were no serious injuries and no fatalities (Bessason et al. 2012, 2014; Bessason and Bjarnason 2016).

Since 1983, 15 Icelandic bridges have been base isolated for seismic protection. Lead-rubber bearings (LRB) have been used in all the cases (Bessason 1992; Naeim and Kelly 1999; Bessason and Haflidason 2004; Jónsson et al. 2010). LRB bearings are commonly produced by pressing a pre-casted oversized lead core in an elastomeric bearing which is prefabricated with a hole for the plug. For the Icelandic bridges, a non-traditional low-cost method has been used for most of the bridges. A hole is drilled into a conventional elastomeric bearing and then the lead plug is cast directly into the hole. The performance of these bearings has been dynamically tested in a laboratory (Bessason 1992). The main finding was that the effective yield shear stress of the lead plug was about 20% lower ($\sigma_{yp} = 8.0$ MPa) than commonly reported in other documents where an oversized plug is pressed into a hole.

Nine of the base isolated bridges are located in South Iceland (Fig. 10.1). Of these, one was built after the 2000 earthquake (New Thjórsá Bridge) and one after the 2008 event (Hvítá Bridge). More details of each of them are given in Table 10.1. At least four of the bridges were subjected to strong ground motion with site-to-fault

Table 10.1 Base isolated bridges in South Iceland (Fig. 10.1)

River/Bridge	Year	Main structural system	Spans (m)	No. bearings[a]
Sog	1983	RC beam	16 + 42 + 16	8
Stóra Laxá	1985	RC beam	36 + 48 + 36	6
Ölfusárós[b]	1988	RC beam	36 + 6x48 + 36	20
Old Thjórsa[b]	1950/1991[c]	Steel arch	12 + 83 + 12	4
Bæjarháls	1994	RC beam	28 + 28	4
Þverá	2002	RC beam	27 + 27	4
New Thjórsá	2003	Composed steel-concrete arch	Arc: 78, total: 170	18
Brúará	2008	Composed steel-concrete beam	22 + 18	6
Hvítá	2010	RC beam	25 + 4x55 + 25	14

[a]Not all the bearings are LRB
[b]Instrumented with strong motion accelerometers
[c]Built 1950 and retrofitted with base isolation (LRB) in 1991

distance less than 15 km. None of the bridges collapsed or were severely damaged and all of them were open for traffic immediately after these events. Detailed post-earthquake analysis of the performance of two of the bridges exist (Bessason and Haflidason 2004; Jónsson et al. 2010) but no for the other bridges.

The overall aim of the work presented here is to summarise and give an overview of the performance of base isolated bridges in Iceland which were subject to the South Iceland earthquakes of June 2000 and May 2008, including what was learned and what could have been done differently if they were to be designed again today. The main focus will be on the four bridges which are located closest to the faults, i.e. the Stóra Laxá Bridge, the Old Thjórsá Bridge, the Ölfusárós Bridge and the Sog Bridge. For all these sites the computed Peak Ground Acceleration (PGA) exceeded 0.3 g (Fig. 10.2).

10.2 Computed and Recorded PGA at the Bridge Sites

Strong motion acceleration meters are installed in the Old Thjórsá Bridge and the Ölfusárós Bridge (Table 10.1). Recorded strong motion data from the South Iceland earthquakes of June 2000 exist for both these bridges. For the May 2008 earthquake only data for the Thjórsá Bridge were recorded, since the accelerometers in the Ölfusárós Bridge were unfortunately out of order. Recorded time series from those events are available from the Internet-Site for European Strong Motion Data (ISESD) (Ambraseys et al. 2002). To estimate the ground motion intensity at the other bridge sites it was necessary to rely on ground motion prediction equations (GMPEs). Only a few models exist which

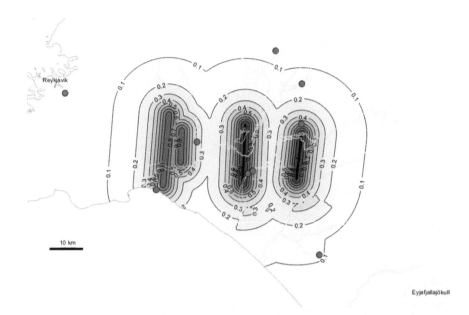

Fig. 10.2 Computed PGA for the three South Iceland Earthquakes of June 2000 and May 2008. Red dots show locations of base isolated bridges in the area (Fig. 10.1). The two South Iceland Earthquakes of June 17th and June 21st are to the right on the map, whilst the May 2008 earthquake (two faults) is to the left (Fig. 10.1)

are based on Icelandic strong motion data. In this study it was decided to use the GMPE provided by Rupakhety and Sigbjörnsson (2009):

$$\log_{10}(PGA) = -1.038 + 0.387 M_W - 1.159 \log_{10}(\sqrt{H^2 + 2.6^2})$$
$$+ 0.123\ S + 0.287\ P(m/s^2) \qquad (10.1)$$

where H is the distance to surface trace of the fault in km, S is a site factor which takes the value 0 for rock sites, and 1 for stiff soil sites. The last term is an error/scatter term where P follows a standard normal distribution, i.e. P∈N(0.1). The model is based on using both the horizontal peak components from each station. Most of the strong motion recordings used in constructing Eq. (10.1) are from Icelandic earthquakes but they were also augmented by records from continental Europe and the Middle East (Rupakhety and Sigbjörnsson 2009). The main characteristic of GMPE given by Eq. (10.1) is that it predicts a relatively high PGA in the near-fault area, whilst the attenuation with distance is more than generally found with a GMPE of similar form. Subsurface fault mapping based on micro-earthquakes exist for all three South Iceland earthquakes of June 2000 and May 2008 (Vogfjörð et al. 2013). By using Eq. (10.1) and geographical fault rupture data (see also Bessason and Bjarnason 2016) it was possible to construct a scenario hazard map based on PGA for these three events and show the locations of the base isolated bridges on the same map (Fig. 10.2). The contours for the 0.2 g line are stretched to the south-east because of sandy soil sites in the flat coastal areas (S = 1 in Eq. (10.1)).

The locations of these soil sites are based on a geological map of South Iceland (Jóhannesson et al. 1982). Three of the bridges are situated in areas where the computed PGA is greater than 0.5 g, i.e. the Stóra Laxá Bridge, the Old Thjórsár Bridge and the Ölfusárós Bridge (Figs. 10.1 and 10.2). The New Thjorsa Bridge (Fig. 10.1) was built after the 2000 earthquakes and therefore only affected by the May 2008 event, and for that event the PGA was <0.10 g at the bridge site.

10.3 Performance of The Base Isolated Bridges

10.3.1 The Stóra Laxá Bridge

The Stóra Laxá Bridge is a 120 m long three-span RC beam bridge with a single lane (Fig. 10.3). It was constructed in 1983. The bridge has three LRB on each abutment and two rubber bearings on each of the two piers. The LRBs are rectangular with the plan size 350 × 450 mm, total rubber thickness 77 mm and lead core diameter of 125 mm. The ratio of total yield force of the LRBs to the weight of the bridge superstructure is $F_{y,LRB}/W \approx 5.5\%$. The rubber bearings at the piers are also rectangular with the plan size 400 × 500 mm and rubber thickness of 49 mm. The bridge is located approximately 3.0 km north of the fault rupture of the June 17, 2000 Mw6.5 earthquake (Figs. 10.1 and 10.2). The longitudinal axis of the bridge is in a NE-SW direction. A location of structures with long natural periods at the end of a fault rupture is generally considered unfavourable with respect to near-fault effects (Somerville et al. 1997). Studies of the recorded ground motion time histories from the South Iceland earthquakes of June 2000 showed that many of them included near-fault low-frequency velocity pulses (Halldórsson et al. 2007). No strong motion record exists for the bridge site, but computed PGA based on the GMPE given by Eq. 10.1 is estimated at 0.61 g. The bridge is believed to have been exposed to strong ground motion during the June 17, 2000 earthquake and there was evidence that the superstructure had moved ±5 cm. There may have been some minor damage of the LRB bearings, but they were not replaced and the original bearings are still serving the bridge.

10.3.2 The Old Thjórsá Bridge

The old Thjórsá Bridge was built in 1951 but seismically retrofitted in 1991 (Fig. 10.4). The approach spans are 12 m long and the main span 83 m. It only has one traffic lane. In the retrofit process, the old original bearings consisting of two fixed steel hinge bearings on the west side and two roller bearings on the east side

Fig. 10.3 The three span 120 m long Stóra Laxá Bridge (Photo: Einar Haflidason)

Fig. 10.4 The old Thjórsá Bridge with main span of 83 m. Lava on west side (left) and bedrock on the east side (Photo: Einar Haflidason)

were replaced with four rectangular LRB, i.e. two on each side. In addition, vertical steel bars (tie bars) were also installed in order to work as a secondary system in the transversal direction to keep the superstructure in place in case of large seismic forces overruling the capacity of the LRB. The plan dimension of the new LRB bearings was 350 × 450 mm, total rubber thickness 77 mm and lead core diameter of 124 mm, i.e. the same bearing size as for the Stóra Laxá Bridge. The mass of the superstructure is 415 tons and the ratio of total yield force of the four LRBs to the weight of the superstructure is $F_{y,LRB}/W \approx 9.6\%$. In 1999 and 2000 it was instrumented with two triaxial accelerometers in the foundations at each side and three single axis accelerometers in the superstructure (Fig. 10.5).

The bridge is located approximately 3.5 km from the fault rupture of the June 21, 2000 earthquake and was subjected to high seismic loads (Fig. 10.1). The computed

10 Performance of Base Isolated Bridges in Recent South Iceland Earthquakes

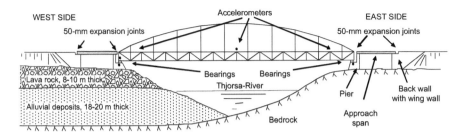

Fig. 10.5 Schematic drawing of the Thjórsá River Bridge showing main parts and the soil profile. The black dots (•) show the location of accelerometers

PGA at the bridge site was close to 0.65 g on the scenario map (Fig. 10.2) and can be compared to recorded values. On the east side the mean PGA value of the two recorded horizontal components was 0.43 g and on the west side the mean peak value was 0.68 g (Bessason and Haflidason 2004). In fact, the highest recorded single component PGA value during the South Iceland earthquakes of June 2000 was measured at the bridge site as 0.84 g on the west side (Thórarinsson et al. 2002).

Studies of peak values, response spectra and other intensity parameters based on the recorded time histories from both sides of the river canyon have shown that there was a significant difference in the ground motion characteristics on each side. In all cases the intensity was greater on the west side independent of the location of the epicentre and fault rupture. This observed difference and amplification could later be explained by unlike site conditions on each side of the river, which was mapped with geotechnical investigations (Bessason and Kaynia 2002). On the west side there is an 8–10 m thick layer of lava rock on top of 18-20 m thick alluvial deposits on older bedrock. On the east side no such sediments exist and the pier and back wall are founded on bedrock (Fig. 10.5).

These observations affected the site selection of the new Thjórsár Bridge, which could be founded on bedrock on both sides of the river canyon by placing it 700 m downstream from the old bridge. The original plan was to locate it much closer. Recent studies from other sites in South Iceland have shown similar site effects, i.e. layers of lava and softer sediments can create site effects that influence the surface ground motion and can be amplified at certain frequency bands (Rahpeyma et al. 2016).

Another important observation from the recorded time histories at the old Thjórsá Brigde site was that they included a near-fault effect that is not always expected for an earthquake of magnitude 6.5 or less (Fig. 10.6) (see also Halldórsson et al. 2007). The main low-frequency velocity pulse was in the longitudinal direction of the bridge with a PGV amplitude of 89 cm/s and a pulse period of 1.90s (Fig. 10.6c). The longitudinal axis of the bridge is almost in the north-south direction (N350°) and therefore nearly parallel with the strike of the June 21 earthquake. In the transverse direction the velocity pulse is not so obvious, neither with respect to amplitude nor to the pulse period. Inspection of the bridge after the June 21, 2000 earthquake showed that the superstructure of the main span and the end spans collided with the piers.

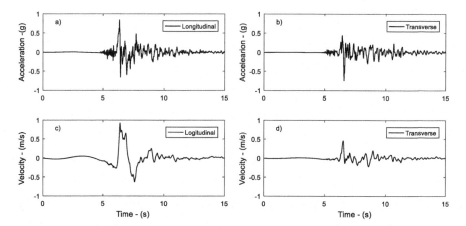

Fig. 10.6 Recorded acceleration time series and integrated velocity time series on the west side of the old Thjórsá Bridge during the South Iceland earthquake of June 21, 2000, (**a**) and (**b**) acceleration in longitudinal and transverse direction, (**c**) and (**d**) velocity in longitudinal and transverse direction

Computational model analysis of the bridge showed that without seismic base isolation the mounting of the old bearings would have been broken and the superstructure then possibly shifted off the concrete foundations into the river in the June 21, 2000 earthquake. Instead the bridge was open for traffic immediately after inspection, after the earthquake.

10.3.3 The Ölfusárós Bridge

The Ölfusárós Bridge is a 370 m long continuous post-tensioned, two-beam concrete structure, built in eight spans with seven piers and two abutments (Fig. 10.7). It was constructed in 1988. At the abutments and on the first pier at each end (i.e. piers 1 and 7) the bridge deck is supported by two friction bearings at each location. At each of the five middle piers (i.e. piers 2, 3, 4, 5 and 6) four rectangular LRB are used, along with vertical steel bars (tie bars) between the piers and the superstructure for earthquake protection of the bridge. The plan dimensions of LRBs are 400x500m, the total rubber thickness 88 mm, and the diameter of the lead cores 125 mm. The total weight of the superstructure is 44MN and the ratio of shear yield force in the LRBs to the weight of the superstructure is $F_{y,LRB}/W \approx 4.5\%$. The bridge's concrete beams have increased thickness over the seven intermittent piers, which stretch 8 m into the span at each side of the piers. On the top of each of the seven piers, concrete blocks were cast in the space between the longitudinal beams in order to limit the horizontal transversal displacements in case of high seismic loads. The gaps between the beams and the blocks on the piers are 70 mm on each side. If the displacements are larger than this, a collision will occur. The limits are stricter at

Fig. 10.7 The base-isolated 370 m long Ölfusárós Bridge, constructed in 1988

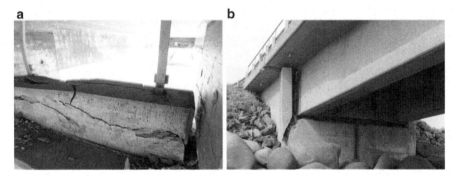

Fig. 10.8 (a) Cracks and smash-up of concrete blocks at top of piers, (b) Broken wing walls at the west abutment, large open crack (250 mm) on the left wall but more closed crack (~10 mm) on the right side but clearly visible

the abutments where horizontal movements of the bridge deck are limited to 50 mm with wing walls on both sides of the deck. For the longitudinal directions expansion joints and back wall with a 50 mm distance were used to limit displacement in this direction.

The bridge was exposed to severe earthquake loads during the South Iceland earthquake on May 29, 2008. Like the Stóra Láxá Bridge, the Ölfusárós Bridge is unfavourably located with respect to the fault rupture, as it is placed approximately 2 km from the south end of the main fault that moved in the earthquake (Jónsson et al. 2010) (Fig. 10.1).

The bridge was damaged during the earthquake and was closed for traffic for a couple of hours whilst inspection was carried out. The concrete blocks on all the piers were hit by transverse motion of the longitudinal beams and severely damaged (Fig. 10.8a). Large cracks formed and big pieces of some blocks were broken and fell away. The tension bars had permanent displacement, either due to yielding of the steel or because they had a bond failure and slipped from the concrete. The wing walls at the abutments were also hit by the superstructure due to the transverse displacement. The impact created a 250 mm wide crack in the south wing wall on the west side of the bridge and nearly dismantled it (Fig. 10.8b). Most of the reinforcement's steel bars in the crack ruptured. The other three wing walls cracked at the bottom of the recess, though the crack width in these was only 1 cm. In the longitudinal direction the superstructure collided with the abutments. Again, the west abutment was more damaged but it was small compared to the damage due to transversal displacements. No damage was observed on the bearings. Observing the damage to the concrete blocks and the wing walls clearly revealed the large displacements of the superstructure. At the abutments the displacements were more than 50 mm in both transversal and longitudinal directions. It is evident from the large gap in the west abutment's south wing wall that the transverse displacement could have measured up to 300 mm. The transverse movements over the piers were at least 70 mm, since the blocks were broken at all piers. After the earthquake the stoppers and the wing walls were repaired. The cost of the repair was estimated to be around 1% of the construction cost of a new identical bridge.

Earthquake response analysis indicated that the bridge was exposed to a near-fault pulse, which most likely was the main reason for large displacements. The computational model was also used to test different design alternatives. This analysis showed that with improved design of the bearing system the bridge should have been able to withstand the earthquake loads without any damage. It should be pointed out that at the time the bridge was designed the then-current information on the earthquake hazard in the area was not known. The improvements were based on replacing friction bearings with LRB and using them at all piers and abutments. The bearings were made higher and the clearance to wing walls and stoppers increased. Furthermore, the stiffness and yield forces in the LRB were chosen to be proportional to the vertical dead load reaction force at each pier and abutment. With this scaling of the bearings the superstructure would mainly behave as a rigid body when moving in the transverse direction as well as in the longitudinal direction (Jónsson et al. 2010).

10.3.4 The Sog Bridge

The Sog Bridge is a 74 m long RC girder bridge in 3 spans with a main span of 42 m and end spans of 16 m (Fig. 10.9). It has 2 lanes and a narrow sidewalk with a total width of the bridge deck of 10.1 m. The bridge has 4 LRB on each abutment and two rubber bearings ($500 \times 500 \times 119$) on each of the two piers. The LRBs are rectangular with dimensions of 300×400, total rubber thickness of 88 mm and lead core diameter of 125 mm. The bridge was located approximately 7 km NW of the fault rupture of the May 2008 Mw 6.3 earthquake (Figs. 10.1 and 10.2). The longitudinal axis of the bridge is in a NW direction.

No strong motion record exists for the bridge site, but computed PGA based on the GMPE given by Eq. (10.1) is estimated at 0.3–0.4 g. Although the bridge is

Fig. 10.9 The base-isolated 72 m long Sog Bridge, constructed in 1983 (Photo: Einar Haflidason)

Fig. 10.10 The Sog Bridge. (**a**) Displacement of the bearings at the south abutment. (**b**) Displacement of bearings at the north abutment

believed to have been exposed to strong ground motion during the 2008 earthquake, no damage was observed after the event and no repair was necessary.

Inspections of the bridge bearings in 2014 indicated some degradation and displacement of the LRB, as can be seen in Fig. 10.10. The displacement was likely the result of the 2008 earthquake.

10.4 Conclusions

At least four base isolated bridges were subjected to strong ground motion during the two equal-size $M_w = 6.5$ South Iceland earthquakes of June 2000 and during the May 2008 South Iceland $M_w = 6.3$ earthquake. All these bridges can be considered to be large on an Icelandic bridge scale. The computed or recorded PGA was above 0.3 g for one of the bridges and for three of them above 0.5 g. Inspection of the bridges after the earthquakes showed in all cases evidence of large motions in expansion joints, pounding of superstructure to abutments and wing walls, tie bars broken, etc. However, for all them the damage was negligible to minor and only one bridge had to be fixed after one of the earthquakes where the repair cost was approximately 1% of the cost of a new bridge.

Nonlinear computational models of two of the bridges (the other two were not modelled) were capable of predicting structural response that correlated fairly well with the observed damage. At least two of the bridges were exposed to near-fault low frequency velocity pulses that caused much larger displacement than if the excitation had been without them. It can be concluded that near-fault pulses must be considered in a design process of new base isolated bridges as well as other types of bridges in South Iceland where the characteristic distance from fault to site is short.

Three of the above bridges were constructed in the 1980s and one retrofitted in 1991. At this time the seismic loads in South Iceland were not well defined. Damage inspection of the bridges after the earthquakes as well as response analysis using recorded time histories from the South Iceland earthquakes of June 2000 and May 2008 indicate that an LRB with greater rubber thicknesses and more clearance between superstructure and abutments, wing walls and other stoppers is preferable. Additional damping devices, e.g. viscous dampers, may be preferable to limit displacements.

References

Ambraseys N, Smit P, Sigbjörnsson R, Suhadolc P, Margaris B (2002) Internet-Site for European Strong-Motion Data (ISESD). European Commission, Research-Directorate General, Environment and Climate Programme

Bessason B (1992) Assessment of earthquake loading and response of seismically isolated Bridges. Ph.D. thesis, Division for Marine Structures, The Norwegian Institute of Technology

Bessason B, Bjarnason JÖ (2016) Seismic vulnerability of low-rise residential buildings based on damage data from three earthquakes (Mw6.5, 6.5 and 6.3). Eng Struct 111:64–79

Bessason B, Haflidason E (2004) Recorded and numerical strong motion response of a base-isolated bridge. Earthquake Spectra 20(2):309–332

Bessason B, Kaynia AM (2002) Site amplification in lava rock on soft sediments. Soil Dyn Earthq Eng 22(7):525–540

Bessason B, Bjarnason JÖ, Gudmundsson A, Sólnes J, Steedman S (2012) Probabilistic earthquake damage curves for low-rise buildings based on field data. Earthq Spectra 28(4):1353–1378

Bessason B, Bjarnason JÖ, Guðmundsson A, Sólnes J, Steedman S (2014) Analysis of damage data of low-rise buildings subjected to a shallow Mw6.3 earthquake. Soil Dyn Earthq Eng 66:89–101

Einarsson P (1991) Earthquakes and present-day tectonics in Iceland. Tectonophysics 189 (1–4):261–279

Halldórsson B, Ólafsson S, Sigbjörnsson R (2007) A fast and efficient simulation of the far-fault and near-fault earthquake ground motions associated with the June 17 and 21, 2000, earthquakes in South Iceland. J Earthq Eng 11(3):343–370

Jóhannesson H, Jakobsson SP, Sæmundsson K (1982) Geological map of Iceland, sheet 6, S-Iceland, 2nd edn. Icelandic Museum of Natural history and Iceland Geodetic Survey, Reykjavik

Jónsson MH, Bessason B, Haflidason E (2010) Earthquake response of a base-isolated bridge subjected to strong near-fault ground motion. Soil Dyn Earthq Eng 30:447–455

Naeim F, Kelly JM (1999) Design of Seismic Isolation Structures, Wiley, New York

Rahpeyma S, Halldorsson B, Olivera C, Green RA, Jónsson S (2016) Detailed site effect estimation in the presence of strong velocity reversals within a small-aperture strong-motion array in Iceland. Soil Dyn Earthq Eng 89:136–151

Rupakhety R, Sigbjörnsson R (2009) Ground-Motion Prediction Equations (GMPEs) for inelastic response and structural behavior factors. Bull Earthq Eng 7(3):637–659

Somerville PG, Smith NF, Graves RW, Abrahamson NA (1997) Modification of empirical strong motion attenuation relations to include the amplitude and duration effect of rupture directivity. Seismol Res Lett 68(1):199–222

Thórarinsson O, Bessason B, Snæbjörnsson JTh, Ólafsson S, Baldvinsson G, Sigbjörnsson R (2002) The South Iceland earthquakes of June 2000: strong motion measurements. In: 12th European conference on earthquake engineering, paper no. 321

Vogfjörð K, Sigbjörnsson R, Snæbjörnsson JT, Halldórsson B, Sólnes J, Stefánsson R (2013) The South Iceland earthquakes 2000 and 2008. In: Sólnes J, Sigmundsson F, Bessason B (eds) Natural hazard in Iceland, volcanic eruptions and earthquakes. University of Iceland Press and Iceland Catastrophe Insurance; 2013. 789 p [in Icelandic]

Chapter 11
Cyclic Capacity of Dowel Connections

Tatjana Isakovic, B. Zoubek, and M. Fischinger

Abstract Dowel connections are most frequently used to join different types of precast elements in RC precast industrial buildings. Such connections are typically subjected to the following types of potential failure mechanism: (a) local failure characterized by the simultaneous yielding of the dowel and crushing of the surrounding concrete, and (b) global failure, characterized by spalling of the concrete between the dowel and the edge of the column or the beam. The strength corresponding to these two types of failure has been estimated using different procedures available in the literature with significantly different accuracy. While the strength corresponding to the local failure is estimated quite precisely, the strength corresponding to the global failure is typically considerably underestimated. The new procedure for the estimation of the global failure is proposed and described in the paper. It was evaluated based on the strut-and-tie model.

Keywords Strength of dowel connections · Cyclic response · Seismic response · Global failure of dowel connections · Strut and tie model

11.1 Introduction

In Europe, the most commonly used system for the construction of precast industrial buildings consists of an assemblage of slender cantilever columns, which are tied together by beams. The majority of industrial facilities in many European countries are housed in such buildings. Recently they have been more frequently used for multi-storey apartment buildings and shopping centres which are used simultaneously by thousands of people. The potential seismic risk involved is therefore high. However, due to the complicated seismic behaviour of these buildings our knowledge is still limited, and the design practice and codes need to be improved.

T. Isakovic (✉) · B. Zoubek · M. Fischinger
Faculty of Civil and Geodetic Engineering, University of Ljubljana, Ljubljana, Slovenia
e-mail: tatjana.isakovic@fgg.uni-lj.si; Matej.Fischinger@fgg.uni-lj.si

The behaviour of precast systems clearly depends on the performance of the specific connections between precast elements. So far, knowledge about the highly complex inelastic seismic behaviour of such connections has been very limited. It was, however, investigated within the FP7 project SAFECAST (Toniolo 2012), where full-scale experiments were performed on specific connections as well as on prototype structures, and the behaviour of different types of precast structures was studied. The study related to the response of dowel connections is presented in this paper.

Typical dowel connections are subject to the following two types of potential failure mechanism (Fig. 11.1): (a) local failure characterized by the simultaneous yielding of the dowel and crushing of the surrounding concrete (Vintzeleou and Tassios 1986; Tanaka and Murakoshi 2011; Zoubek et al. 2013, Magliulo et al. 2014) and (b) global failure, characterized by spalling of the concrete between the dowel and the edge of the column or the beam (Fuchs et al. 1995; Vintzeleou and Tassios 1986; Psycharis and Mouzakis 2012; Capozzi et al. 2012).

A local failure mechanism will typically occur if the distance of the dowel from the edge is large enough (e.g. about six diameters of the dowel or more). In majority of cases this type of failure is ductile. If, however, the dowel is placed closer to the edge of the column or the beam, global failure is more probable, since spalling of the concrete between the dowel and the edge is likely to occur. When there are no stirrups in the critical region, this failure will be brittle, since the capacity of the connection is governed by the tensile failure of the concrete between the dowel and the edge of the column or beam.

Local failure has been the subject of many studies and can be predicted with great accuracy using the procedures, proposed in the literature (e.g. in the references previously cited). Contrary, the studies of the global failure are only few (e.g. experimental

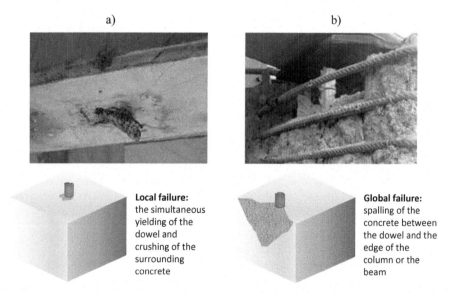

Fig. 11.1 Two types of failure: (**a**) local failure, (**b**) global failure

studies by Fuchs et al. 1995 and analytical studies by Vintzeleou and Tassios 1986). Thus the procedures, proposed to estimate this type of failure, are considerably less precise and typically unacceptably conservative. The reason is mostly related to the inadequately defined role of the stirrups around the dowel. They can significantly influence (increase) the strength and change the type of failure from brittle to ductile.

In this paper only the less investigated global type of failure is analysed. A new procedure for estimation of the corresponding strength is proposed. It explicitly takes into account the contribution of stirrups in the critical region around the dowel employing the strut and tie model. The procedure is presented in Sect. 11.2.

The proposed procedures were evaluated by means of experiments that were performed within the scope of the SAFECAST project (Performance of Innovative Mechanical Connections in Precast Building Structures under Seismic Conditions) at the University of Ljubljana (UL) and the National Technical University of Athens. These experiments are described briefly in Sect. 11.3. The experimental and analytical results are compared in Sect. 11.4.

11.2 Global Failure of Dowel Connections

Dowel connections in which the dowel is placed close to the edge of the concrete elements are susceptible to splitting of the concrete between the dowel and the edge of the element. When there are no stirrups in the critical region around the dowel, failure typically occurs due to excessive tensile stresses (see solid line in Fig 11.1). The failure is brittle.

In the critical region around the dowel, RC elements may often be reinforced by quite compact transverse reinforcement. Such reinforcement changes the stress field and typically changes the type of failure of the connection from brittle to ductile (compare the response designated by the solid line with that represented by the dashed and dotted lines in Fig. 11.2).

The influence of stirrups on the strength of the dowel connection depends on their diameter and spacing. If the precast elements are reinforced by a relatively large quantity of stirrups, the strength provided by them will be typically greater than the tensile strength of the concrete itself. In such cases the strength of the connection increases after cracking of the concrete (indicated by the dashed line in Fig. 11.2). However, if fewer stirrups are provided, the tensile strength of concrete can be larger than the strength of stirrups. In such cases the strength of the connection is typically reduced after cracking of the concrete (see the dotted line in Fig. 11.2).

In the study presented in this paper, it is considered that the global strength of the dowel connection is provided by stirrups after cracking of the concrete (the contribution of the concrete to the strength is neglected). As has already been discussed, the strength defined in this way can be larger or smaller than the strength provided by the tensile strength of the concrete (see Fig. 11.2 for more details).

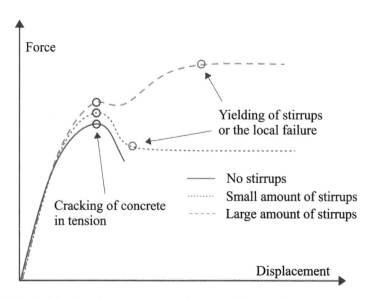

Fig. 11.2 Typical force/displacement response of a eccentric beam-column dowel connection

11.3 The New Procedure for Estimation of the Strength Corresponding to the Global Failure

Taking into account the crucial role of stirrups, a different approach from that presented in the previous studies was implemented for the estimation of the strength of dowel connections in precast buildings. The stirrups were considered explicitly, employing a strut and tie model, as is illustrated in Fig. 11.3.

Strut and tie models are already well established, and have been widely used in different codes (ACI 318–08, CEN 2004a, NZS 3101) mainly to solve those problems where Bernoulli's hypothesis about a linear distribution of strains cannot be applied. Generally, strut and tie models permit designers to choose the way in which the load is transferred, selecting certain arrangement of stirrups. This arrangement defines the configuration of an equivalent truss, where the compressive stresses in the concrete are in equilibrium with the tensile stresses in the stirrups.

In the first column of Fig. 11.3 typical configurations of the connections are presented. The related strut and tie model is shown in the second column. The third column presents the stresses, calculated by the FEM analysis (Zoubek et al. 2013). In the last column of Fig. 11.3, a closed expression of the strength capacity of the dowel connections is given. This strength is defined as a force corresponding to the yielding of the first layer of the stirrups. The complete utilization of the compression struts is connected to the local ductile failure mechanism (see Zoubek et al. 2015 for more details), and is not considered in Figs 11.3 and 11.4.

Let us now examine the strength capacity of the connections in more detail on the example of a single eccentric dowel and perimeter hoops (CASE 1 in Fig. 11.3). In

11 Cyclic Capacity of Dowel Connections

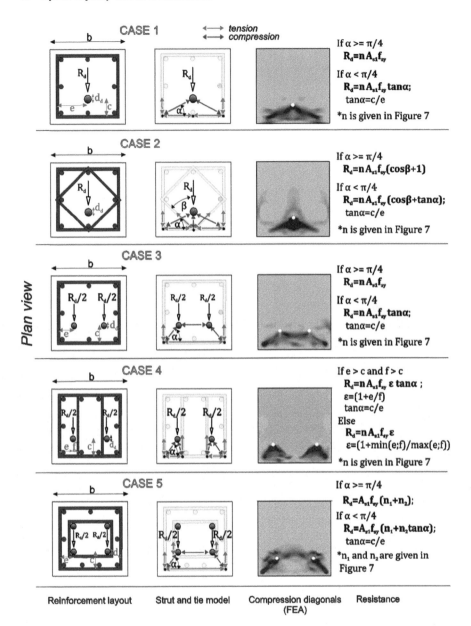

Fig. 11.3 Proposal for the calculation of the resistance of eccentric dowel connection for different reinforcement layouts which are most frequently used in practice

this case the equivalent truss consists of two compression diagonals (blue lines) and stirrups (red lines). If the dowel is placed relatively close to the edge of the concrete section (i.e. if the angle α in Fig. 11.3 is smaller than $\pi/4$), yielding will occur in the

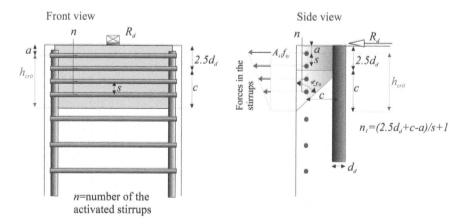

Fig. 11.4 Schematic presentation of the assumed distribution of stresses in the confining bars and the number of activated stirrups n for different arrangements of the reinforcement (Fig. 11.3)

legs of the stirrups which are perpendicular to the direction of loading. In this case the strength can be expressed as:

$$F_{max} = 2A_{s1} f_{sy} \tan \alpha \tag{11.1}$$

where A_{s1} is the cross-section of one leg of the perimeter hoop, and f_{sy} is the yield strength of the steel.

If the distance of the dowel from the edge is larger (i.e. if the angle α in Fig. 11.3 is greater than $\pi/4$ and $tg\alpha > 1$) yielding will occur in the leg of the stirrups which is parallel to the loading direction. The strength of the connection can then be expressed as:

$$F_{max} = 2A_{s1} f_{sy} \tag{11.2}$$

The critical region, where rupture of the concrete is typically observed, is not limited to one cross-section of the column or beam, but it is spread along a certain length of the dowel. Consequently, more than one layer of stirrups may be activated. All these stirrups influence the strength of the connection. Based on the FEM analysis (Zoubek et al. 2013) and based on the experimental data, it has been observed that the height of the critical region can be defined as is illustrated in Fig. 11.4.

Taking into account the typical shape of the ruptured concrete it can be assumed that the width of this region is approximately constant (see Fig. 11.4, left). Its depth, however, changes as is illustrated in Fig. 11.4 (right). At the top of the column all over the height of 2.5 d_d (d_d is the diameter of the dowel) this depth is equal to the distance between the dowel and the axes of the stirrups. It is then gradually reduced. Taking into account the results of the FEM analyses and the experimental data, it has

11 Cyclic Capacity of Dowel Connections

been observed that the depth is reduced almost linearly at an angle of 45° (see Fig. 11.4 for more details).

The number of the engaged stirrups n can be defined, taking into account the height of the critical region h_{crit} (see Fig. 11.4) and vertical distance between the stirrups s (see Fig. 11.4) as:

$$n = h_{crit}/s + 1 \qquad (11.3)$$

The strength of the dowel connection R_d is defined as the force F (see Eqs. 11.1 and 11.2) which is applied to the dowel when yielding of the first layer of stirrups (i.e. of the top stirrup) occurs. Stresses in the other stirrups in the critical region are linearly reduced, as is illustrated in Fig. 11.4. The total resistance of all the stirrups can be expressed considering the average stress σ_{avg} in the stirrups as:

$$R_d = F_{max} = 2\ n\ A_{s1}\ \sigma_{avg}\ \tan\alpha = 2\ n\ A_{s1}\ (f_{sy}/2)\ \tan\alpha$$
$$= n\ A_{s1}\ f_{sy}\ \tan\alpha \qquad (11.4)$$

and

$$R_d = F_{max} = 2\ T_1 = 2\ n\ A_{s1}\ \sigma_{avg} = 2\ n\ A_{s1}\ (f_{sy}/2) = n\ A_{s1}\ f_{sy} \qquad (11.5)$$

for the case $\alpha \leq \pi/4$ and $\alpha > \pi/4$, respectively.

In the previous equations, A_{s1} is the cross section of one stirrup's leg; f_{sy} is the yield strength of the steel; n is the number of activated stirrups; and α is the angle marked in Fig. 11.3.

11.4 Evaluation of the Proposed Analytical Procedure

11.4.1 Short Description of the Experiments

Experimental investigations were used to evaluate the proposed analytical procedures. Within the scope of the SAFECAST project, quite extensive experimental work was performed on dowel connections at the University of Ljubljana (UL) and at the National Technical University of Athens (NTUA). A detailed description of these experiments can be found in (Fischinger et al. 2012) for the test performed at UL, and in (Psycharis and Mouzakis 2012) for the tests performed at NTUA. Two examples of the test specimens are presented in Fig. 11.5. They consist of a part of the beam and column, having dimensions that are typical for precast industrial buildings built in Europe. The beam and the column are connected together by means of one or two dowels.

In all tests the horizontal loading was applied to the beam by means of a hydraulic actuator. The tests performed at NTUA were essentially pure shear tests (Fig. 11.5,

Fig. 11.5 Beam-to-column dowel connection specimens tested at (left) the University of Ljubljana (UL) and (right) the National Technical University of Athens (NTUA)

right). No vertical loading was applied. In the case of experiments performed at UL (Fig. 11.5, left) a vertical loading of 100 kN was applied at the mid-point of the beam span by means of a vertical actuator.

As opposed to the tests performed at NTUA, which were essentially shear tests, in case of the tests performed at UL large relative rotations between the beams and the columns were imposed. Actually, the strength reduction of the dowel connections due to these rotations was one of the main points of interest at UL.

For the details of the tested specimens, please see (Fischinger et al. 2012) and (Psycharis and Mouzakis 2012). They vary with regard to the number of dowels, the diameter of the dowels, their distance from the edge of the columns or beams, and the amount of longitudinal and transverse reinforcement in the columns and beams. The strength of the concrete and steel also varies.

11.4.2 *Comparison between the Experimental and Analytical Results*

The analytical procedure proposed in Sect. 11.3 was evaluated by means of the experiments which are described in Sect. 11.4.1. The strength of the dowel connection was defined as the lower of the strength values corresponding to the global and the local failure. The local failure was estimated using the similar procedure as those reported in the literature (for more details see Zoubek et al. 2015).

The very good correlation which was observed between the analytical and experimental values is illustrated graphically in Fig. 11.6, using the resistance ratio, which is defined as r = calculated resistance/actual resistance. Two groups of results are presented. The grey diamonds and circles represent the resistance

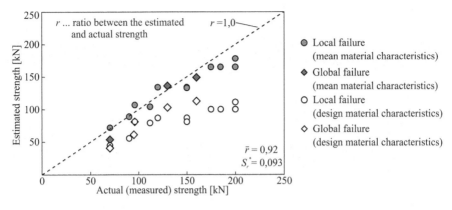

Fig. 11.6 Comparison of the calculated and actual resistance of the dowel connections

ratios, which correspond to the analytical values, obtained by taking into account the mean values of the concrete and steel strengths. The circles indicate local failure, whereas the diamonds indicate global failure. The mean material characteristics are defined based on the results of uniaxial compressive tests of concrete cylinders and uniaxial tensile test of steel bars. The prediction of the proposed procedure for the mean material characteristics agrees very well with the experiments (Fig. 11.6). The estimate of the mean of the resistance ratio is $\bar{r} = 0.92$ and standard deviation 0.093. When the mean values of strengths are reduced to the corresponding design values (as defined according to the Eurocode 2 (CEN 2004a) and Eurocode 8 (CEN 2004b) standards, the resistance ratios presented by white diamonds and circles are obtained.

11.5 Conclusions

The failure of the dowel connections can occur due to the two reasons:
(a) due to the simultaneous yielding of the dowel and crushing of the surrounding concrete when the compression capacity of the concrete is exceeded,
(b) due to the spalling of the concrete between the dowel and the edge of the column or the beam when the tensile stresses of the concrete are exceeded or the yielding of the stirrups occurs.

The first type of failure is local, since it is localized to the limited area around the dowel. The second type of failure is global since it affects the larger area between the dowel and the edge of the concrete elements. The local failure occurs, when the dowel is well embedded into the surrounding concrete. The global failure typically occurs when the dowel is placed closer to the edge of the concrete cross-section, typically at the distance smaller than six diameters of the dowel.

While the local failure mechanism has been relatively well investigated, and presented in several studies, the procedures for the estimation of the global failure have been only few, and they have been unacceptably conservative, since they have not recognized the role of the stirrups around the dowel properly.

Thus the new procedure, which can be used to estimate the strength corresponding to the global failure of dowel connections is proposed. Taking into account an appropriate strut and tie model, the influence of the stirrups on the achievable resistance as well as on the type of the failure of dowel connections is taken into account explicitly and accurately. The comparison with experimental results revealed that the accuracy of the proposed method is good. Since the method is quite simple it can be easily applied in the design practice.

Acknowledgements The presented research was supported by the SAFECAST project "Performance of Innovative Mechanical Connections in Precast Building Structures under Seismic Conditions" (Grant agreement no. 218417-2) within the framework of the Seventh Framework Programme (FP7) of the European Commission. The experiments performed by UL were realized in cooperation with the Slovenian National Building and Civil Engineering Insti-tute (ZAG). The specimens were constructed at the Primorje d.d. company. The research was partly supported by the Ministry of Education, Science and Sport of Republic of Slovenia.

References

ACI 318-08 (2008) Building code requirements for structural concrete and commentary. American Concrete Institute, Farmington Hills, Michigan, USA

Capozzi V, Magliulo G, Manfredi G (2012) Nonlinear Mechanical Model of Seismic Behaviour of Beam-Column Pin Connections. In: 15th world conference on earthquake engineering, Portugal, Lisbon, 24–28th September, 2012

CEN 2004a (2004) Eurocode 2, Design of concrete structures, general rules and rules for buildings. CEN, Brussels

CEN. 2004b (2004) Eurocode 8: design of structures for earthquake resistance – part 1: general rules, seismic actions and rules for buildings. CEN, Brussels

Fuchs W, Eligehausen R, Breen JE (1995) Concrete Capacity Design (CCD) approach for fastening to concrete. ACI Struct J 92(1):73–94

Fischinger M, Zoubek B, Kramar M, Isakovic T (2012) Cyclic Response of Dowel Connections in Precast Structures. In: 15th world conference on earthquake engineering, Portugal, Lisbon, 24–28th September, 2012

Magliulo G, Ercolino M, Cimmino M, Capozzi V, Manfredi G (2014) FEM analysis of the strength of RC beam-to-column dowel connections under monotonic actions. Constr Build Mater 69:271–284

NZS 3101: 2006 (2006) Concrete structures standard. Part 1 – the design of concrete structures and part 2 – commentary. Standards New Zealand, Wellington

Psycharis IN, Mouzakis HP (2012) Shear resistance of pinned connections of precast members to monotonic and cyclic loading. Eng Struct 41:413–427

Tanaka Y, Murakoshi J (2011) Reexamination of dowel behavior of steel bars embedded in concrete. ACI Struct J 108(6):659–668

Toniolo G (2012) SAFECAST Project: European research on seismic behavior of the connections of precast structures. In: 15th world conference on earthquake engineering, Portugal, Lisbon, 24–28th September, 2012

Vintzeleou EN, Tassios TP (1986) Mathematical model for dowel action under monotonic and cyclic conditions. Mag Concr Res 38:13–22

Zoubek B, Fahjan J, Isakovic T, Fischinger M (2013) Cyclic failure analysis of the beam-to-column dowel connections in precast industrial buildings engineering structures. Eng Struct 52:179–191

Zoubek B, Fischinger M, Isakovic T (2015) Estimation of the cyclic capacity of beam-to-column dowel connections in precast industrial buildings. Bull Earthq Eng (7, 7):2145–2168

Chapter 12
Ductile Knee-Braced Frames for Seismic Applications

Sutat Leelataviwat, P. Doung, E. Junda, and W. Chan-anan

Abstract This paper presents the behavior and design concept of efficient structural steel systems based on innovative applications of knee braces. Advantages of knee-braced frames (KBF) include relatively simple connections for ease of construction and reparability after an earthquake and less obstruction as compared to conventional bracing systems. Various configurations of KBFs can be designed and detailed for different levels of strength, stiffness, and ductility. KBFs are designed so that all inelastic activities are confined to the knee braces and designated yielding elements only. Key design concepts to ensure ductile behavior of KBFs are first summarized. Finally, results from experimental and analytical studies into the behavior of KBFs are briefly presented. The results show that KBFs can provide viable alternatives to conventional structural systems.

Keywords Knee braces · Knee-braced frames · Seismic resistant steel structures · Cyclic tests

12.1 Introduction

Knee braces were widely used in the past for wind-resistant design. However, the application of knee braces for seismic resistant structures is still limited. In the past several years, the authors have conducted extensive experimental and analytical studies to develop ductile knee-braced frames for seismic applications (Srechai 2007; Suksen 2007; Leelataviwat et al. 2011a, b; Junda 2011;

S. Leelataviwat (✉) · P. Doung
Department of Civil Engineering, King Mongkut's University of Technology Thonburi, Bangkok, Thailand
e-mail: sutat.lee@kmutt.ac.th

E. Junda · W. Chan-anan
Department of Civil and Environmental Engineering, Nakhon Pathom Rajabhat University, Nakhon Pathom, Thailand

Wongpakdee et al. 2014; Doung 2015). Knee-braced frames (KBFs) utilize relatively simple connections for ease of construction and reparability after an earthquake. More importantly, the knee braces provide much less obstruction as compared to the braces of conventional systems, making this system architecturally attractive. With a slight modification in the brace connections, knee braces can also be utilized in the seismic strengthening of existing steel frames. Various configurations of KBFs can be designed and detailed for different levels of strength, stiffness, and ductility. Figure 12.1 shows various KBF configurations that have been investigated by the authors to date.

In Fig. 12.1a, knee braces are combined with a moment frame to provide largest strength and stiffness. This system can utilize either conventional buckling braces or buckling restrained braces (BRBs) for higher ductility. For this system, the frame is designed so that the knee braces will yield under seismic loads followed by plastic hinging of beams at the ends of the beam segments outside the knee portions. In Fig. 12.1b, partially restrained (PR) connections are used instead of rigid connections. PR connections such as bolted top and seat angle with double web angle connections can be used for beam-column connections. These PR connections make the erection of the frame relatively simple and allow the frame to be repaired after an earthquake. Properly detailed, these PR connections can exhibit considerable ductility and energy dissipation capacity. In Fig. 12.1c, simple or shear connections are used. For this system, the beam is designed to be fully elastic under the largest forces generated by the knee braces. This system is most efficient in terms of ease of construction and reparability after an earthquake. Hence, it ranks highly on ductility and resiliency but may lack the strength exhibited by the back-up moment frames. In Fig. 12.1d, open-web truss frames replaces the solid beams used in other systems. This system is suitable for long-span applications.

The design of all the KBF systems above can be carried out based on a capacity design concept that results in ductile behavior. The frames are designed so that the knee braces will yield. All inelastic activities are confined to designated yielding elements and are directed away from the critical areas, decreasing the dependence of the performance on the material and quality of workmanship.

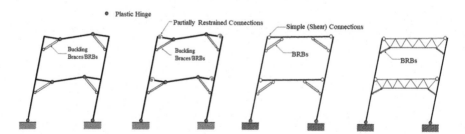

Fig. 12.1 Various Knee-braced systems: (**a**) Knee-braced moment frame; (**b**) Knee-braced moment frame with partially restrained connections; (**c**) Knee-braced frames; and (**d**) Knee-braced truss moment frame

In this paper, the key concepts for the design of ductile KBF systems are summarized first. Examples of the response from cyclic tests of selected systems are presented. Finally, an example of the dynamic response of a selected system is provided. This paper provides a comprehensive overview of the design concept and behavior of viable seismic resistant structural systems based on knee bracing concept.

12.2 Design Concept of Knee-Braced Frames

Based on the past experimental and analytical studies, the ductile behavior of KBFs hinges on two important design considerations, controlling the deformation demands on the knee braces and the design of the columns to resist the forces induced by the knee braces. The following sections elaborate on these two key aspects. It should be noted that the following discussions focus on a KBF with simple connections. However, the concepts are also applicable to all the systems presented in Fig. 12.1.

12.2.1 Knee Braces Design

For the systems shown in Fig. 12.1, knee braces are the primary designated yielding elements. Hence, they are expected to deform well into the inelastic range. For this reason, BRBs are more suitable than conventional braces. Compared to the braces in a conventional braced frame, knee braces may experience larger axial strain demand for a given frame drift. Fig. 12.2 compares the deformation and the strain demand of the braces in a KBF and a conventional concentrically braced frame (CBF). The strain demands shown in Fig. 12.2 were computed assuming rigid beams and columns. Fig. 12.3 shows the brace strain versus drift angle plots for the KBF and CBF. As can be seen, comparing to the diagonal brace in the CBF, knee braces in the

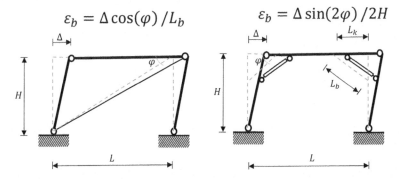

Fig. 12.2 Deformation and the strain demand of the braces in a KBF and a conventional CBF

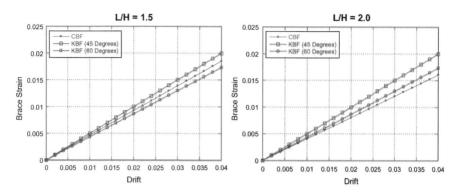

Fig. 12.3 Brace strain in a KBF and CBF

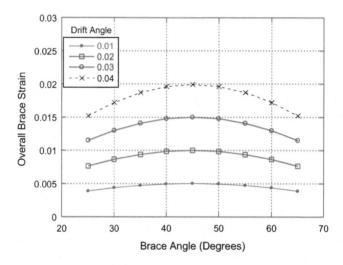

Fig. 12.4 Brace strain as a function of brace angle for different frame drifts

KBF can experience larger of smaller deformation demand depending of the frame configuration and brace angle (φ). For shorter braces like knee braces, the deformation capacity are generally smaller than that of a conventional brace because the deformation can only distribute over a short length. Therefore, one of the most important aspects in the design of a KBF is to ensure that the deformation or strain demand can be safely accommodated by the knee braces. Longer knee braces (with large brace angle, φ) can generally accommodate larger deformation demand but longer braces may also obstruct the passage in the bay.

In a KBF, the size and brace angle must be chosen based on the balance between function, deformation demand, and brace ductility. Figure 12.4 shows the brace strain demands as a function of knee brace angle (φ) for different drift angles. The strain varies depending primarily on the brace angle and becomes largest for $\varphi = 45$ degrees. Based on Fig. 12.4, the type of braces and brace angle can be chosen according to the expected level of frame drift.

As mentioned earlier, one of the most important design considerations is the controlling of the deformation demands on the knee braces. For this reason, KBF is most suited for a displacement-based design procedure. For a given brace angle, a plot such as Fig. 12.4 can facilitate the selection of a target drift of the frame. Once the target drift is chosen, a displacement-based design method can be used to obtained the require frame strength to ensure that frame drift remains within the target. Any displacement-based design procedures can be used for this purpose. One displacement-based design method that has been successfully used by the authors is called Performance-based Plastic Design (PBPD) method (Goel and Chao 2008).

In the PBPD method, the design base shear for a selected hazard level and a target drift is calculated using energy balance concept. The required frame strength is computed by equating the work needed to push the structure monotonically up to the target drift to that required by an equivalent elastic-plastic single degree of freedom system to achieve the same state. For KBFs, a target deformation can be selected (based on Fig. 12.4) and the required base shear strength can be computed. The sizes of the BRBs or knee braces can then be chosen based on the required frame strength. It has been found that the PBPD method is very effective in controlling the deformation of frame and the braces to within the target. The details of the PBPD method as applied to KBFs can be found elsewhere (Srechai 2007; Goel and Chao 2008; Wongpakdee et al. 2014).

12.2.2 Column Design

One of the concerns regarding the use of KBFs is that the knee braces may induce large flexural moments in the columns leading to a soft-story type mechanism under seismic excitations. However, recent developments in the design and assessment of structural systems under seismic excitation allow KBFs to be designed with a high degree of accuracy and confidence. For KBF systems, the columns should be designed to remain fully elastic (except at the bases) under the maximum forces induced by fully strain-hardened braces. In order to achieve this objective, the columns can be designed based on the capacity design concept.

One approach that has been used successfully is to apply the concept of plastic design with the corresponding yield mechanism shown in Fig. 12.1. To remain fully elastic, the columns must be designed to resist the knee brace forces adjusted to fully-yielded and strain-hardened conditions. Based on the PBPD approach, a capacity design method that considers the equilibrium of the entire column subjected to all forces can be carried out. This method is sometimes referred to as "column tree" analysis. Figure 12.5 shows the example of a column tree analysis. The forces associated with the beams and BRBs are applied to the column tree. The columns can also be designed using pushover analysis. Once the brace sizes have been determined, the frame can be "pushed" up to the target drift level by assuming

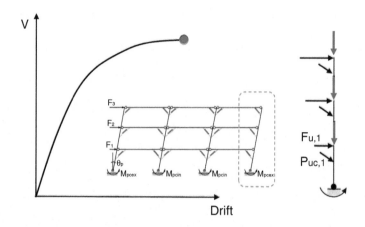

Fig. 12.5 Column tree design based on pushover analysis

elastic columns except at the bases where plastic hinges are allowed to form. The moment and axial force diagrams of the elastic columns are then used to design the member sizes. This process can be done iteratively until the column sizes satisfy the elastic behavior objective.

12.3 Experimental Results

Large-scaled experiments have been carried out to assess the performance of KBFs with different configurations. Specimens in the form of a portal frame and a T-shaped sub-assemblage have been tested under cyclic loading. Some selected results from the tests are reviewed in this paper. Figure 12.6 shows a test specimen of a knee-braced moment frame with partially restrained connections. For this specimen, the beam-to-column joints consisted of top and seat angle connections. Angles were also used to connect the beam web to the column flange. Regular braces made of a hollow circular section were used for this specimen. The knee braces connected to beams and columns by bolted connections. In the test, the specimen was subjected to quasi-static, cyclic loading until failure.

The hysteretic loops are shown in Fig. 12.6c. The results show a stable hysteretic response through the entire loading history. The pinching behavior which is the characteristic of a frame with partially restrained connections is also apparent. The pinching is mainly due to the combination of the opening and closing of gaps and slippage at bolt holes. The specimen was able to deform upto 4% drift when the fracture initiated in the braces.

Figure 12.7 shows one of the test specimens for a KBF with BRBs (Fig. 12.1d). A T-shaped sub-assemblage representing half of a beam and half of a column was tested. Single plate shear connection was used at the beam-to-column connection. The column was mounted to the strong floor. The cyclic load was applied at the end

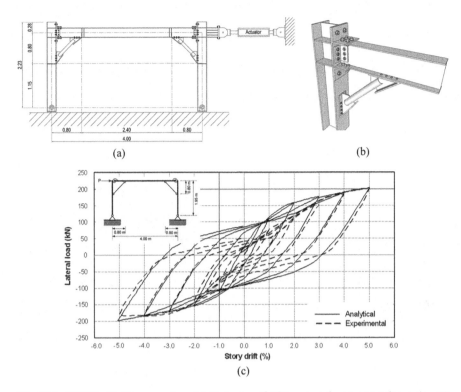

Fig. 12.6 KBMF with PR connections (**a**) Test set-up (**b**) Beam-to-column connection region (**c**) Cyclic test result

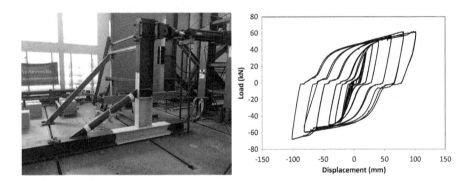

Fig. 12.7 KBF with single plate shear connections

of the beam. The BRB was oriented at a 60 degree angle to minimize the strain as shown in Fig. 12.4. The specimen shows very ductile behavior with full and stable hysteretic loops. The test was stopped due to fracture of one of the bolts at the single plate shear connection. However, even with the loss of one of the bolts at the shear connection, there was only a minor decrease in the lateral load resistance of the frame. The specimen was able to deform more than 5%. The connection was designed primarily to carry the shear force and was unable to accommodate the rotation of the beam.

12.4 Dynamic Response

A 3-story building shown in Fig. 12.8 was selected as an example to illustrate the dynamic response of KBFs. The building was assumed to be an office building. In this study, the frame was designed as KBF with BRBs and single plate shear connections. The study frame was designed in accordance with the PBPD method described above. The structural system was designed for a Design Category D with $S_1 = 0.6$ g and $S_s = 1.5$ g following ASCE 7-10 (2010). The design base shear of the study frame was evaluated at two hazard levels, DBE and MCE, for 2% and 3% target drifts respectively. Nonlinear static pushover and dynamic time history analyses were conducted using a set of 44 ground motions based on FEMA P695 (2009).

Sample results from pushover and time history analyses are shown in Figs. 12.9 and 12.10. As can be seen, the inelastic activities occurred only at the designated locations. The frame performed as intended in the design. The story drifts were within the target limits for both DBE and MCE levels.

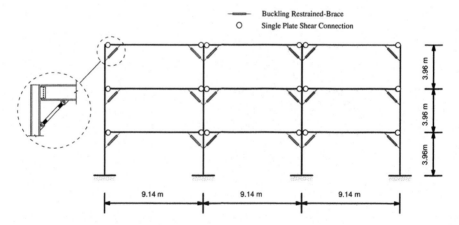

Fig. 12.8 Example 3-story KBF with single plate shear connections

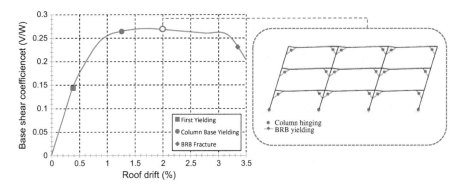

Fig. 12.9 Pushover analysis results

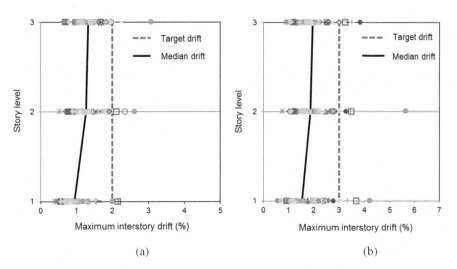

Fig. 12.10 Time history analysis results (**a**) DBE level (**b**) MCE level

12.5 Summary

The behavior and design concept of efficient structural systems based on innovative applications of knee braces are presented in this paper. Selected experimental and analytical study results are reviewed and discussed. Two key design issues to achieve ductile behavior for KBFs consist of limiting the deformation demand in the knee braces and designing the columns to remain elastic. Design approaches to control the deformation demands of the knee braces and the design of the columns are presented. Based on research carried out thus far, it was found that KBFs with various configurations represent viable alternatives to conventional structural systems. Various configurations of KBFs can be designed and detailed for different levels of strength, stiffness, and ductility.

Acknowledgements The research presented in this paper has been supported by different agencies over a number of years. The authors would like to acknowledge funding from the Thailand Research Fund (TRF), National Research Council of Thailand, and King Mongkut's University of Technology, Thailand.

References

ASCE 7-10 (2010) Minimum design loads for buildings and other structures. American Society of Civil Engineers (ASCE), ASCE/SEI 7-10, Virginia

Doung P (2015) Seismic collapse evaluation of buckling-restrained Knee-Braced frames with single plate shear connections. M.Eng. thesis, Department of Civil Engineering, King Mongkut's University of Technology Thonburi, Bangkok, Thailand

FEMA P695 (2009) Quantification of building system performance factor. Federal Emergency Management Agency, Redwood City

Goel S-C, Chao S-H (2008) Performance-based plastic design: earthquake-resistant steel structures. International Code Council, Country Club Hills

Junda E (2011) Dynamic response of Knee-Braced moment Frames with partially restrained connections under earthquake excitations, M.Eng. thesis, Department of Civil Engineering, King Mongkut's University of Technology Thonburi, Bangkok, Thailand

Leelataviwat S, Srechai J, Suksan B, Warnitchai P (2011a) Performance-based design approach for ductile knee-braced moment frames. In: Proceeding of the 9th U.S. National and 10th Canadian conference on earthquake engineering. No. 487

Leelataviwat S, Suksan B, Srechai J, Warnitchai P (2011b) Seismic design and behavior of ductile knee-braced moment frames. J Struct Eng ASCE 137(5):579–588

Srechai J (2007) Dynamic response of knee braced moment frames under earthquake excitations, M.Eng. thesis, Department of Civil Engineering, King Mongkut's University of Technology Thonburi, Bangkok, Thailand

Suksen B (2007) Cyclic testing of Knee Braced moment Frames, M.Eng. thesis, Department of Civil Engineering, King Mongkut's University of Technology Thonburi, Bangkok, Thailand

Wongpakdee N, Leelataviwat S, Goel SC, Liao W-C (2014) Performance-based design and collapse evaluation of buckling restrained knee braced truss moment frames. Eng Struct 60:23–31

Chapter 13
Seismic Capacity Reduction Factors for a RC Beam and Two RC Columns

Pablo Mariano Barlek Mendoza, Daniela Micaela Scotta, and Enrique Emilio Galíndez

Abstract Many building structures can be damaged and even collapse during a severe earthquake. For this reason it is important to take immediate decisions about the safety of damaged structures in order to avoid possible human losses in case of aftershocks. Therefore, a quantitative damage assessment should be made to estimate residual seismic capacity. Reinforced Concrete (RC) Frames are one of the most common earthquake resistant elements used in Argentina. Consequently, it was judged necessary to study the residual seismic capacity of the basic components of this structural type. Experimental results and numerical models of RC beams and columns were considered to establish the reduction factors for different damage classes. A beam tested by the authors and two columns reported in the bibliography were studied. The results were compared with the values suggested by the revised Japanese Guideline for Post-Earthquake Damage Evaluation and Rehabilitation (2014) and a satisfactory agreement was found between them.

Keywords Residual seismic capacity · RC beams · RC columns · Crack width

13.1 Introduction

After a seismic event it is necessary to carry out a post-earthquake safety assessment in order to identify safe buildings from those that are not. In a subsequent stage a detailed damage assessment is performed on structures to determine the need for repair, reconstruction or rehabilitation. Consequently, it is important to develop a technical guide to help engineers perform these tasks with a unified criterion.

Many evaluation methodologies were proposed in different countries over the years. The Japanese post-earthquake damage assessment methodology is one of the most widespread. Seismic capacity reduction factors are fundamental in this

P. M. Barlek Mendoza (✉) · D. M. Scotta · E. E. Galíndez
Instituto de Estructuras "Ing. Arturo M. Guzmán", Universidad Nacional de Tucumán, San Miguel de Tucumán, Argentina

methodology. These factors are obtained from experimental tests on structural members. In this paper seismic capacity reduction factors are calculated for a RC beam and two columns. The results are compared with values recommended by the Japanese guideline.

13.2 Post-Earthquake Damage Assessment Based on Residual Seismic Capacity

In Japan, the Guideline for Post-Earthquake Damage Evaluation and Rehabilitation (JBDPA 1991) was originally developed in 1991 and revised in 2001 and 2014. The main objective of the guideline is to provide rational criteria in order to identify and rate building damage quantitatively. Residual seismic capacity of reinforced concrete (RC) buildings is evaluated through the R-index. This index is defined as the ratio of post-earthquake seismic capacity to the structure's original capacity. Mathematically,

$$R\,[\%] = \frac{_D I_S^D}{I_S \cdot 100} \tag{13.1}$$

Where:

$_D I_S$: Seismic Performance Index of the structure after earthquake damage.
I_S: Seismic Performance Index of the structure before earthquake damage.

Experience in post-earthquake damage assessment has shown that $R = 95\%$, the limit between slight and light damage, is a limit state of serviceability. Thus, structures with slight damage do not need to be repaired and can continue to operate after the seismic event. Additionally, $R = 60\%$, the limit between moderate and severe damage, is a limit state of reparability. Therefore, most of the buildings found with severe damage were demolished and rebuilt. Maeda et al. (2014).

Seismic performance index I_S is widely applied in the assessment of existing RC buildings in Japan. Its definition can be found in the Standard for Seismic Evaluation of Existing Reinforced Concrete Buildings (1977). This index takes into account such factors as strength, ductility, irregularities in structural configuration and the ageing process of materials. There are three different procedures to determine I_S depending on the refinement level required for assessment. However in post-earthquake safety assessment for a structure with storey collapse mechanism: $I_S = \sum Q_{ui}$, where Q_{ui} is the lateral strength capacity of a vertical structural member of the most damaged storey. In addition, for a structure with beam-sway collapse mechanism: $I_S = \sum M_{ui}$, where M_{ui} is the ultimate flexural moment at a yielding hinge in the mechanism.

Post-earthquake performance index $_D I_S$ can be evaluated using seismic capacity reduction factors η (Eq.2). These last factors account for the loss of residual energy dissipation capacity corresponding to the damage state of each structural member. In post-earthquake damage assessment the expressions used to calculate $_D I_S$ are as follows: For a structure with storey collapse mechanism $_D I_S = \sum Q_{ui} \cdot \eta_i$, and for a structure with beam-sway mechanism $_D I_S = \sum M_{ui} \cdot \eta_i$.

The Damage Evaluation Guideline (JBDPA 2001) recognizes five different damage classes for structural members (Table 13.1). These categories are classified according to the maximum residual crack width ($\max W_0$) measured on the member.

Figure 13.1 shows two hypothetical lateral force – displacement curves: one for a ductile member and the other for a brittle member. During an earthquake when a member reaches the peak displacement (δ_p), a residual displacement (δ_0) occurs. The areas corresponding to E_d and E_r are the dissipated energy during the earthquake and the residual energy dissipation capacity of the structural member after the event, respectively. The η factor can be defined as the ratio of residual energy dissipation capacity (E_r) to the member's original energy dissipation capacity ($E_T = E_d + E_r$). Hence,

Table 13.1 Damage Classes of Structural Members (JBDPA 2001)

Damage class	Observed damage
I	Some cracks are found. $\max W_0 \leq 0.20 mm$
II	$0.20 mm \leq \max W_0 \leq 1.00 mm$
III	Heavy cracks are found. Some spalling of concrete is observed. $1.00 mm \leq \max W_0 \leq 2.00 mm$
IV	Many heavy cracks are found. Reinforcing bars exposed due to spalling of the covering concrete. $\max W_0 > 2.00 mm$
V	Buckling of reinforcement. Crushing of concrete and vertical deformation of columns and/or shear walls. Side-sway, subsidence of upper floors and/or fracture of reinforcing bars. Member collapse.

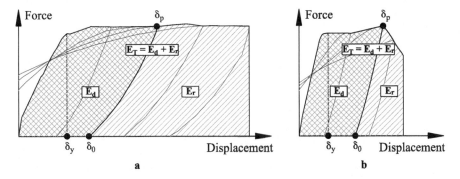

Fig. 13.1 Dissipated and residual energies used to define seismic capacity reduction factor η. (**a**) Ductile member, (**b**) Brittle member

Table 13.2 Seismic Capacity Reduction Factors η (JBDPA 2014 Revision)

Damage class	Beams		Columns		
	Ductile	Brittle	Ductile	Quasi-Ductile	Brittle
I	0.95	0.95	0.95	0.95	0.95
II	0.75	0.70	0.75	0.70	0.60
III	0.50	0.40	0.50	0.40	0.30
IV	0.20	0.10	0.20	0.10	0.00
V	0.00	0.00	0.00	0.00	0.00

$$\eta = \frac{E_r}{E_T} = \frac{E_r}{E_d + E_r} \qquad (13.2)$$

The 2014 revision of the Guideline is based on information and knowledge gathered from the 2011 East Japan Earthquake. The two main aspects of this revision include the re-evaluation of the η factors and the proposal of a methodology to calculate the R-index for structures with beam-sway collapse mechanism.

The new values for η factors suggested by the JBDPA were enhanced by more experimental data on beams and a new category for quasi-ductile columns was added. These values can be seen on Table 13.2. Definitions of ductile, quasi-ductile and brittle members can be found on Maeda et al. (2014). As shown on Table 13.2, seismic capacity reduction factors for a structural member decrease as damage class increases. On the other hand, for elements with the same damage class η factors decrease as ductility capacity decreases.

Previous versions of the Guideline only considered storey collapse mechanism on RC frames. In this type of mechanism the damage is prominent on vertical members, i.e. columns. Present day seismic codes try to avoid storey collapse mechanism and recommend beam-sway mechanism instead. Ductile failure patterns such as beam yielding are observed in the beam-sway mechanism. This has led to the development of a new evaluation method for structures that exhibit this kind of behaviour. As a result, there is a renewed interest in seismic reduction factors for RC beams.

13.3 RC Beam

13.3.1 Experimental Study

A pin supported RC beam with a central stub was tested by Scotta et al. (2012). The objective was to determine seismic capacity reduction factors for a ductile beam. The detail of the specimen is showed in Fig. 13.2. Symmetrical longitudinal

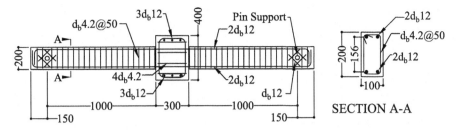

Fig. 13.2 RC beam tested by Scotta et al. (2012). Dimensions in mm

Table 13.3 Material Properties of RC Beam (see Sect. 13.3) and RC Columns (see Sect. 13.4)

Property	Scotta et al. (2012) RC beam [MPa]	U1 – Tanaka (1990) RC column [MPa]	U8 – Zahn (1986) RC column [MPa]
Concrete compressive strength (f'_c)	42.1	25.6	40.1
Concrete Elastic Modulus (E_c)	29,761	25,298	31,662
Concrete tensile strength ($f_t = 0.1 f'_c$)	4.21	2.56	4.01
Steel strength at yield point (f_y)	563	474	440
Steel strength at ultimate point (f_{su})	668	721	674
Steel Elastic Modulus (E_s)	216,000	200,000	200,000
Post-yielding steel Young's modulus (E_{sh})	5200	1800	2000

reinforcement was adopted. Sufficient transverse reinforcement was provided to ensure adequate deformation capacity in the hinge region. Material parameters are presented in Table 13.3.

The beam was subjected to a series of controlled pseudo-static cyclic vertical displacements. The load was applied on the central stub. Vertical displacement history imposed consisted of two cycles for each ductility ratio (μ) ranging from 1 to 5. Test configuration did not allow displacements higher than 52.5 mm. However, at this displacement value the specimen still exhibited capacity to resist loads. Load – displacement hysteresis loops were obtained from this test.

Flexural crack widths were measured in the hinge area, Fig. 13.3a. The cracks were measured using a 10 mm total length ruler with 0.1 mm precision. Measurements were carried out at the peak of each cycle and when the vertical force was unloaded. The relationship between maximum residual crack widths and peak displacements is shown in Fig. 13.3b.

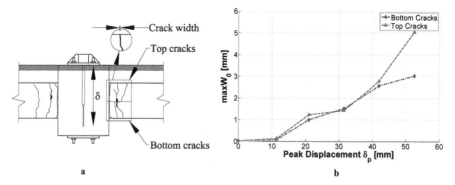

Fig. 13.3 RC beam test. (**a**) Crack width measurement, (**b**) Peak displacement – maximum crack width relationship

13.3.2 Analytical Models

A nonlinear model of the beam presented in Sect. 13.3.1 was made to reproduce test results, Scotta and Galíndez (2012). A fiber element model implemented in OpenSees (Open System for Earthquake Engineering Simulation) was used. Bar slip due to strain penetration of the longitudinal reinforcement in the central stub was taken into account. In addition, this model was also used to study seismic reduction factors η for beams with different ductility capacity.

Figure 13.4 shows the discretization and main characteristics of the numerical model. Half of the beam was simulated due to symmetrical geometry. Five elements were employed. Element 1 was defined with null length to apply the bar slip model. From element 2 to 5, five Gauss Points were defined. Popovics (1973) uniaxial model was considered for concrete and a modified Chang and Mander (1994) model was used for steel. The material parameters were the same as the ones specified on Table 13.3.

Figure 13.5 presents a comparison between numerical and experimental load-displacement hysteresis loops. Ultimate displacement was assumed as the displacement when concrete reaches a compressive strain of 0.003 in fiber F4. Good agreement between both curves was observed.

The tested beam presented an ultimate ductility ratio of 6.00. To obtain beams with less ductility capacity, longitudinal reinforcement in the numerical models was increased. 326, 339 and 452 mm^2 were adopted to obtain ductility ratios of 5.00, 4.00 and 3.00, respectively.

The relationship between maximum residual crack width (maxW_0), and the seismic capacity reduction factor (η), was determined by a simple analytical model, Scotta et al. (2012). This model was used to determine maxW_0 from residual displacements δ_0 obtained in the numerical models. A ductile beam was assumed to only have bending deformation. If the beam is idealized as a rigid body, the flexural deformation can be represented by the rotation of the rigid body, Fig. 13.6. This assumption gives an estimation of flexural deformation (R_{0f}), due to total flexural residual crack widths (ΣW_{0f}).

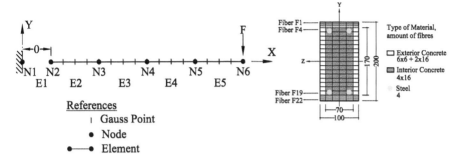

Fig. 13.4 OpenSees Numerical Model of RC Beam tested by Scotta et al. (2012)

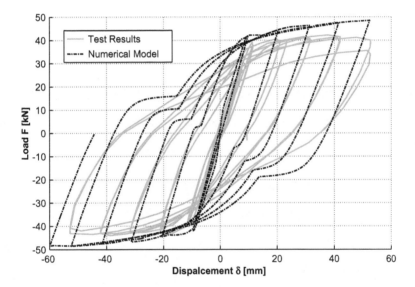

Fig. 13.5 Load – displacement hysteresis loops for RC beam (Scotta et al. 2012)

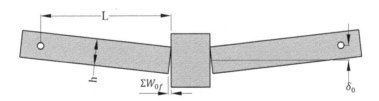

Fig. 13.6 Idealization of residual flexural deformation on RC beam

$$R_{0f} = \frac{\Sigma W_{0f}}{h} = \frac{\delta_0}{L} \tag{13.3}$$

Rearranging Eq.3 leads to Eq.4:

$$\max W_0 = \delta_0 \cdot \frac{h}{L \cdot n_f} \tag{13.4}$$

Where:

$$n_f = \frac{\Sigma W_{0f}}{\max W_{0f}} \tag{13.5}$$

From experimental results, $n_f = 2$.

13.4 RC Columns

13.4.1 Experimental Studies

In addition to the tested beam, studies were carried out on two columns tested by other authors, i.e. Unit 1 (U1) by Tanaka (1990) and Unit 8 (U8) by Zahn (1986). The former had low axial load ratio $P/(A_g \cdot f'_c) = 0.20$ and exhibited ductile behaviour, while the latter had high axial load ratio $P/(A_g \cdot f'_c) = 0.39$ and exhibited a more fragile behaviour. The main mechanical properties of these specimens can also be found on Table 13.3. Geometrical properties and reinforcement details are shown on Fig. 13.7.

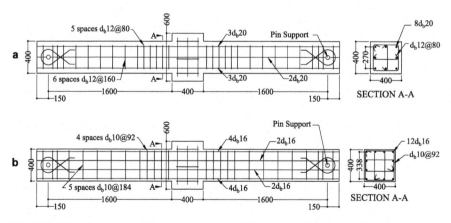

Fig. 13.7 RC columns tested by other authors: (**a**) U1 – Tanaka (1990), (**b**) U8 – Zahn (1986). Dimensions in mm

Unit 1 Tanaka (1990) had an axial load $P = 819,00 kN$ that remained constant during the test. This unit was subjected to displacement cycles that represented the following ductility ratios: 0.75, 2.00, 4.00, 6.00 and 8.00. The length of the plastic hinge formed near the stub was 400 mm. Flexural failure occurred at a peak displacement of 120 mm.

Unit 8 Zahn (1986) had a much higher axial load, $P = 2502,00 kN$. The displacement history imposed over this column included cycles with ductility ratios 0.75, 2.00, 4.00 and 6.00. In this case, the length of the plastic hinge was 310 mm. Collapse was observed at a maximum displacement of 50 mm when the longitudinal reinforcement started to exhibit prominent buckling. This specimen also underwent flexural failure.

Crack widths were not measured in any of the column tests. As a consequence, in order to establish the relationship between maximum residual crack widths and residual displacements the expression by Choi et al. (2006) was adopted. In both cases shear deformation can be neglected because the two specimens exhibited flexural failure. Therefore, the expression can be written as follows.

$$\delta_0 = \frac{\max W_{0f} \cdot n_f}{(h_c - x)} \cdot L \qquad (13.6)$$

Where:

δ_0: Total residual displacement.

$$n_f = \frac{\sum W_{0f}}{\max W_{0f}} \cong 2.00$$

$h_c = 400 mm$: Column height, measured perpendicularly to the flexural axis.
x: Distance between the most compressed concrete fiber and the neutral axis.
$L = 1600\ mm$: Column Equivalent Length.

It is important to notice that if x is approximately equal to zero, then Eq. 13.6 for columns becomes Eq. 13.4 for beams. That is to say that in the case of beams, the neutral axis is close to the most compressed concrete fiber and most of the section is tensioned.

13.4.2 Analytical Models

Two numerical models with fiber discretization, one for each column, were made using PERFORM-3D nonlinear analysis software, Barlek Mendoza et al. (2014). Three different materials were considered for these models: cover concrete, confined concrete and rebar steel. Each model consisted on two frame elements. The first

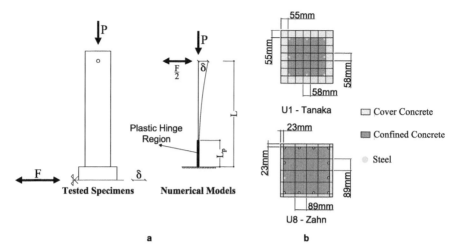

Fig. 13.8 (a) Tested RC columns vs. Analytical models, (b) Fiber discretization used for each column

element extended over the plastic hinge region. Its cross section was subdivided in fibers. The second element remained elastic and extended over the remaining length of the column, see Fig. 13.8a.

U1 – Tanaka (1990) cross section was subdivided in 57 fibers. Additionally, 48 fibers were used to discretize U8 – Zahn (1986) cross section. Figure 13.8 shows the meshing considered for each column.

Material properties were in accordance with Table 13.3. Tensile stress collaborations were neglected for both cover concrete and confined concrete. For U1 – Tanaka (1990), confined concrete stress – strain relationship was established using a Modified Kent and Park model (1971). In contrast, Mander model (1988) for confined concrete was used for U8 – Zahn (1986). Trilineal stress – strain relationships with no strength loss were assumed for reinforcing steel. Cyclic degradation was also considered using energy degradation factors.

Load – displacement hysteresis loops for U1 – Tanaka (1990) and U8 – Zahn (1986) are shown in Fig. 13.9a, b, respectively. Good adjustment between test results and numerical models was observed in both cases. The numerical model of U1 – Tanaka (1990) was able to reproduce higher displacement cycles with great accuracy. Contrarily, U8 – Zahn (1986) analytical model was better than U1 model for lower ductility ratio cycles.

Maximum residual crack widths can be inferred using residual displacements (δ_0) obtained from numerical models, see Eq. 13.6. As a consequence, it is possible to establish a relationship between maximum residual crack widths and seismic capacity reduction factors. This relationship requires both the simplified model (Choi et al. 2006) and numerical models.

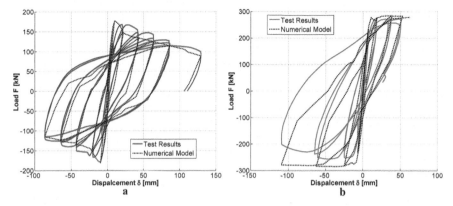

Fig. 13.9 Load – Displacement hysteresis loops for RC columns. (**a**) U1 – Tanaka (1990), (**b**) U8 – Zahn (1986)

13.5 Results and Discussion

Seismic capacity reduction factors were evaluated using Eq. 13.2. Total absorbable energy (E_T) of each member was calculated with the corresponding load – displacement envelope curve. Residual energy dissipation capacity (E_r) was determined as the difference between dissipated energy and E_T. Values obtained for RC beams and RC columns are summarised in Table 13.4. Additionally, in Fig. 13.10 these factors are compared to the values specified in the 2014 revision of the guideline.

13.5.1 RC Beam

The η factors obtained for RC beams are shown in Fig. 13.10a. These results include experimental and numerical analysis. Vertiacal and horizontal Dashed lines indicate the values specified in the guideline for ductile and brittle beams, respectively.

Experimental results for both top and bottom cracks were also plotted in Fig. 13.10a. Notice how these experimental curves are approximately matched by the curve that represents the reduction factors for the numerical model with an ultimate ductility ratio of 6.00. In general, the values obtained in this study for ductile beams are slightly greater than the ones suggested by the guideline.

The curves determined for different ductility ratios appear to have a linear tendency. However, it can be verified that for small values of $\max W_0$ there is a hardly noticeable change in their slopes. In addition, it can be seen that the greater the ductility ratio, the less steep the slope of these curves is.

Analytical models with ductility ratios lower than 4.00 were closer to the values recommended by the JBDPA for brittle beams. The numerical model of the beam with ultimate ductility ratio of 3.00 tends to be below the suggested limits. On the other hand, the curve corresponding to a ductility capacity of 5.00

Table 13.4 Seismic Capacity Reduction Factors η

Damage class	RC beams				RC columns	
	$\mu = 6.00$	$\mu = 5.00$	$\mu = 4.00$	$\mu = 3.00$	U1 – Tanaka (1990)	U8 – Zahn (1986)
I	0.95	0.94	0.92	0.87	0.93	0.87
II	0.79	0.72	0.64	0.50	0.81	0.43
III	0.56	0.42	0.26	0.00	0.67	0.33
IV	0.10	0.00	0.00	0.00	0.06	0.00
V	0.00	0.00	0.00	0.00	0.00	0.00

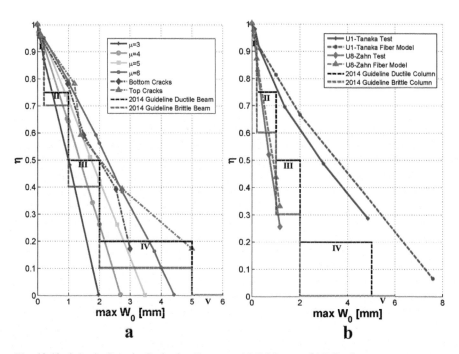

Fig. 13.10 Seismic Capacity Reduction Factors η. (**a**) RC beams, (**b**) RC columns

has higher reduction factors, which can be compared to the suggested values for ductile beams.

13.5.2 RC Columns

Fig. 13.10b shows the seismic capacity reduction factors for the columns analysed in this paper. In addition, the values recommended in the guideline for different damage classes are plotted for both ductile and brittle columns.

Curves corresponding to test results and analytical models were very similar, especially for U8 – Zahn (1986). This can be attributed to the fact that numerical models hysteresis loops closely matched the ones measured experimentally. For U1 – Tanaka (1990), the analytical model tended to overestimate the η factors. This tendency is notorious for high values of maximum residual crack widths.

It is important to notice that in the case of U1 – Tanaka (1990) the reduction factors η are greater than the ones suggested for ductile columns by the JBDPA. This last statement is true for both experimental and analytical results.

For U8 – Zahn (1986), crack widths were much narrower because of the high axial load. In this sense, there were no crack widths corresponding to Damage Class IV. This indicates that the column passed directly from Damage Class III to Damage Class V (collapse) because of the buckling of the longitudinal reinforcement. Additionally, it can be seen that there are values of η below the horizontal dashed lines with guideline recommendations for brittle columns. This means that the element rapidly lost energy dissipation capacity as maximum crack widths grew.

13.6 Conclusions

In this study, seismic capacity reduction factors η for a RC beam and two RC columns were determined. Test results and numerical models were used to this end. The results were compared with the reference values specified in the 2014 revision of the JBDPA Guideline.

Even though the total number of cases analysed in this paper is limited, the results obtained suggest that it would be possible to determine reasonable reduction factor values combining numerical models and simplified models that relate maximum residual crack widths to residual displacements.

For the ductile elements (both beams and columns), seismic capacity reduction factors were higher than the values recommended in the guideline. However, in the case of brittle members, the reduction factors suggested by the code were sometimes greater than the ones calculated.

Numerical studies such as the one carried out on RC beams with different ductility ratios could be used in order to define when an element can be considered ductile, quasi-ductile or brittle.

It is important to continue studying the different type of failures that structural members can exhibit and their relationship with seismic capacity reduction factors. In this sense, the next step in the research project is to consider shear and flexure interaction, especially on columns with moderate to high axial load ratio.

References

Barlek Mendoza P, Galíndez E, Pavoni S (2014) Modelación Numérica de Columnas de Hormigón Armado Sometidas a Cargas Cíclicas (in Spanish). In: 23° Jornadas Argentinas de Ingeniería Estructural, Buenos Aires, Argentina, September 2014

Chang G, Mander J (1994) Seismic energy based fatigue damage analysis of bridge Columns: part I – evaluation of seismic capacity. Northwestern center for engineering education research, technical report 94–0006

Choi H, Nakano Y, Takahashi N (2006) Residual seismic performance of R.C. frames with unreinforced block wall based on crack widths. In: First European conference on earthquake engineering and seismology. Geneva, Switzerland, 3–8 September 2006

Kent D, Park R (1971) Flexural members with confined concrete. J Struct Div ASCE 97(7):1969–1990

Maeda M, Matsukawa K, Ito Y (2014) Revision of guideline for post-earthquake damage evaluation of RC buildings in Japan. In: Tenth U.S. national conference on earthquake engineering, frontiers of earthquake engineering. Anchorage, Alaska, 21–25 July 2014

Mander J, Priestley N, Park R (1988) Theoretical stress – strain model for confined concrete. J Struct Eng ASCE 114(8):1804–1825

Popovics S (1973) A numerical approach to the complete stress strain curve for concrete. Cem Concr Res 3(5):583–599

Scotta D, Galíndez E (2012) Simulación Numérica de Elementos de Hormigón Armado Sometidos a Cargas Cíclicas Reversibles (in Spanish). In: VIII Jornadas de Ciencia y Tecnología de Facultades de Ingeniería del NOA (VIII JCTNOA). San Miguel de Tucumán, Argentina

Scotta D, Galíndez E, Pavoni S (2012) Residual seismic capacity of reinforced concrete beams under reversible flexural cyclic loads (in Spanish). In: XXXV Jornadas Sul Americanas de Engenharia Estrutural. Rio de Janeiro, Brazil, 19–21 September 2012

Tanaka H (1990) Effect of lateral confining reinforcement on the ductile behaviour of reinforced concrete columns. PhD dissertation, University of Canterbury, New Zealand

The Japan Building Disaster Prevention Association (JBDPA) (1991, revised in 2001 and 2014) Guideline for post-earthquake damage evaluation and rehabilitation (in Japanese)

The Japan Building Disaster Prevention Association (JBDPA) (1977, revised in 1990 and 2001) Standard for seismic evaluation of existing reinforced concrete buildings (in Japanese)

Zahn F (1986) Design of reinforced concrete bridge columns for strength and ductility. PhD dissertation, University of Canterbury, New Zealand

Chapter 14
Single-Degree-of-Freedom Analytical Predictive Models for Seismic Isolators

Todor Zhelyazov

Abstract Single-degree-of-freedom constitutive relations aimed at modeling the mechanical response of seismic isolators, specifically lead-core bearing devices and elastomeric bearings, are discussed in this contribution. Two constitutive models are considered. For the lead-core bearings, a model that defines the relation between the shear displacement and the generated shear force is postulated. This model involves a hysteresis parameter the evolution of which is separately defined. The mechanical behavior of the elastomeric bearings I s modeled assuming a constitutive law that takes into account several interacting mechanisms including the mechanical damage accumulation. Numerical procedures are developed for both constitutive relations. These numerical procedures are to be implemented in detailed finite element models and accurate finite element analysis of base-isolated structures.

Keywords Lead core bearing device · Constitutive model · Analytical models · Material degradation

14.1 Introduction

Seismic isolation and energy dissipation systems are relatively new techniques that require a special attention in the design phase, i.e., an accurate and detailed modeling of the dynamic response of the base-isolated structure.

The seismic isolation can be used for both new structures and refurbishment and rehabilitation of existing buildings. In the later case, it might reduce the need of measures for stiffening and strengthening of the superstructure.

In the recent past, seismic isolation technology was mainly used for large structures and buildings that house displacement sensitive and acceleration sensitive

---

T. Zhelyazov (✉)
Technical University of Sofia, Sofia, Bulgaria
e-mail: todor.zhelyazov@tu-sofia.bg

Fig. 14.1 Schematization of a lead-core bearing device

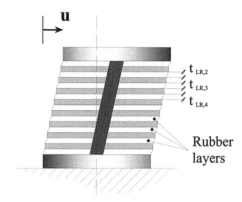

equipment. With the progress in the technology the field of application of seismic isolation is extended to housing and commercial buildings.

The most widely accepted techniques for seismic isolation are the elastomeric bearing and friction pendulum systems. The scope of this study is confined to elastomeric bearings and to their modification - the lead-core rubber bearings.

Single-Degree-of Freedom (SDOF) predictive models for lead-core bearing (LCB) devices typically formulate a relation between the shear displacement (or shear strain) and the shear force (or the shear stress). The shear strain is generally defined as the ratio between the shear displacement and the total height of the rubber layers in the bearing device (Fig. 14.1):

$$\gamma = \frac{u}{\sum_{i=1}^{n} t_{LR,i}} \qquad (14.1)$$

LCB devices typically consist of components possessing different mechanical properties: lead-core, rubber and steel elements. In this context SDOF models should provide a macro-characteristic representative for the mechanical response of this multiple-component system on the macro scale. Constitutive models for LCB devices have evolved from the bilinear model, in which pre-yield stiffness is simply replaced by the post-yield stiffness after the yielding of the lead, to more sophisticated models involving differential equations. These are the Bouc-Wen model and a differential equation model in which a damage variable is introduced.

Two constitutive models are discussed in this contribution - the Bouc-Wen model (i) and a relation involving a damage variable (ii). The Bouc-Wen model consists of an equation that defines a "skeleton" curve (a relation between shear displacement and shear force) and another equation which specifies the evolution of the hysteresis parameter. In the constitutive model accounting for damage, the total shear stress acting in LCB device is split into several terms that quantify the contributions of the different mechanisms. The above models are implemented in numerical procedures created using Python software. Upon appropriate model parameter calibration, these numerical tools can be used in designing structures which contain seismic isolators such as LCB devices. Only analytical and numerical aspects of the discussed constitutive models are considered here.

This contribution appears thanks to a continuous collaboration, started in 2013 between the author and the Earthquake Engineering Research Centre, a part of the University of Iceland.

14.2 Constitutive Models

The uniaxial and the biaxial responses of elastomeric bearing have been modelled by (Yasaka et al. 1988) using the multiple spring model. Way and Jeng (1988) have proposed a plasticity-based nonlinear model. Nowadays, a widely accepted strategy is to model the mechanical behaviour of seismic isolation defining a set of equations that postulate a skeleton curve (i.e., the one defined by Eq. (14.2)) which is accompanied by a relation introducing a hysteresis parameter (see Eq. (14.3)) (Bouc 1967,;1971; Wen 1976; Constantinou and Tadjbakhsh 1985; Song and Der Kiureghian 2006) as follows:

$$Q = \alpha \frac{Q_y}{d_y} u + (1 - \alpha) Q_y Z \qquad (14.2)$$

$$d_y \frac{dZ}{dt} = -\gamma \left| \frac{du}{dt} \right| Z |Z|^{(\eta-1)} - \beta \frac{du}{dt} |Z|^\eta + A \frac{du}{dt} \qquad (14.3)$$

In Eq. (14.2) u stands for the current displacement and Z for a dimensionless hysteresis component which should satisfy Eq. (14.3). The other material parameters in (14.2) are defined as follows: α is the post-yielding to pre-yielding stiffness ratio, d_y is the displacement for which the yielding in the lead core takes place, and Q_y is the shear force, corresponding to the yielding. In Eq. (14.3) β, γ and A are dimensionless model parameters and η controls the transition phase at the yielding of the lead core.

Further, the modified mid-point method (see for example Press et al. 2007) is implemented to integrate numerically the ordinary first-order differential Eq. (14.3) (see Fig. 14.2):

$$\frac{dZ}{dt} = f(Z(u), \dot{u}) \qquad (14.4)$$

$$z_0 = Z(u_i) \qquad (14.5)$$

$$z_1 = z_0 + hf(Z(u_i), \dot{u}) \qquad (14.6)$$

$$z_{k+1} = z_{k-1} + 2hf(Z(u_i + kh), \dot{u})(k = 1 \ldots n - 1) \qquad (14.7)$$

$$Z(u_j) = \frac{1}{2}[z_n + z_{n-1} + hf(Z(u_i + H), \dot{u})] \qquad (14.8)$$

This is an algorithm to calculate the value of the function Z at point u_j, provided the value at point u_i is known. In Eq. (14.8) $H = u_j - u_i$; n is the number of substeps into which the interval H is divided. Overdot denotes a derivative with respect to time.

Fig. 14.2 Procedure for numerical integration

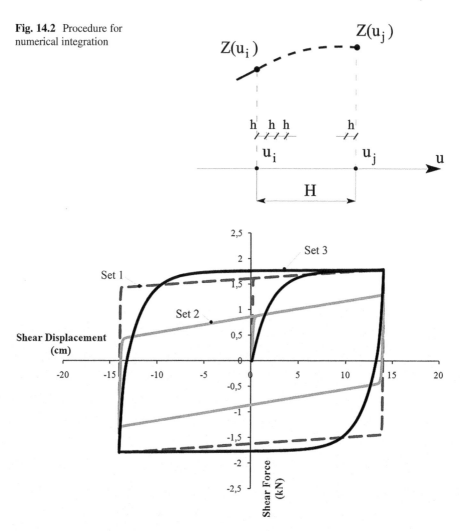

Fig. 14.3 Mechanical response of a lead-core bearing in terms of shear displacement- shear force curve obtained by using numerical procedure based on the Bouc-Wen model

To solve Eqs. (14.2) and (14.3) a numerical procedure is created in Python software. The algorithm contains a subroutine (Eq. 14.4, 14.5, 14.6, 14.7, and 14.8) for numerical integration using the modified mid-point method.

Several shear force – shear displacement relationships are shown in Fig. 14.3.

Table 14.1 Bouc-Wen model: mechanical constants

	γ	η	β	α	A
Set 1	0.3	1	0,9	0.01	2
Set 2	0.3	1	0.1	0.12	2
Set 3	0.25	1	−0.15	0.05	2

Model parameters used to obtain this constitutive relation are summarized in Table 14.1.

14.3 The Case of Multiaxial Loading

It should be pointed out that in the discussion thus far a uniaxial loading has been presumed. For the case of a multiaxial loading, the displacement vector **u** is resolved into its components u_x and u_y and Eq. (14.2) should be replaced by the following system of equations (see Nagarajaiah et al. 1991):

$$\left| \begin{array}{l} F_x = \alpha \dfrac{Q_y}{d_y} u_x + (1-\alpha) Q_y Z_x \\ F_y = \alpha \dfrac{Q_y}{d_y} u_y + (1-\alpha) Q_y Z_y \end{array} \right. \quad (14.9)$$

In Eqs. (14.9) F_x and F_y are the components of the shear force generated in the bearing device. The biaxial interaction is further modelled by the following set of equations (Park et al. 1986; Nagarajaiah et al. 1991):

$$\begin{pmatrix} \dfrac{dZ_x}{dt} d_y \\ \dfrac{dZ_y}{dt} d_y \end{pmatrix} = \begin{pmatrix} A \dfrac{du_x}{dt} \\ A \dfrac{dU_y}{dt} \end{pmatrix} - \begin{bmatrix} Z_x^2 \left[\gamma.sign\left(\dfrac{du_x}{dt} Z_x\right) + \beta \right] & Z_x Z_y \left[\gamma.sign\left(\dfrac{du_y}{dt} Z_y\right) + \beta \right] \\ Z_x Z_y \left[\gamma.sign\left(\dfrac{du_x}{dt} Z_x\right) + \beta \right] & Z_y^2 \left[\gamma.sign\left(\dfrac{du_y}{dt} Z_y\right) + \beta \right] \end{bmatrix} \begin{pmatrix} \dfrac{du_x}{dt} \\ \dfrac{du_y}{dt} \end{pmatrix}$$

(14.10)

Herein, $\frac{du_x}{dt}$, $\frac{du_y}{dt}$ are velocity components in the X and Y directions. As in Eq. (14.3) d_y stands for the yield displacement, Z_x and Z_y are dimensionless hysteresis parameters, and A, β and γ are model parameters controlling the shape of the hysteresis loop.

14.4 Stress-Strain Constitutive Relation Accounting for Material Degradation

An alternative approach, compared to the model discussed in the previous section can be found in (Dall'Astra and Ragni 2006; Govindjee and Simo 1992; Haupt and Sedlan 2001). This type of constitutive relations can be situated in line with models that are compatible with thermodynamics fundamentals (Fabrizio and Morro 1992).

An internal variable is introduced to assess mechanical degradation. Total shear stress is split into three terms: a term which represents an elastic contribution (τ_e), and two terms describing overstresses relaxing in time (τ_1 and τ_2):

$$\tau = \tau_e + \tau_1 + \tau_2 \tag{14.11}$$

The elastic contribution is assessed as:

$$\tau_e = F(\gamma) \tag{14.12}$$

with F(γ) being a polynomial depending on the shear strain γ. Overstresses are defined as functions of the shear strain, model parameters (E_1, E_2) and internal variables (γ_1, γ_2):

$$\tau_1 = F(E_1, \gamma, \gamma_1) \tag{14.13}$$
$$\tau_2 = F(E_2, \gamma, \gamma_2) \tag{14.14}$$

Evolutions of internal variables $\gamma_{v,1}$ and $\gamma_{v,2}$ are defined as follows (Dall'Astra and Ragni 2006):

$$\dot{\gamma}_1 = \left(\frac{|\dot{\gamma}|}{\eta_1(\gamma, \dot{\gamma}, q_e)} + \nu_1\right)\tau_{v,1} \tag{14.15}$$

$$\dot{\gamma}_2 = \left(\frac{H(-\gamma\dot{\gamma})}{\eta_2} + \nu_2\right)\tau_{v,2} \tag{14.16}$$

ν_1, η_2 and ν_2 in the above equations are model parameters. Generally they are identified through curve fitting on the basis of acquired experimental data.

In Eq. (14.16) H stands for the Heaviside function:

$$H(q) = \begin{cases} 1 & \text{if } q > 0 \\ 0 & \text{if } q \leq 0 \end{cases} \tag{14.17}$$

As it can be seen in Eq. (14.15), it is presumed that the quantity η_1 is a function of the shear strain γ, shear strain rate $\dot{\gamma}$ and a damage parameter- q_e. The damage parameter supplies information about material degradation. Upon appropriate calibration on the basis of the current value of the damage parameter, degrading phenomena in material can be rationally estimated. In the present study, the evolution of the damage parameter is defined by using a governing equation proposed by (Dall'Astra and Ragni 2006):

$$\dot{q}_e = \begin{cases} \zeta|\dot{\gamma}|(0.5\gamma - q_e) & if \quad q_e \leq 0.5|\gamma| \\ 0 & if \quad 0.5|\gamma \leq q_e \leq 1| \end{cases} \quad (14.18)$$

As already stated, damage parameter should enable the evaluation of the current state of mechanical damage after a given period of exploitation. The loading history or the loading path which has led to the current state of mechanical degradation is taken into account through the variable γ - the shear strain and its time rate of change $\dot{\gamma}$ - see Eq. (14.18). In the present study, only numerical simulations of typical identification test e.g. repetitive loading paths as the one shown in Fig. 14.4 are considered.

The evolutions of the damage parameter obtained by integrating Eq. (14.18) for different shear strain rates are depicted in Fig. 14.5. Only the first quarter of one loading cycle (as shown in Fig. 14.4) is considered - variation of the shear strain from zero to *1.5*. It can be seen that, according to the adopted constitutive relation, the increase in shear strain rate results in the decrease of the accumulated damage.

It should be noted that in order to make the damage parameter more informative, the model should be calibrated against experimental data and the degrading phenomena - assessed through direct observation if appropriate.

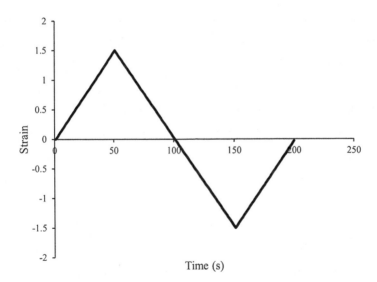

Fig. 14.4 A typical loading path

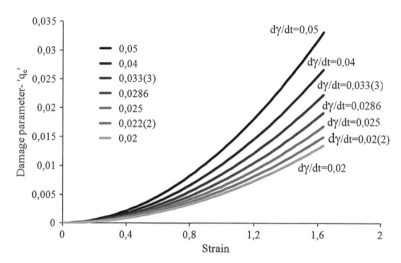

Fig. 14.5 Evolutions of the material degradation in function of the shear strain

14.5 Identification of the Model Parameters

The identification procedure discussed in this section is based on genetic algorithms (Goldberg 1989).

The procedure is inherently an algorithm for calibrating the model constants involved in the chosen material model. Such procedures are often referred to as "curve fitting". The curve fitting consists in finding a set of model parameters for which the model output fits best a reference set of data points. Generally, the target set of data points is an experimentally obtained relationship. The author believes that the target set of data points could be alternatively obtained by finite element modeling (i.e. by simulating a typical identification test see Fig. 14.6).

Two consecutive steps of the iterative procedure are schematized in Fig. 14.7.

The procedure starts by defining an array that can be visualized as a $p \times n$ 2-D matrix, where p is the number of the model constants that are to be identified by means of the curve fitting procedure (i.e., number of rows in the above – defined matrix) and n is the number of sets.

A typical iteration contains three operations: "analysis" module; "comparison" module and "modification" module.

In the "analysis" module each set of model constants (a total of n outputs) is used to produce a "shear displacement – shear force" relationship (plotted in grey – see Fig. 14.7). In the "comparison" module, the output obtained from a given set of model constants is compared with the target set of data points (plotted in black – see Fig. 14.7). In Fig. 14.7, only the best fit within a given iteration is depicted.

In order to tune the model constants, the components of the data array are modified in the "modifications" module before the next iteration. A *TF* (Test Function) is associated with each set of model constants (i.e., with each column of

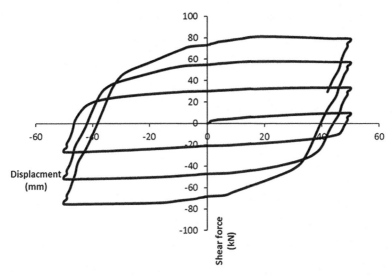

Fig. 14.6 Evolution of a macroscopic characteristic of the bearing device, specifically the "Shear displacement-Shear force" relationship, obtained by finite element analysis. (Reported by Zhelyazov et al. 2016)

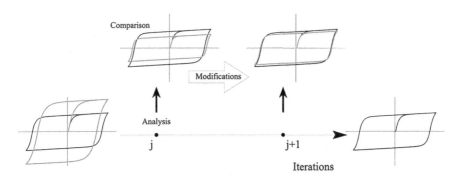

Fig. 14.7 Implementation of the curve fitting algorithm: starting from a given set of parameters (that produce an output plotted in grey) and a target curve (plotted in black), through iterations, the output of the model is getting closer to the target curve and eventually, at the end of the procedure, a set of model constants giving a best fit with the target relationship is defined

the array). *TF* is used to assess the output obtained using a given set of model constants. With an appropriate normalization *TF=1* means perfect match with the target set of data points. An illustrative example is given in Fig. 14.8.

In the k^{-th} iteration, two sets of model constants $S_k^{(a)}$ and $S_k^{(b)}$ (plotted in grey in Fig. 14.8a and in Fig. 14.8b respectively) are to be compared with a given reference curve (Fig. 14.8 – plotted in black). By applying the procedure outlined in Kwok et al. (2007) it is obtained that

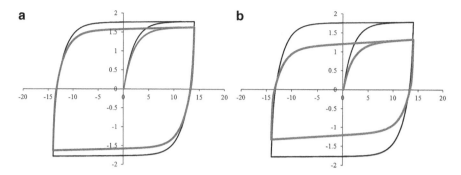

Fig. 14.8 (a) Comparison between the output of the adopted constitutive relation obtained by using a set of model constants conventionally labeled $S^{(a)}$ with a reference set of data points, (b) Comparison between an output obtained by using a set of constants conventionally labeled $S^{(b)}$ and the same reference curve

$$TF_k^{(a)} = 0,40589 < 0,109369 = TF_k^{(b)} \tag{14.19}$$

In the next iteration the set $S^{(a)}$ is conserved, whereas the set $S^{(b)}$ is modified as follows:

$$S_{k+1}^{(a)} = S_k^{(a)} \tag{14.20}$$

whereas the set $S^{(b)}$ is modified as follows:

$$S_{k+1}^{(b)} = \xi S_k^{(a)} + (1 - \xi) S_k^{(b)} \tag{14.21}$$

In Eq. (14.21) ξ is a random number and $\xi \in (0, 1)$.

The curve fitting procedure ends either when a predefined maximum number of iterations is reached or a satisfactory fit to the target set of data points is obtained.

14.6 Concluding Remarks and Further Resarch

Analytical models aimed at reproducing the mechanical response of a lead-core bearing devices for passive seismic isolation have been discussed.

Numerical procedures designed to implement the widely accepted Boc-Wen model, a constitutive relation accounting for the material degradation as well as a curve fitting procedure have been presented. A generalization of the Bouc – Wen model aimed at accounting for the case of displacement vector in general position with respect to the adopted reference frame (i.e., displacement vector **u** in general position) has been also discussed.

Within the further research program it is presumed to implement the Bouc – Wen constitutive relation generalized to the case of multiaxial loading in the finite element model of a base isolated structure.

As it was pointed out, the author believes that identification of the model constants involved in the SDOF model could be done by using results obtained by finite element analysis. Such a strategy might significantly optimize the identification procedure by reducing the experimental part. Further refinement of the finite element model by exploring the capabilities to explore the thermo-mechanical response of a bearing device is foreseen.

The algorithms discussed in the contribution can be further implemented in an automated procedure for synthesis of a bearing device on the basis of near-fault strong motion records typical for a given region.

An effort will be dedicated to the refinement and further automation of the numerical procedure for identification of model constants.

References

Bouc R (1967) Forced vibration of mechanical systems with hysteresis. In: Proceedings of the fourth conference on nonlinear oscillation, Prague, Czechoslovakia, p 315

Bouc R (1971) Modèle mathématique d'hystérésis: application aux systèmes à un degré de liberté. Acustica (in French) 24:16–25

Wen YK (1976) Method of Random Vibration of hysteretic system. J Eng Mech Div ASCE 102 (2):249–263

Constantinou MC, Tadjbakhsh M (1985) Hysteretic Dampers in base isolation: random approach. J Struct Eng 111(4):705–721

Song J, Der Kiureghian A (2006) Generalized Bouc–Wen model for highly asymmetric hysteresis. J Eng Mechanics ASCE 132(6):610–618

Press WH, Teukolsky SA, Vetterling WT, Flannery BP (2007) Numerical recipies, the art of scientific computing, 3rd edn. Cambridge University Press, Cambridge

Dall'Astra A, Ragni L (2006) Experimental tests and analytical model of high damping rubber dissipating devices. Eng Struct 28(13):1874–1884

Govindjee S, Simo JC (1992) Mullins effect and the strain amplitude dependence of the storage modulus. Int J Solids Struct 29:1737–1751

Haupt P, Sedlan H (2001) Viscoplasticity of elastomeric materials: experimental facts and constitutive modelling. Arch Appl Mech 71:89–109

Fabrizio M, Morro A (1992) Mathematical problems in linear viscoelasticity. SIAM—Studies In Applied Mathematics, Philadelphia

Goldberg DE (1989) Genetic algorithms in search, optimization and machine learning. Addison-Wesley, Reading

Kwok NM, Ha QP, Nguyen MT, Li J, Samali B (2007) Bouc- Wen model parameter identification for a MR fluid damper using computationally efficient GA. ISA Trans 46(2):167–179

Nagarajaiah S, Reinhorn AM, Constantinou MC (1991) Nonlinear dynamic analysis of 3-D-base-isolated structures. J Struct Eng 117:2035–2054

Park YJ, Wen YK, Ang AHS (1986) Random vibrations of hysteretic systems under bidirectional ground motions. Earthq Eng Struct Dyn 14(4):543–557

Way D, Jeng V (1988) NPAD – a computer program for the analysis of base isolated structures. Proc ASME Pressure Vessels and Pipng Conference, American Society of Mechanical Engineers 147:65–69

Yasaka A, Mizukoshi K, Izuka M, Takenaka Y, Maeda S, Fujimoto N (1988) Bilinear hysteresis model for base isolation devices. In: Summaries of technical papers of annual meeting, Architectural Inst. Of Japan, Tokyo, Japan, 1:395–400

Zhelyazov T, Rupakhety R, Ólafsson S (2016) Modeling the mechanical response of a lead- core bearing device: damage mechanics approach. In: Papadrakakis M, Papadopoulos V, Stefanou G, Plevris V (eds) ECCOMAS Congress, June 2016. VII European Congress on Computational Methods in Applied Sciences and Engineering, Crete Island

Chapter 15
The Evaluation of Nonlinear Seismic Demands of RC Shear Wall Buildings Using a Modified Response Spectrum Analysis Procedure

Fawad Ahmed Najam and Pennung Warnitchai

Abstract In the standard Response Spectrum Analysis (RSA) procedure, the elastic force demands of all significant vibration modes are first combined and then reduced by a response modification factor (R) to get the inelastic design demands. Recent studies, however, have shown that it may not be appropriate to reduce the demand contributions of higher vibration modes by the same factor. In this study, a modified RSA procedure based on equivalent linearization concept is presented. The underlying assumptions are that the nonlinear seismic demands can be approximately obtained by summing up the individual modal responses, and that the responses of each vibration mode can be approximately represented by those of an equivalent linear SDF system. Using three high-rise buildings with RC shear walls (20-, 33- and 44-story high), the accuracy of this procedure is examined. The modified RSA procedure is found to provide reasonably accurate demand estimations for all case study buildings.

Keywords Response spectrum analysis · Response modification factor · Nonlinear model · RC shear wall · Equivalent linear system · High-rise buildings

15.1 Background

The understanding of complex structural behavior is one of the primary goals of structural analysis. Although, recent advancements in technological fields have opened a whole new paradigm dealing with detailed structural modeling techniques and

---

F. A. Najam (✉)
NUST Institute of Civil Engineering (NICE), National University of Sciences and Technology (NUST), Islamabad, Pakistan
e-mail: fawad@nice.nust.edu.pk

P. Warnitchai
School of Engineering and Technology, Asian Institute of Technology, Khlong Nueng, Pathumthani, Thailand

intricate analysis procedures, a clear understanding may not always be guaranteed by complex procedures. In fact, it can be more effectively developed using convenient analysis methods capable of simplifying the complex response in terms of its components. Therefore, practicing engineers are always interested in simple and conceptually elegant procedures which can provide reasonably accurate response estimates in lesser time and computational effort. For example, for a high-rise building project, setting up a full nonlinear structural model—sophisticated enough to capture all important aspects of material and component nonlinearity—may be an onerous task compared to a linear elastic model. Moreover, the latest analysis guidelines require to use a large number of ground motions records representing the anticipated seismic hazard at the building site. An ordinary design office may not have necessary expertise and resources to undergo the complete process of performing the detailed nonlinear response history analysis (NLRHA) for each project. For most practical cases, the linear elastic analysis may serve the purpose of estimating design demands within their required degree of accuracy.

The most commonly used analysis procedure to determine the design forces and displacement demands is the standard Response Spectrum Analysis (RSA) procedure. Although the development of fast numerical solvers, user-friendly software, and significant decrease in computational cost over last two decades have resulted in an increased use of new analysis procedures, RSA is still the most widely applied procedure, owing to its practical convenience. As prescribed in various codes and design guidelines, it assumes that the nonlinear force demands can be estimated by simply reducing the linear elastic force demands (combined from all vibration modes) by a response modification factor (R) (ASCE 7-05, 2006) or a behavior factor, q (EC 8, 2004). The displacements and drifts corresponding to these reduced forces (i.e. after reduction by R or q) are then multiplied with a deflection amplification factor, C_d (ASCE 7-05, 2006) or displacement behavior factor, q_d (EC 8, 2004) to obtain the expected maximum deformations produced by design seismic forces. Being an approximate analysis procedure, the RSA procedure is often subjected to criticism when compared with the detailed NLRHA procedure. The adequacy of seismic design factors (R and C_d) has also remained a subject of immense research for last few decades. While agreeing with some of the inherent limitations of the standard RSA procedure compared to the detailed NLRHA, we still believe that the concept of vibration modes is a valuable tool to simplify and quickly understand the complex dynamic behavior. Seeing the total response as a combination of contributions from a few vibration modes—while the behavior of each mode is represented by a single-degree-of-freedom (SDF) system governed by few parameters—results in a significantly improved physical insight. This study therefore, focuses on a possible improvement in existing practice and the current prevailing use of the RSA procedure for determining the design demands of high-rise buildings.

15.2 Motivation

The use of response modification factor (R), as recommended in the standard RSA procedure, is equivalent to proportionally reducing the demand contribution of every vibration mode with the same factor. However, various studies based on modal decomposition of inelastic dynamic response of buildings have shown that the response of all vibration modes does not experience the same level of nonlinearity under a ground motion. Eible and Keintzel (1988) were among the first to identify that the shear force demand corresponding to each vibration mode of a cantilever wall structure is limited by the yield moment at the base. Based on this observation, they proposed the concept of *"modal limit forces"*, defined as the maximum values that a certain mode's forces in an elastoplastic structure can attain. Later studies (Paret et al. 1996; Sasaki et al. 1998) also confirmed that the amount of inelastic action experienced by different vibration modes can be significantly different. For RC cantilever walls, Priestley (2003) demonstrated that the higher-mode response is not affected by ductility in the same manner as the response of fundamental mode. The inelastic action mostly occurs under the response of fundamental mode, while higher modes tend to undergo lower levels of nonlinearity. Therefore, reducing the forces of all modes by the same reduction factor may result in a significant underestimation of nonlinear demands. Various other studies (Klemencic et al. 2007; Zekioglu et al. 2007) have also identified this limitation of standard RSA procedure by comparing the design demands of RC core wall buildings with the true nonlinear demands obtained from the detailed NLRHA procedure.

More recently, Ahmed and Warnitchai (2012), and Mehmood et al. (2016a, b) applied the Uncoupled Modal Response History Analysis (UMRHA) procedure to various high-rise buildings with RC shear walls and gravity frame systems. This approximate analysis procedure was originally formulated by Chopra and Goel (2002) and it allows to decompose the complex nonlinear dynamic response of buildings in to contributions from few individual vibration modes. These studies have shown that the UMRHA procedure is able to provide reasonably accurate predictions of story shears, story overturning moments, inter-story drifts, and floor accelerations. The current study uses the UMRHA procedure as its starting point and is intended to propose a convenient and practical analysis procedure based on it. Therefore, it is necessary to first briefly review the basic concepts and underlying assumptions of the UMRHA procedure.

15.3 Theoretical Formulation and Basic Assumptions

15.3.1 The Uncoupled Modal Response History Analysis (UMRHA) Procedure

The UMRHA procedure can be viewed as an extended version of the classical modal analysis procedure. In the latter, the complex dynamic responses of a linear multi-degree-of-freedom (MDF) structure are considered as a sum of many independent vibration modes. The response behavior of each mode is essentially similar to that of a SDF system, which is governed by a few modal properties, making it easier to understand. Furthermore, only a few vibration modes can accurately describe the complex structural responses in most practical cases. This classical modal analysis procedure is applicable to any linear elastic structures. However, when the responses exceed the elastic limits, the governing equations of motion become nonlinear and consequentially, the theoretical basis for modal analysis becomes invalid. Despite this, the UMRHA procedure assumes that even for inelastic responses, the vibration modes still exist, and the complex inelastic responses can be approximately expressed as a sum of these modal responses. In this procedure, the behavior of each vibration mode is represented by a nonlinear SDF system governed by Eq. (15.1).

$$\ddot{D}_i(t) + 2\xi_i \omega_i \dot{D}_i(t) + F_{si}(D_i, \dot{D}_i)/L_i = -\ddot{x}_g(t) \qquad (15.1)$$

Where L_i is the product of modal participation factor (Γ_i) and modal mass (M_i) of any i^{th} mode, while ω_i and ξ_i are the natural circular frequency and the damping ratio of the ith mode, respectively. To compute the response time history of $D_i(t)$ against a ground acceleration vector ($\ddot{x}_g(t)$), one needs to know the nonlinear force-deformation function $F_{si}(D_i, \dot{D}_i)$ which describes the restoring force produced by the structure under a deformed shape of ith mode. This restoring force function is a combined result of lateral response contributions from all structural components, and is usually selected by idealizing the actual modal cyclic pushover curve for each mode with a suitable hysteretic model. Further details of this procedure and its validation can be seen in Chopra and Goel (2002), Ahmed and Warnitchai (2012), and Mehmood et al. (2016a, b).

The UMRHA procedure provides an opportunity to clearly understand the individual modal behavior as well as their relative contributions to total response. Using this effective tool, recent studies reconfirmed that the response of each contributing vibration mode exhibits a different level of inelastic action under the same ground motion. Therefore, reducing each mode's force demands with the same response modification factor—as recommended in the standard RSA procedure—may not be appropriate. For a relatively more accurate determination of nonlinear seismic demands, the behavior of each mode should actually correspond to its close-to-real nonlinear state, instead of scaling down the response from linear elastic behavior.

15.3.2 *The Basic Concept of the Modified Response Spectrum Analysis (MRSA) Procedure*

The underlying assumption of the UMRHA procedure—that the response of complex nonlinear structure can be approximately represented by a summation of responses from a few nonlinear modal SDF systems—further leads to an idea that properly-tuned *"equivalent linear"* SDF systems can represent these nonlinear modal SDF systems. The basic concept is that a fairly accurate estimate of nonlinear response can be obtained by analyzing a hypothetical *equivalent linear* system with properties such that the peak displacement response of both nonlinear and *equivalent linear* systems is same. This approach—referred in literature as the Equivalent Linearization, EL—has remained a subject of immense research over past few decades, with a large number of studies dealing with the conversion of nonlinear systems into equivalent linear systems. However, unlike existing EL procedures where the full structure is idealized as a single equivalent linear SDF system, here the concept is applied separately to each significant vibration mode of a structure by converting its representative nonlinear SDF system into its equivalent linear counterpart. For any i^{th} vibration mode, the governing equation of motion of inelastic SDF system (Eq. 15.1) can be approximately replaced by the following equation.

$$\ddot{D}_i(t) + 2\xi_{eq,i}\omega_{eq,i}\dot{D}_i(t) + \omega_{eq,i}^2 D_i(t) = -\ddot{x}_g(t) \tag{15.2}$$

Where $\xi_{eq,\,i}$ and $\omega_{eq,\,i}$ are the damping coefficient and circular frequency of an equivalent linear system. Their values should be selected such that the difference in dynamic responses of actual nonlinear system and the equivalent linear system is minimum. If such optimum values of $\xi_{eq,\,i}$ and $\omega_{eq,\,i}$ are known, Eq. (15.2) can be readily solved to determine $D_i(t)$ without the need of nonlinear function $F_{si}(D_i,\dot{D}_i)$. The equivalent linear frequency ($\omega_{eq,\,i}$) can also be converted to equivalent natural time period ($T_{eq,\,i}$). In this study, these two parameters, $T_{eq,\,i}$ and $\xi_{eq,\,i}$, will be referred onwards as the *"equivalent linear properties"*. Figure 15.1 shows the conversion of an ith mode nonlinear SDF system governed by Eq. (15.1), in to an equivalent linear SDF system governed by Eq. (15.2).

If the equivalent linear properties for each vibration mode are the best representative of their corresponding nonlinear SDF systems, the modal superposition of responses from these *equivalent linear* SDF systems is expected to provide a fairly accurate estimate of nonlinear demands, compared to the direct scaling-down of

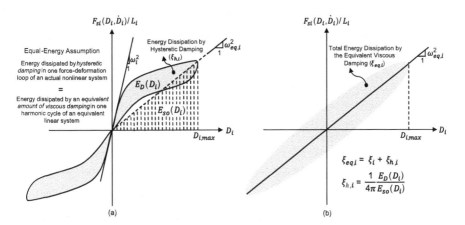

Fig. 15.1 Replacing an ith mode nonlinear SDF system with the corresponding equivalent linear system (**a**) A nonlinear SDF system with initial circular natural frequency ω_i, and with initial inherent viscous damping ξ_i (**b**) An *"equivalent linear"* SDF system with an equivalent circular natural frequency ω_{eq}, and with an equivalent viscous damping $\xi_{eq,\,i}$

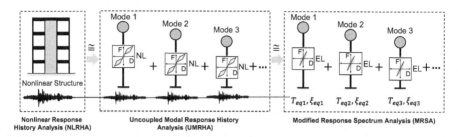

Fig. 15.2 The basic concept of the proposed Modified Response Spectrum Analysis (MRSA) procedure

elastic force demands—as assumed in the standard RSA procedure. This scheme—referred onwards as the Modified Response Spectrum Analysis, MRSA—although involves an additional step (i.e. estimation of equivalent linear properties), it can offer a significant reduction in effort and time compared to the nonlinear dynamic analysis of an inelastic model. Figure 15.2 presents the basic concept of the proposed MRSA procedure, as guided by the UMRHA procedure.

A large number of the EL procedures were proposed in last few decades, dealing with an accurate estimation of equivalent linear properties (T_{eq} and ξ_{eq}). A comprehensive review and comparison of various EL procedures can be found in Liu et al. (2014), and Lin and Miranda (2009). The intent of this study is to test the concept (i.e. the application of equivalent linearization approach for the RSA procedure); it is not intended to propose any refinement in the existing EL methods. Therefore, it is more rational to stick to the simplest and more conventional EL approach, instead of going after empirical and relatively more ambitious efforts to determine T_{eq} and ξ_{eq}. In this study, for an ith vibration mode, the secant natural period ($T_{sec,\,i}$) at maximum displacement amplitude ($D_{i,\,max}$ in Fig. 15.1) is opted as the equivalent linear period for proposing and evaluating the MRSA procedure. The total equivalent viscous damping for an ith vibration mode ($\xi_{eq,\,i}$), is determined as follows.

$$\xi_{eq,i} = \xi_i + \xi_{h,i} \text{ where } \xi_{h,i} = \frac{1}{4\pi} \frac{E_D(D_i)}{E_{so}(D_i)} \qquad (15.3, 15.4)$$

ξ_i is the initial viscous damping ratio and $\xi_{h,\,i}$ is the additional hysteretic damping ratio of an ith vibration mode. In this study, the amount of $\xi_{h,\,i}$ is estimated by using the equal-energy dissipation principle. In this approach, the energy dissipated by the inelastic action in the original nonlinear SDF system is equated to the energy dissipated by viscous damping of an equivalent linear system against a sinusoidal excitation. Based on this equality, $\xi_{h,\,i}$ can be determined using Eq. (15.4).

This selected scheme to determine the equivalent linear properties for any i^{th} vibration mode may not be the most accurate among all the existing EL procedures. However, it retains the theoretical background and conceptual simplicity, and therefore, can be considered reasonable for evaluating the proposed MRSA procedure. Using this scheme, the equivalent linear properties ($T_{eq,\,i}$ and $\xi_{eq,\,i}$) can be determined as a function of D_i (or the peak roof drift ratio, x_i^r/H—the ratio of peak roof displacement to the total height of building), for any nonlinear modal SDF system. In this study, these relationships (relating the equivalent linear properties with the roof drift ratio) are developed which are specific only to the selected case study buildings with RC shear walls and gravity frames (explained in Sect. 15.3). An iterative scheme is proposed (and explained in Sect. 15.5) to determine the $T_{eq,\,i}$ and $\xi_{eq,\,i}$ for any i^{th} vibration mode, using developed relationships. However, once the efficiency of the proposed MRSA procedure in predicting nonlinear demands is established, the generalized relationships can be developed as a function of ductility ratio ($\mu = D_{i,\,max}/D_y$, where D_y is the yield displacement) for different structural systems in future studies. The equivalent linear properties can then be obtained directly, without the need of determining the modal hysteretic behavior $F_{si}(D_i, \dot{D}_i)$. The proposed MRSA procedure can, therefore, be conveniently performed without the need of making a nonlinear structural model.

15.4 Case Study Buildings and Ground Motions

In this study, the accuracy of the MRSA procedure is examined by using three existing high-rise case study buildings. These buildings (20-, 33- and 44-story high, denoted as B1, B2 and B3, respectively) are located in Bangkok, the capital city of Thailand, and are only designed for gravity and wind loads. They are selected to represent a range of typical existing RC shear wall buildings in many countries around the world. All three case study buildings have a podium (for first few stories) and a tower continued up to the roof level. The primary gravity-load-carrying system in B1 is RC beam-column frame with RC slabs, while B2 and B3 have the flat plate system (RC column-slab frame). The lateral load in all three buildings is mainly resisted by a number of RC walls and cores. All three buildings have mat foundation resting on piles. Masonry infill walls are also extensively used in these buildings. Salient structural and architectural features of these buildings are given in Table 15.1.

The proposed MRSA procedure will be applied to these case study buildings to evaluate their nonlinear seismic demands. For this purpose, the linear elastic models were created in ETABS (2011). The initial natural periods (along with an initial inherent damping ratio of 2.5%) of elastic models are used to determine the $T_{eq,\,i}$ and $\xi_{eq,\,i}$ for any i^{th} vibration mode, using the adopted EL scheme. The detailed NLRHA procedure will also be carried out using a suite of compatible ground motion records. The seismic demands obtained from the detailed NLRHA procedure will be used as a benchmark to gauge the accuracy of the proposed MRSA procedure. In order to further understand and evaluate the composition of seismic demands, the UMRHA procedure will also be carried out to obtain modal contributions of various key response quantities.

For the detailed NLRHA procedure and the cyclic modal pushover analysis, full 3D inelastic finite element models of case study buildings were created in Perform

Table 15.1 Basic geometry and characteristics of case study buildings

Building			B1	B2	B3
Height (m)			60	116	152
No. of stories			20	33	44
Typical story height (m)			2.8	3.5	3.5
Height of podium (m)			14	22	43
Natural periods of first three translational vibration modes (sec)	x direction	T_1	1.44	2.81	2.79
		T_2	0.38	0.60	0.71
		T_3	0.17	0.31	0.33
	y direction	T_1	2.12	3.21	3.61
		T_2	0.63	0.97	1.12
		T_3	0.21	0.47	0.31
RC wall section area/total building footprint area (%)			0.40	1.22	0.99
RC column section area/total building footprint area (%)			1.20	2.20	1.80

3D (2006). Each RC wall is modeled by nonlinear concrete and steel fiber elements over the entire height since flexural cracking and yielding may occur at any location due to the higher-mode effects. Each RC column is modeled by a combination of a linear elastic beam-column element with nonlinear plastic zones at its two ends. The concrete slabs are assumed to remain elastic and are modeled by elastic thin shell elements. Each masonry infill wall is modeled by two equivalent compression-only diagonal struts following the FEMA 356 (2000) guidelines. The inherent modal damping ratio of 2.5% is assigned to every significant vibration mode.

A set of seven ground motion records (M_w 6.5–7.5) are selected from the PEER strong ground motion database. They are recorded on relatively near-source sites (<50 Km) with site class D. The ground motions are scaled and adjusted by spectral matching (Hancock et al. 2006) to match with a 5% damped design basis earthquake (DBE-level) target response spectrum prescribed in ASCE 7-05, corresponding to a building site with $S_s = 1.5$ g and $S_1 = 0.75$ g. Figure 15.3 shows the target and matched acceleration response spectra of the selected ground motions.

15.5 Nonlinear Responses Using the UMRHA and NLRHA Procedures

The nonlinear models of case study buildings are subjected to the detailed NLRHA procedure to compute the benchmark seismic demands for comparison with the MRSA procedure. To obtain further insight to the composition of these demands, the UMRHA procedure is also carried out. For this purpose, the nonlinear models were first subjected to the cyclic pushover analyses using the modal inertia force distribution pattern. The resulting cyclic pushover curves (e.g. in the form of base shear coefficient V_{bi}/W versus the cyclic roof drift ratio x_i^r/H) are used to identify the suitable nonlinear function $F_{si}(D_i, \dot{D}_i)$ for the UMRHA procedure. They are also

15 The Evaluation of Nonlinear Seismic Demands of RC Shear Wall... 193

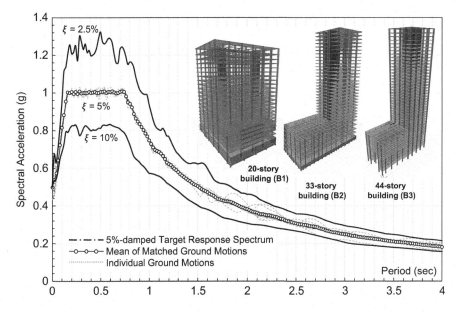

Fig. 15.3 The acceleration response spectra of ground motions, and 3D views of three case study buildings

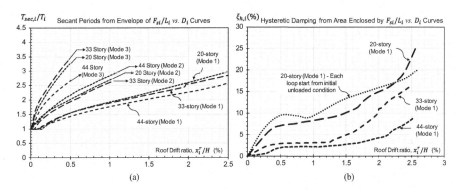

Fig. 15.4 The equivalent linear properties determined from the cyclic pushover analyses (**a**) $T_{sec,i}/T_i$ vs. Roof Drift Ratio (x_i^r/H) (**b**) $\xi_{h,i}$ vs. Roof Drift Ratio (x_i^r/H)

used to develop the $T_{eq,i}/T_i$ vs. x_i^r/H, and $\xi_{h,i}$ vs. x_i^r/H relationships for computing the equivalent linear properties in the MRSA procedure. The selected EL scheme is applied to the cyclic pushover curves for first three translational modes of case study buildings in their x directions. The resulting $T_{sec,i}/T_i$ vs. x_i^r/H relationships as shown in Fig. 15.4a, where T_i denotes the initial time period for an ith vibration mode. A similar trend of equivalent natural period indicates that the representative curves for equivalent natural period can be developed in future studies for the structural systems with similar nonlinear function $F_{si}(D_i, \dot{D}_i)$. The hysteretic damping $\xi_{h,i}$

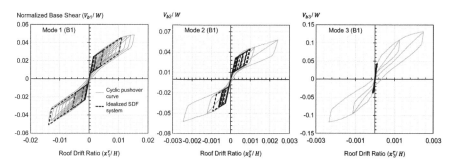

Fig. 15.5 The cyclic modal pushover curves vs. the hysteretic response of idealized nonlinear SDF systems (representing an ith vibration mode) under one ground motion

is determined using the equal-energy dissipation principle applied to the actual cyclic pushover curves. The point-by-point application of Eq. (15.4) resulted in $\xi_{h,\,i}$ vs. x_i^r/H relationships for any i^{th} vibration mode. For first mode in x directions of three case study buildings, these relationships are shown in Fig. 15.4b. In the MRSA procedure, the total equivalent viscous damping (ξ_{eq}) is determined by adding $\xi_{h,\,i}$ to the initial modal viscous damping (ξ_i) (Eq. (15.3)).

For the UMRHA procedure, the F_{si}/L_i vs. D_i relationships for each significant i^{th} mode are idealized by the self-centering type bilinear flag-shaped behavior, and its controlling parameters are tuned to get optimal matching. The modal nonlinear SDF systems (governed by Eq. (15.1)) were subjected to the selected ground motion set to determine $D_i(t)$ and $F_{si}(t)$, which were then converted to various displacement and force related response quantities, respectively. For the x direction of 44-story building (B3), Fig. 15.5 shows the comparison between the actual cyclic modal pushover curve and the response of idealized modal nonlinear SDF systems, in the form of V_{bi}/W vs. x_i^r/H relationships for first three vibration modes. It can be seen that first-mode response is experiencing a reasonable inelastic action as indicated by the significant reduction in stiffness compared to the initial uncracked state. The second mode however, is experiencing a relatively lower level of nonlinearity as compared to the first mode, while the response of third mode remained approximately in the linear elastic range. This observation reconfirms that the use of same response modification factor (R) for each vibration mode, as prescribed in the standard RSA procedure, is not appropriate.

Figure 15.6 shows the envelopes of individual modal contributions to story displacements, inter-story drift ratios, story shears and story overturning moments obtained from the UMRHA procedure, for the same example case (B3). It also compares the combined response envelopes obtained from the UMRHA procedure with those obtained from the NLRHA procedure. A reasonable matching shows the applicability of the UMRHA procedure to decompose the complex nonlinear responses into contributions from a few significant vibration modes. Given that the combined response envelopes obtained from the UMRHA procedure match well with those obtained from the NLRHA procedure,

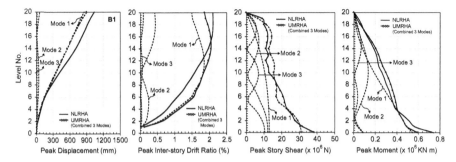

Fig. 15.6 The individual modal contributions, and the comparison of combined seismic demands obtained from the UMRHA procedure with those obtained from the NLRHA procedure

the individual modal contributions (shown in Fig. 15.6) can be compared with the corresponding equivalent linear modal demands to evaluate the accuracy of the MRSA procedure at individual mode level.

15.6 Evaluation of the MRSA Procedure for Case Study Buildings

The seismic demands of all three case study buildings are determined using the MRSA procedure. The practical implementation of this procedure is similar to the standard RSA procedure. It is equivalent to mode-by-mode subjecting a structural model having elongated natural period ($T_{sec,\ i}$) of any i^{th} vibration mode, to a response spectrum constructed for a higher damping ratio ($\xi_{eq,\ i}$). Using the initial modal properties and the relationships shown in Fig. 15.4, an iterative procedure is proposed in the MRSA procedure for the determination of equivalent linear properties. The initial elastic roof drift ratio (x_i^r/H) for an ith vibration mode can be determined using the displacement response spectrum corresponding to its initial natural period (T_i) and initial viscous damping (ξ_i). This can be used to pick the trial values of $T_{sec,\ i}$ and $\xi_{eq,\ i}$ from the developed $T_{sec,\ i}/T_i$ vs. x_i^r/H and $\xi_{h,\ i}$ vs. x_i^r/H relationships. The value of roof drift ratio (x_i^r/H) is then updated by again determining the spectral displacement corresponding to the trial $T_{sec,\ i}$ and $\xi_{eq,\ i}$. This process can be repeated for any i^{th} mode until the starting value of x_i^r/H for an iteration converges to the resulting x_i^r/H at final equivalent linear properties. The seismic demands for each significant vibration mode (corresponding to its final equivalent linear properties) can then be determined by modifying the results of the standard RSA procedure. The modifying factors for force and displacement demands can be determined as the ratios of spectral acceleration (S_a) and spectral displacement (S_d) at equivalent linear properties to those at initial properties, respectively. Since, most of the commercial software provide the facility to apply user-defined scale factors or modifiers to load cases, the MRSA procedure can be

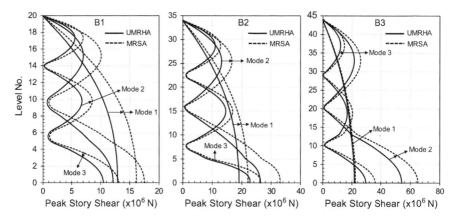

Fig. 15.7 The comparison of modal story shears obtained from the UMRHA and MRSA procedures

automated in already available software capable of performing the standard RSA procedure. This automation also allows to conveniently determine all the local response quantities similar to the standard RSA procedure.

Figure 15.7 shows the comparison of peak modal story shear obtained from the MRSA and the UMRHA procedures for first three translational modes of all three case study buildings. The results from the UMRHA procedure show that for B1, the base shear is mainly dominated by the first vibration mode, while for B2 and B3, it is dominated by the second mode. This is well predicted by the MRSA procedure (Fig. 15.7). A reasonable match for all three case study buildings shows that the MRSA procedure is able to provide an accurate prediction of nonlinear seismic shear demands for every important mode.

Figure 15.8 shows a detailed comparison between the seismic demands computed by the MRSA and the NLRHA procedures for all three case study buildings. The seismic demands computed from the standard RSA procedure are also included in the presented comparison. The response modification factor (R) and the deflection amplification factor (C_d) are taken as 4.5 and 4, respectively in the standard RSA, in accordance with Table 12.2-1 of ASCE 7-05 (2006). It can be seen that the shear force demands obtained from the standard RSA procedure (i.e. reduced by $R = 4.5$) are significantly lower than those from the NLRHA procedure over the entire height of case study buildings. The displacements and inter-story drift ratios (after amplification by C_d) are also underestimating the corresponding true nonlinear demands. The underestimation in force demands is mainly due to the use of same response modification factor ($R = 4.5$) to reduce the elastic seismic demands of second and third vibration modes, as the first mode. Both the UMRHA (Fig. 15.5) and the MRSA (Fig. 15.7) procedures showed that the second mode experienced a lower level of nonlinearity as compared to the first mode, and therefore, the same R factor would result in significantly non-conservative estimates of story shear forces and overturning moments. Similar is true for third mode which remained approximately

15 The Evaluation of Nonlinear Seismic Demands of RC Shear Wall... 197

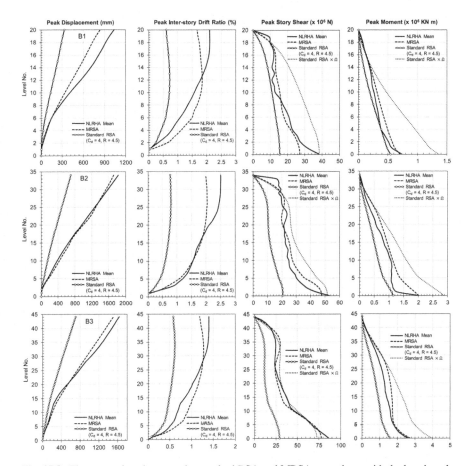

Fig. 15.8 The comparison between the standard RSA and MRSA procedures with the benchmark NLRHA values for all three case study buildings

linear as suggested by both the UMRHA and MRSA procedures, but was reduced by R as required in the standard RSA procedure. This underestimation may result in unsafe design of new buildings or inaccurate performance assessment of existing buildings (which may also give a false sense of safety in some cases).

The MRSA procedure on the other hand, is consistently providing the accuracy comparable to the UMRHA procedure. The displacements and inter-story drift ratios are matching with the corresponding demands obtained from the NLRHA procedure, within a reasonable degree of accuracy. A slight overestimation in overturning moment demands is observed. However, considering the ease offered by the MRSA procedure compared to the detailed NLRHA procedure, a small error in estimation on the conservative side may be tolerated. This satisfactory performance of the MRSA procedure encourages us that it can be considered (and developed further for general use) as a simplified analysis option in cases where it is not practical to perform the detailed NLRHA procedure.

15.7 Conclusions

This study presents a modified version of the response spectrum analysis based on the equivalent linearization approach. The underlying idea is that a properly tuned linear elastic SDF model (with elongated natural period and with additional damping) can approximately represent the nonlinear behavior of a vibration mode, and hence can provide a reasonable estimate of nonlinear seismic demands of that mode. This modified response spectrum analysis (MRSA) procedure is applied to three case study high-rise buildings with shear walls. It is shown that the MRSA procedure can estimate nonlinear seismic demands of these buildings with reasonable accuracy, either for those of individual vibration modes or for their sum (total demands). The MRSA procedure also retains the convenience offered by the standard RSA procedure for practicing engineers; it does not require nonlinear analysis nor nonlinear modeling. This study is only a first step towards the development of a more versatile MRSA procedure. Considering the impact this idea may have on common design office practice, several improvements can be made in the future to make it applicable to buildings and structures of various types, configurations, and materials used.

References

Ahmed M, Warnitchai P (2012) The cause of unproportionately large higher mode contributions in the inelastic seismic responses of high-rise core-wall buildings. Earthq Eng Struct Dyn 41(15):2195–2214

ASCE 7-05 (2006) Minimum design loads for buildings and other structures. ASCE/SEI 7-05, Reston, Va

Chopra AK, Goel RK (2002) A modal pushover analysis procedure for estimating seismic demands for buildings. Earthq Eng Struct Dyn 31(3):561–582

CSI (2006) Perform 3D, version 4. Computers and Structures, Inc., Berkeley

CSI (2011) ETABS nonlinear v9.7.4, version 9.7.4. Computers and Structures, Inc., Berkeley

Eible J, Keintzel E (1988) Seismic shear forces in RC cantilever walls. In: Proceeding of ninth world conference on earthquake engineering, vol. VI, Tokyo, Kyoto, Japan

Eurocode EC 8 (2004) Design of structures for earthquake resistance part 1: general rules, seismic actions and rules for buildings, EN 1998-1. European Committee for Standardization (CEN), Brussels, Belgium

FEMA 356 (2000) Pre-standard and commentary for the seismic rehabilitation of buildings. Federal Emergency Management Agency, Washington

Hancock J, Watson-Lamprey J, Abrahamson NA, Bommer JJ, Markatis A, McCOY E, Mendis R (2006) An improved method of matching response spectra of recorded earthquake ground motion using wavelets. J Earthq Eng 10(Special Issue 1):67–89

Klemencic R, Fry A, Hooper JD, Morgen BG (2007) Performance based design of ductile concrete core wall buildings – issues to consider before detail analysis. Struct Design Tall Spec Build 16:599–614

Lin YY, Miranda E (2009) Evaluation of equivalent linear methods for estimating target displacements of existing structures. Eng Struct 31(12):3080–3089. ISSN 0141-0296

Liu T, Zordan T, Briseghella B, Zhang Q (2014) Evaluation of equivalent linearization analysis methods for seismically isolated buildings characterized by SDOF systems. Eng Struct 59 (2014):619–634

Mehmood T, Warnitchai P, Ahmed M, Qureshi MI (2016a) Alternative approach to compute shear amplification in high-rise reinforced concrete core wall buildings using uncoupled modal response history analysis procedure. Struct Design Tall Spec Build 26. https://doi.org/10.1002/tal.1314

Mehmood T, Warnitchai P, Suwansaya P (2016b) Uncoupled modal response history analysis procedure for seismic evaluation of tall buildings. J Earthq Eng, 2016 (Accepted)

Paret TF, Sasaki KK, Eilbeck DH, Freeman SA (1996) Approximate inelastic procedures to identify failure mechanisms from higher mode effects. In: Proceeding of the 11th WCEE, Acapulco, Mexico, paper 966

Priestley MJN (2003) Does capacity design do the job? An examination of higher mode effects in cantilever walls. Bull NZ Nat Soc Earthq Eng 36(4)

Sasaki KK, Freeman SA, Paret TF (1998) Multi-mode pushover procedure (MMP)—a method to identify the effect of higher modes in a pushover analysis. In: Proceedings of the 6th US national conference on earthquake engineering, Seattle, USA

Zekioglu A, Wilford M, Jin L, Melek M (2007) Case study using the Los Angeles tall buildings structural guidelines: 40-storey concrete core wall building. Struct Design Tall Spec Build 16:583–597

Chapter 16
Seismic Fragility Assessment of Reinforced Concrete High-Rise Buildings Using the Uncoupled Modal Response History Analysis (UMRHA)

Muhammad Zain, Naveed Anwar, Fawad Ahmed Najam, and Tahir Mehmood

Abstract In this study, a simplified approach for the analytical development of fragility curves of high-rise RC buildings is presented. It is based on an approximate modal decomposition procedure known as the Uncoupled Modal Response History Analysis (UMRHA). Using an example of a 55-story case study building, the fragility relationships are developed using the presented approach. Fifteen earthquake ground motions (categorized into 3 groups corresponding to combinations of small or large magnitude and source-to-site distances) are considered for this example. These ground motion histories are scaled for 3 intensity measures (peak ground acceleration, spectral acceleration at 0.2 s and spectral acceleration at 1 s) varying from 0.25 to 2 g. The presented approach resulted in a significant reduction of computational time compared to the detailed Nonlinear Response History Analysis (NLRHA) procedure, and can be applied to assess the seismic vulnerability of complex-natured, higher mode-dominating tall reinforced concrete buildings.

Keywords Seismic risk assessment · UMRHA · NLRHA · Fragility relationships · High-rise RC buildings

---

M. Zain · F. A. Najam
NUST Institute of Civil Engineering (NICE), National University of Sciences and Technology (NUST), Islamabad, Pakistan
e-mail: fawad@nice.nust.edu.pk

N. Anwar (✉)
Asian Institute of Technology (AIT), Bangkok, Thailand
e-mail: nanwar@ait.asia

T. Mehmood
COMSATS Institute of Information Technology (CIIT), Wah Cantt, Pakistan

16.1 Introduction

Due to social and business needs, most of the population migrates towards the urban areas resulting in an increased risk associated with structural collapse/failures. The need and complexities of high-rise buildings are also rapidly increasing in densely populated urban areas. New approaches are emerging to tackle the risks associated with design and construction of structures, such as Consequence-Based Engineering (CBE). CBE is gaining popularity in structural engineering community which enables the engineers to explicitly account for the uncertainty and risk aspects in their practice by employing the probabilistic safety assessments. The ability of CBE to cover more than case-by-case scenarios to mitigate future losses is one of the major attractions offered by this approach. Vulnerability (usually described in terms of fragility relationships) demonstrates the probability that at certain level of intensity of ground motion damage will occur. Therefore, fragility assessment is one of the integrated portions of CBE and it represents the vulnerability information in the form conditional probability of exceedance of particular damage states for particular given seismic intensity, as shown in Eq. 16.1. Fragility relationships can be employed to perform both pre-earthquake planning, as well as for the post-earthquake loss estimation.

$$P(fragility) = P[LS|IM = x] \qquad (16.1)$$

The fragility relationships (also known as hazard-vulnerability relationships) also serve as one of the best available means to select the most suitable and appropriate retrofitting strategies for structural systems. Different approaches are employed by various researchers to derive these relationships which can be classified into four major categories: (1) Empirical Fragility relationships: these are generated by employing the statistical analysis of the data obtained from previous earthquakes. (2) Judgmental Fragility Curves: such curves are primarily based upon the experts' opinions. (3) Analytical Fragility Curves: these curves can provide the most reliable results if sufficient data from past earthquakes is available. However, there is still a margin of uncertainty involved due to limitations and uncertainties in modelling the nonlinear response of RC structures. Since these curves are developed by simulating the expected nonlinear behaviour of the structure itself, they demand a huge computational effort and resources. (4) Hybrid Fragility Curves: these curves are made through the combination of the other two or three types of fragility curves. The process of fragility derivation consists of thorough evaluation of the uncertainties which are classified into two major categories, aleatory and epistemic. The former one indicates the uncertainties associated intrinsically with the system i.e. the uncertainties involved in the ground motions, while the later one characterizes the uncertainties that are mostly due to the deficiency and lack of knowledge and data about the structural properties and behaviour (Gruenwald 2008).

Various researches have proposed simplified methodologies for developing the analytical fragility relationships, but most of the studies have focused on low-rise and mid-rise structures. Relatively less work has been carried out to develop fragility curves for high-rise structures. Jun Ji et al. (2007) presented a new methodology for developing the fragility curves for high-rise buildings by making a calibrated and efficient 2D model of the original structure using genetic algorithms, considering the PGA, Sa at 0.2 s, and Sa at 1.0 s as the seismic intensity indicators. High-rise buildings are very complex structural systems, composed of many structural and non-structural components. The seismic response of high-rise structures remains convoluted as the many vibration modes other than the fundamental mode participate significantly in their seismic response. Among various numerical analysis procedures for evaluating seismic performance of high-rise buildings, the Nonlinear Response History Analysis (NLRHA) procedure has been widely considered and accepted as the most reliable and accurate one. However, the procedure is computationally very expensive, and it does not provide much physical insight into the complex inelastic responses of the structure. Nonlinear Static procedures (NSP) on the other hand, accounts only for the response contribution of fundamental vibration mode and are not considered suitable for evaluating higher-mode dominating structures. In this study, a simplified procedure called the "Uncoupled Modal Response History Analysis (UMRHA)" (Chopra 2007) procedure is used as tool to develop the fragility curves for high-rise buildings. In UMRHA, the nonlinear response contribution of individual vibration modes are computed and combined into the total response as explained in next section.

16.2 Basic Concept of the Uncoupled Modal Response History Analysis (UMRHA) Procedure

The UMRHA procedure (Chopra 2007) can be viewed as an extended version of the classical modal analysis procedure in which the overall complex dynamic response of a linear Multi-Degree-of-Freedom (MDOF) structure is considered as a sum of contributions from only few independent vibration modes. The response behavior of each mode is essentially similar to that of a Single-Degree-of-Freedom (SDOF) system governed by a few modal properties. Strictly speaking, this classical modal analysis procedure is applicable to only linear elastic structures. When the responses exceed the elastic limits, the governing equations of motion become nonlinear and consequently, the theoretical basis for modal analysis becomes invalid. Despite this, the UMRHA procedure assumes that even for inelastic responses, the complex dynamic response can be approximately expressed as a sum of individual modal contributions (assuming them uncoupled). The number of modes included in the analysis are selected based on cumulative modal mass participation ratio of more than 90%. The governing equation of motion of SDOF subjected to a horizontal ground motion $\ddot{x}_g(t)$ can be written as:

$$\ddot{D}_i + 2\xi_i \omega_i \dot{D}_i + F_{si}(D_i, \dot{D}_i)/L_i = -\ddot{x}_g(t) \tag{16.2}$$

Where $L_i = M_i \Gamma_i$ and $M_i = \boldsymbol{\phi}_i^T \mathbf{M} \boldsymbol{\phi}_i$; ω_i and ξ_i are the natural vibration frequency and the damping ratio of the ith mode, respectively. Eq. (16.2) is a standard governing equation of motion for inelastic SDOF systems. To compute the response time history of $D_i(t)$ from this equation, one needs to know the nonlinear function $F_{si}(D_i, \dot{D}_i)$. A reversed cyclic pushover analysis is performed for each important mode to identify this nonlinear function $F_{si}(D_i, \dot{D}_i)$. The cyclic pushover analysis for the ith mode can be carried out by applying a force vector with the ith modal inertia force pattern $\mathbf{s}_i^* = \mathbf{M} \boldsymbol{\phi}_i$ (where $\mathbf{M}$ is the mass matrix of the building and $\boldsymbol{\phi}_i$ is the ith natural vibration mode of the building in its linear range. The relationship between roof displacement, obtained from the cyclic pushover (denoted by x_i^r) and D_i is approximately given by

$$D_i = x_i^r / (\Gamma_i \phi_i^r) \tag{16.3}$$

where ϕ_i^r is the value of $\boldsymbol{\phi}_i$ at the roof level. The relationship between the base shear V_{bi} and F_{si} under this modal inertia force distribution pattern is given by

$$F_{si}/L_i = V_{bi}/\Gamma_i L_i \tag{16.4}$$

By this way, the results from the cyclic pushover analysis are first presented in the form of cyclic base shear (V_{bi})—roof displacement (x_i^r) relationship, and then transformed into the required $F_{si} - D_i$ relationship. At this stage, a suitable nonlinear hysteretic model can be selected, and its parameters can be tuned to match with this $F_{si} - D_i$ relationship. The response time history of $D_i(t)$ as well as $F_{si}(t)$ can then be calculated from the nonlinear governing Eq. (16.2). The response of each mode belongs to the ith vibration mode and can be generally represented by $r_i(t)$. By summing the contributions from all significant modes, the total response history $r(t)$ is obtained as follows.

$$r(t) = \sum_{i=1}^{m} r_i(t) \tag{16.5}$$

where m is the number of significant vibration modes.

16.3 Methodology and Structural Modeling

A UMRHA-based methodology is intended to include higher mode effects in the process of developing the fragility relationships. Computational effort is always considered as one of the major concerns in seismic fragility analysis. The current methodology focuses towards the reduction of the computational effort by proposing a theoretically close, yet simpler method (UMRHA) to replace full nonlinear response history analysis (NLRHA). A 55 story core-wall high-rise building, located in a seismically active area (Manila, Philippines) is selected for application of presented methodology. Full 3D nonlinear finite element model (shown in Fig. 16.1) was created in PERFORM 3D (CSI, 2000) employing the concepts of capacity-based design (all the primary structural members are not allowed to undergo shear failure). The whole core wall and link beams are modeled as nonlinear, while the other components are kept linear. Complete hysteretic behaviors (F-D Relationships) were assigned to all nonlinear structural components as well as materials to explicitly account for stiffness degradation and hysteretic damping. Table 16.1 and Fig. 16.1 show the major characteristics of the selected building in terms of geometry and material properties of key structural elements, respectively. Figure 16.2 shows the first four mode shapes of the building in its weaker direction.

Figure 16.3 shows the proposed methodology in the form of a flow chart. The procedure requires to perform both monotonic and reversed cyclic pushover analyses to identify strong/weak direction and damage states of building, and to develop complete hysteretic behaviors for equivalent single-degree-of-freedom (SDOF) systems, respectively. For case study building, the procedure is validated by comparing displacement histories obtained from UMRHA and NLRHA (presented in Sect. 16.6). Selected ground motions are applied to full analytical model as well as to

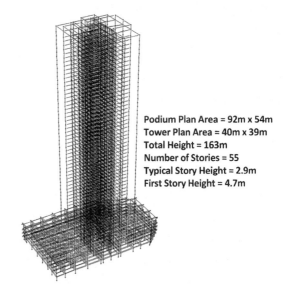

Fig. 16.1 The 3D analytical model of a 55-story case study building (CSI, 2000)

Podium Plan Area = 92m x 54m
Tower Plan Area = 40m x 39m
Total Height = 163m
Number of Stories = 55
Typical Story Height = 2.9m
First Story Height = 4.7m

Table 16.1 Material properties of key structural elements

Member	Nominal concrete strength, psi (MPa)
Columns and shear walls	
Lower basement to 11th level	10,000 (69)
12th level to 21st level	8500 (59)
21st level to roof deck level	7000 (48)
Beams, girders, and slabs	
Foundation to 40th level	6000 (41)
40th level to roof deck level	5000 (34)

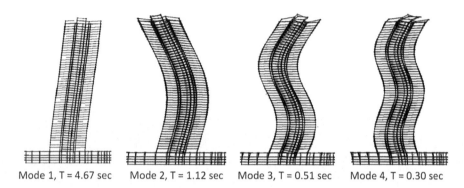

Mode 1, T = 4.67 sec   Mode 2, T = 1.12 sec   Mode 3, T = 0.51 sec   Mode 4, T = 0.30 sec

Fig. 16.2 First 4 mode shapes of the considered building in weaker direction

all equivalent SDOF systems. The SDOF displacement histories (obtained in terms of spectral displacement) are converted back to actual displacements and are added linearly to obtain a complete displacement history.

For the case study structure, equivalent SDOF systems are created for the first four modes (shown in Fig. 16.3) providing the modal mass participation ratio of more than 90%. A computer program RUAUMOKO 2D (Carr 2004), developed at University of Canterbury, is used for the solving SDOF systems. It provides a convenient interface and allows user to assign and control a reasonably large number of hysteretic behaviors to various nonlinear components.

16.4 Uncertainty Treatment

Probabilistic nature of seismic fragilities is greatly influenced by the uncertainties (aleatory or epistemic) and assumptions involved in the process. Wen et al. (2003) describes the aleatory uncertainty as the one which can be explicitly recognized by a stochastic model, whereas those which exist in the model itself and its parameters,

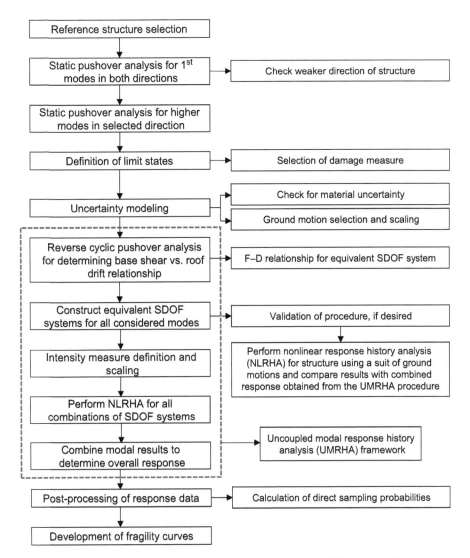

Fig. 16.3 Proposed methodology for analytical fragility assessment of high-rise buildings

are epistemic. In this case, aleatory uncertainties represent the uncertainty associated with the ground motions (existing intrinsically in the earthquakes), while epistemic uncertainties represent the existence of ambiguity in the structural capacity itself (mainly due to the lack of knowledge and possible variation in construction process). The current study focuses towards the consideration of both types of uncertainties involved in the seismic fragility assessment by considering 15 ground motions and varying the materials' strengths.

16.4.1 Material Uncertainty

The intrinsic variability of material strengths is also one of the major sources of uncertainty involved in the seismic fragility assessment. In this study, the compressive and tensile strengths of concrete, as well as, the yield strength of steel are considered as random variables. The studies by Ghobarah et al. (1998) and Elnashai et al. (2004) employed statistical distributions to define the uncertainty involved in the yield strength of steel. There seem to be a consent in terms of employing normal and log-normal distributions to elaborate the variability in the yield strength of steel with coefficient of variation (COV) ranging from 4% to 12%. Bournonville et al. (2004) evaluated the variability of properties of reinforcing bars, produced by more than 34 mills in U.S. and Canada. The current study employs the research outcome from Bournonville et al. (2004). The mean and COV for the yield point are 480 MPa and 7% respectively, while the mean value and COV for the ultimate strengths are 728 MPa and 6% respectively. The intrinsic randomness concrete strength can be captured using experimental data. Hueste et al. (2004) studied the variable nature of concrete strength, including the experimental results from the testing of higher strength concretes. The current study utilizes the results reported by Hueste et al. (2004).

16.4.2 Selection of Ground Motions Accelerograms

Bazzurro and Cornell (1994) suggested that seven ground motions are sufficient for covering the aspect of uncertainty from the earthquakes while the recent Tall Building Initiatives (TBI) Guidelines (2010) also recommends the same number of ground motions. Some researchers also have demonstrated the use of simulated ground motion histories e.g. Shinozuka et al. (2000b) used the simulated ground motions by Hwang and Huo (1996) for developing the analytical fragility curves for bridges. Jun Ji et al. (2009) presented new criteria for ground motion selection based on earthquake magnitude, soil conditions and source-to-site distance.

In this study, 15 ground motion records (organized in to 3 categories) are selected following the same criteria as proposed by Jun Ji et al. (2009) i.e. earthquake magnitude, site soil conditions and source-to-site distance. The larger magnitude earthquake histories usually contain several peaks compared to moderate and small earthquakes mostly causing the structure to undergo a significant extent of nonlinearity. The source-to-site distance influences the filtration of frequency fractions during the process of wave propagation, while soil conditions are mainly held responsible for the amplification or dissipation of seismic waves. Based on these considerations, the selected ground motions are divided among 3 categories each corresponding to an adequate variation in the magnitude, distance to source, and soil

Table 16.2 Selected ground motion records

Category	Earthquake	Magnitude	Distance to Rupture	Soil at Site
1	Chi-chi, Taiwan	7.6	7.30	Stiff
	Imperial Valley	6.5	2.50	Soft
	Kobe, Japan	6.9	1.20	Soft
	Loma Prieta, USA	6.9	5.10	Stiff
	Northridge	6.7	17.5	Stiff
2	Aftershock of Friuli EQ, Italy	5.7	10.0	Soft
	Alkion, Greece	6.1	25.0	Soft
	Anza (horse Cany)	4.9	20.0	Soft
	Caolinga	5.0	12.6	Stiff
	Dinar, Turkey	6.0	1.02	Soft
3	Chi-chi, Taiwan	7.6	39.3	Soft
	Kobe, Japan	6.9	89.3	Stiff
	Kocaeli, Turkey	7.4	76.1	Stiff
	Kocaeli, Turkey	7.4	78.9	Soft
	Northridge	6.7	64.6	Soft

conditions i.e. Near-source and Large-magnitude, Near-source and Moderate-magnitude, and Distant-source and Large-magnitude. The selected ground motions and their properties, with reference to their categories, are enlisted in the Table 16.2.

16.5 Definition of Limit States and Seismic Intensity Indicators

The definition of limit states of the structure is a fundamental component in seismic fragility assessment. For high-rise buildings, there is no universally acceptable and consistently applicable criterion to develop a relationship between damage and various demand quantities. Several researchers have proposed different performance limit states of buildings, usually classified in two major categories (qualitative and quantitative). In qualitative terms, HAZUS (1999) provides four limit states of building structures (Slight, Moderate, Major, and Collapse). Smyth et al. (2004) and Kircher et al. (1997) also used four damage states; slight, moderate, extensive, and collapse. Whereas, the quantitative approach describes the damage states in terms of mathematical representations of damage, depending upon some designated and specific structural responses. Different researchers have employed different damage indicators for representing damage at local and global levels. Shinozuka et al. (2000a) used ductility demands as damage indicators for prescribed damage states. Guneyisi and Altay (2008) used inter-story drift (ISD) ratio to develop the fragility curves. Although many others have used damage indicators related to

Table 16.3 Definitions of considered limit states

Level	Limit state	Definition
Limit state 1 (LS 1)	Damage control	The very first yield of longitudinal steel reinforcement, or the formation of first plastic hinge.
Limit state 2 (LS 2)	Collapse prevention	Ultimate strength/capacity of main load resisting system.

energy and forces, but ISD is the most frequently used parameter and can correlate adequately with both the non-structural and structural damage. This study also employs ISD ratios as the seismic response (damage) indicator and defines the two limit states of case study building based on study conducted by Jun Ji et al. (2009). The first one is "Damage Control", and the second is "Collapse Prevention". Qualitative definitions of the considered limit states are provided in Table 16.3.

A nonlinear static pushover analysis for first 4 modes in weaker direction was conducted for full 3D nonlinear model. The definitions of the prescribed limit states are then applied to the results of the pushover analysis to obtain the quantitative definitions of limit states in terms of ISD ratios. It should be noted that limit states of case study building are defined for each mode separately to include the higher modes effects on the selected damage criteria. Another essential step in the fragility analysis is a proper selection of an intensity measure to relate structural performance. An adequate intensity measure would correlate well between the structural response and the associated vulnerability (Wen et al. 2004). In earlier studies, peak ground acceleration (PGA) was one of the frequently used seismic intensity indicator. Other widely used measures involve Modified Mercalli Intensity (MMI), Arias Intensity (AI), and Root Mean Square (RMS) Acceleration (Singhal and Kiremidjian 1997). Some studies include spectral acceleration (Sa) as intensity measure at fundamental period of structure (Kinali and Ellingwood 2007), Sa at 0.2 s and Sa at 1.0 s. In this study, PGA, Sa at 0.2 s and Sa at 1.0 s are considered as seismic intensity measures considering the idea that PGA alone may not serve as an accurate intensity indicator to correlate theoretically computed structural damage with observed performance (Sewell 1989).

16.6 Results and Discussion

This section presents the results obtained from UMRHA and NLRHA with the view to develop fragility curves for the case study tall building. However, first the effect of uncertainties in material strengths will be presented to check the sensitivity of results.

16.6.1 Effect of Material Strength Uncertainties

It is preferable to concentrate on dominant factors that can play an influential role in the probabilistic variation of the response; therefore, the results' sensitivity prior to performing the complete UMRHA is checked, and only that type of uncertainty is considered that can cause a significant variation in the building response. The sensitivity of the results in response of the variation in material strengths requires complete analytical simulations. Ibarra and Krawinkler (2005) described that the uncertainty in ductility capacity and post-capping stiffness generate the principal additional contributions to the dispersion of capacity, especially near the collapse. The former one is more important when P-Δ effects are large. The study further suggested that the record-to-record variability is also the major contributor to total uncertainty.

In the case of high-rise structures, it may not be suitable to conduct Monte Carlo simulation as it requires a large number of analyses for reasonably accurate results, and a run time of each analysis is around 40 h for 3D structural model. In this study, 5 pairs of material strengths (concrete and steel) are selected and the results are presented here based on "worst-case-scenario". Figures 16.4 and 16.5 show the sensitivity of the results in response to the variation of material properties, when 3D analytical model was subjected to the application of a ground motion history and the first modal pushover analysis respectively. The steel strength is mean (μ) + 1 standard deviation (σ), while the concrete strength is mean (μ) − 1 standard deviation (σ). This pair of strength is selected considering that high steel strength attracts more earthquake forces, and the reduction in concrete quality decreases the shear capacity. The results show that even with this much variation in material strengths, the response of building does not vary significantly. Thus, the material uncertainty can be treated as an epistemic uncertainty.

16.6.2 Fragility Derivation

For case study tall building, dynamic response histories were determined using both NLRHA and UMRHA procedures to validate the methodology. Before UMRHA, a

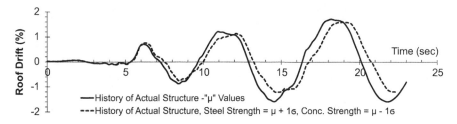

Fig. 16.4 Sensitivity of roof drift (%) to variation in material properties when the model was subjected to a ground motion history

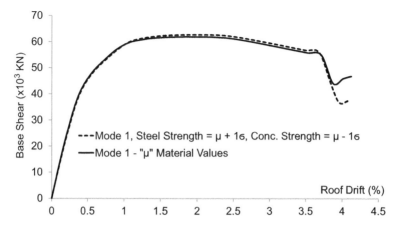

Fig. 16.5 Sensitivity of first modal monotonic pushover curve from variation in material properties

reversed cyclic pushover analysis for all 4 modes was conducted to obtain global hysteretic behavior which was converted later to an idealized force-deformation model to construct equivalent SDOF systems. As an example, Figure 16.6 shows the envelope of cyclic pushover curve for the first mode in weaker direction. Table 16.4 shows some of the important properties of equivalent SDOF systems determined from idealized force-deformation model. Selected ground motions were then applied to each of the SDOF system (representing each mode) and the individual response was linearly added to obtain overall displacement histories. Figure 16.7 shows the comparison of roof drift (%) history obtained from NLRHA of 3D analytical model and from UMRHA. It can be seen that UMRHA response is reasonably matching with NLRHA. After getting a reasonable degree of confidence on UMRHA validation, each ground motion history is scaled to eight intensity levels (PGA, Sa at 0.2 s and Sa at 1.0 s) ranging 0.25–2.0 g, with the increment of 0.25 g. In UMRHA, for each of equivalent SDOF system, 120 dynamic analyses are conducted making total number of analyses as 480 for each seismic intensity measure. A total of 1440 analyses are conducted to consider first 4 modes.

Once the results are obtained, the quantitative definitions of limit states are applied to assess the levels of structure's performance subjected to ground motions with a particular intensity. When a computed value approaches or surpasses the distinctly defined limit states (in any of the considered vibration mode), that particular event is counted in the sample to compute the probabilities. This process of probability calculation is performed for each limit state at each specific level of intensity (determined by dividing the number of ground motions causing that specific limit state by the total number of considered ground motions). Figure 16.8 shows the peak roof drift values for all three ground motion categories against all considered levels of PGA. The random nature of peak dynamic response can be seen for all 15 ground motions highlighting the importance of considering aleatory uncertainties in seismic fragility analysis. Similar relationships were developed for

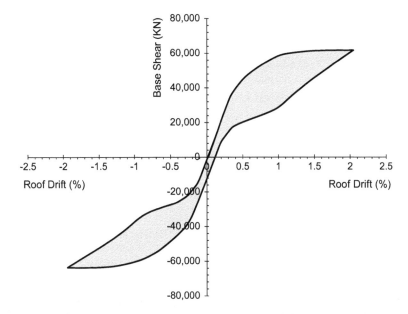

Fig. 16.6 Envelope of reversed cyclic pushover curve for the first mode of the case study building

Table 16.4 Important characteristics of SDOF systems

Properties	Mode 1	Mode 2	Mode 3	Mode 4
Γ_n	1.50286	0.811329	0.604722	0.731589
D_{ny} (mm)	490	395	310	200
$\frac{F_{sny}}{L_n}$ (mm/s^2)	722.495	9073.197	49174.89	72711.33

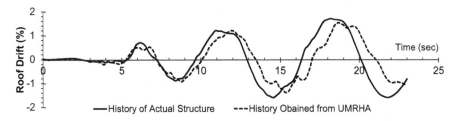

—History of Actual Structure   ---History Obained from UMRHA

Fig. 16.7 Comparison between roof drift history of 3D analytical model from NLRHA and UMRHA

other two intensity measures (Sa at 0.2 s and Sa at 1.0 s) and converted later to generalized fragility curves (as shown in Fig. 16.9). A lognormal distribution is assumed to develop generalized fragility curves governed by eq. 6 below.

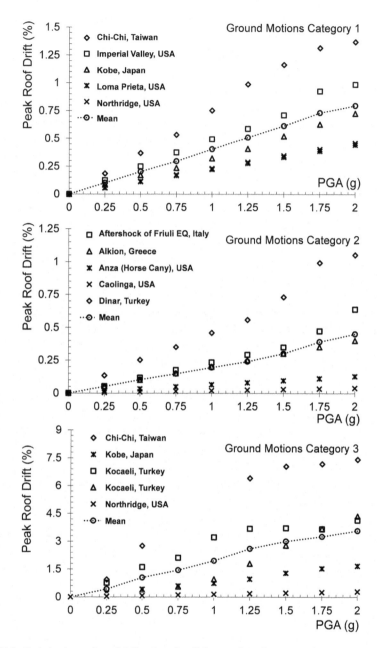

Fig. 16.8 Peak (and mean) roof drift values for all 3 ground motion categories

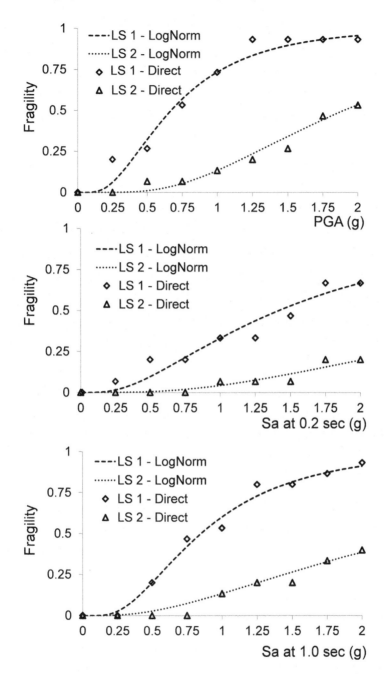

Fig. 16.9 Developed fragility curves for the case study building for defined intensity measures (Solid lines represent the developed lognormal functions, and the dots show the directly calculated sampling probabilities)

Table 16.5 Lognormal distribution parameters for fragility relationships

Limit State	PGA		S_a at 0.2 s		S_a at 1.0 s	
	λ_c	β_c	λ_c	β_c	λ_c	β_c
LS 1	−0.3991	0.6390	0.3490	0.7990	−0.1624	0.6310
LS 2	0.6420	0.5799	1.3754	0.8015	0.9385	0.8450

$$P(LS|IM) = \phi[\ln IM - \lambda_c]/\beta_c \qquad (16.6)$$

Where $P(LS|IM)$ describes the probability of exceedance of a specific limit state of the building at a particular intensity measure with an explicit intensity, whereas ϕ represents the standard normal cumulative distribution function, and β_c and λ_c are the controlling parameters, which represent the slope of the curve and its median, respectively. Nonlinear curve-fitting techniques are utilized to make an optimized estimation of the controlling parameters for each of the fragility curves. Figure 16.9 shows the developed lognormal fragility relationships of the considered building at the intensity measures of PGA, Sa at 0.2 s, and Sa at 1.0 s along with the direct sampling probabilities. The values of controlling parameters are enlisted in Table 16.5.

16.7 Conclusions

In this study, a simplified approach based on the UMRHA procedure is proposed for the development of analytical fragility curves of high-rise RC buildings. The methodology is executed on a 55 story case study building to demonstrate the process. Following conclusions can be drawn from this study:

(a) The damage limit states are generally defined only on the basis of first-mode pushover analysis which are not able to account for the chances of secondary nonlinearity (e.g. secondary plastic hinge developments) for the high-rise structures having significant response contribution from higher vibration modes. The presented approach offers an advantage of defining the limit states, including high-modes of vibrations and is expected to be more reliable compared to simplified methodologies based on the damage limit states definitions for only first vibration mode.
(b) Uncertainties arising from both seismic demand and structural capacity are evaluated and it is concluded that the random nature of ground motion should be given due consideration while developing generalized fragility functions.
(c) Computational effort and cost is always an ever-growing challenge for analytical assessment of seismic risk to existing and new buildings. The NLRHA procedure for one high-rise building subjected to one ground motion record takes around 30 h of computation time of a 3.4 GHz processor and 4.0 GB RAM desktop computer. The processing of computed dynamic responses into the required format takes another 4–5 h. For the UMRHA procedure, it takes around

1 h to complete cyclic pushover analyses for three vibration modes, 10 min for the analysis of three equivalent nonlinear SDOF systems, and 20 min for transforming and processing computed responses into the final format. This low computational effort for the UMRHA (compared to the NLRHA) may allow us to explore the nonlinear responses of tall buildings to a reasonably high number of ground motions in a convenient and practical manner.

References

Bazzurro P, Cornell CA (1994) Seismic hazard analysis of nonlinear structures. I: methodology. J Struct Eng 120(11):3320–3344

Bournonville M, Dahnke J, Darwin D (2004) Statistical analysis of the mechanical properties and weight of reinforcing bars. University of Kansas Report

Carr AJ (2004) Ruaumoko – inelastic dynamic analysis program. Department of Civil Engineering, University of Canterbury, Christchurch

Chopra AK (2007) Dynamics of structures: theory and applications to earthquake engineering. Prentice Hall, Hoboken

Elnashai A, Borzi B, Vlachos S (2004) Deformation-based vulnerability functions for RC bridges. Struct Eng Mech 17(2):215–244

Ghobarah A, Aly N, El-Attar M (1998) Seismic reliability assessment of existing reinforced concrete buildings. J Earthq Eng 2(4):569–592

Gruenwald J (2008) Risk-based structural design: Designing for future aircraft. AE 440 Technical Report

Guneyisi EM, Altay G (2008) Seismic fragility assessment of effectiveness of viscous dampers in RC buildings under scenario earthquakes. Struct Saf 30(5):461–480

Hazus (1999) Earthquake loss estimation methodology technical and user manual. Federal Emergency Management Agency (FEMA), Washington

Hueste MBD, Chompreda P, Trejo D, Cline DB, Keating PB (2004) Mechanical properties of high-strength concrete for prestressed members. ACI Struct J 101(4)

Hwang H, Huo J (1996) Simulation of earthquake acceleration time histories. Center for Earthquake Research and Information, The University of Memphis, Technical Report

Ibarra LF, Krawinkler H (2005) Global collapse of frame structures under seismic excitations. Pacific Earthquake Engineering Research Center, Berkeley

Ji J, Elnashai AS, Kuchma DA (2007) An analytical framework for seismic fragility analysis of RC high-rise buildings. Eng Struct 29(12):3197–3209

Ji J, Elnashai AS, Kuchma DA (2009) Seismic fragility relationships of reinforced concrete high-rise buildings. Struct Design Tall Spec Build 18(3):259–277

Kinali K, Ellingwood BR (2007) Seismic fragility assessment of steel frames for consequence-based engineering: a case study for Memphis. Eng Struct 29(6):1115–1127

Kircher CA, Nassar AA, Kustu O, Holmes WT (1997) Development of building damage functions for earthquake loss estimation. Earthquake Spectra 13(4):663–682

PEER (2010) Guidelines for performance-based seismic design of tall buildings, PEER Report No. 2010/05. University of California, Berkeley

Powell G (2000) Perform-3D CSI. User manual and User guide. CSI Inc.

Sewell RT (1989) Damage effectiveness of earthquake ground motion: characterizations based on the performance of structures and equipment. PhD thesis, Stanford University

Shinozuka M, Feng MQ, Kim HK, Kim SH (2000a) Nonlinear static procedure for fragility curve development. J Eng Mech 126(12):1287–1295

Shinozuka M, Feng MQ, Lee J, Naganuma T (2000b) Statistical analysis of fragility curves. J Eng Mech 126(12):1224–1231

Singhal A, Kiremidjian A (1997) A method for earthquake motion-damage relationships with application to reinforced concrete frames. NCEER-97-0008

Smyth AW, Altay G, Deodatis G, Erdik M, Franco G, Gulkan P, Kunreuther H, Lus H, Mete E, Seeber N (2004) Probabilistic benefit-cost analysis for earthquake damage mitigation: evaluating measures for apartment houses in Turkey. Earthquake Spectra 20(1):171–203

Wen Y, Ellingwood B, Veneziano D, Bracci J (2003) Uncertainty modeling in earthquake engineering. Mid-America earthquake center project FD-2 report

Wen Y, Ellingwood B, Bracci JM (2004) Vulnerability function framework for consequence-based engineering. MAE Center Project DS-4 Report, April 28, 2004

Chapter 17
Ambient Vibration Testing of a Three-Storey Substandard RC Building at Different Levels of Structural Seismic Damage

Pinar Inci, Caglar Goksu, Ugur Demir, and Alper Ilki

Abstract In this paper, the effects of structural damage on the modal characteristics of a substandard full-scale reinforced concrete (RC) building were investigated. The RC building was a representative of large number of existing substandard RC buildings. The building was subjected to different levels of structural damage through quasi-static reversed cyclic lateral loading. Ambient vibration tests were carried out not only before and after quasi-static lateral loading cycles, but also at a certain damage level. The vibration test survey showed that modal frequencies decreased while damping ratios increased with the increasing levels of damage. More importantly, since the structural damage pattern due to quasi-static cyclic lateral loading was similar to ones observed in existing RC structures after earthquakes, determining the changes in dynamic characteristics at different levels of structural damage can be useful for estimating the residual performance of the structure after an earthquake.

Keywords Ambient vibration test · Damage · Damping · Frequency · RC

17.1 Introduction

Structural health monitoring (SHM) has been employed as a routine application in some industries such as aerospace, automotive, manufacturing, etc. for several decades. It has also been considered for civil infrastructure systems, such as dams, bridges, tall buildings in recent years. Eventually, since the existing building stock, which includes huge number of aged RC residential and public structures, represent vulnerability against earthquakes, a great knowledge for SHM has being used extensively for obtaining dynamic characteristics of these civil structures. Various experimental and analytical vibration based techniques have been considered to

P. Inci (✉) · C. Goksu · U. Demir · A. Ilki
Civil Engineering Faculty, Istanbul Technical University, Istanbul, Turkey
e-mail: pinarinci@itu.edu.tr

Fig. 17.1 Test building

identify the dynamic characteristics of structures for model updating issues, which has a key role for seismic performance assessment of concerned structures (Brownjohn and Xia 2000; Foti et al. 2012; Sanayei et al. 2015). Similarly, well-known SHM applications were adopted and developed to identify the change in dynamic characteristics of structures, which have been subjected to seismic actions (Ivanovic et al. 2000; Kusunoki et al. 2012; Vidal et al. 2014). In this study, similar to these condition assessment studies, it was aimed to present the change in dynamic characteristics of a substandard RC building due to damage on structural members. An important point of this study, was i to quantify the influence of different types and extents of structural damages on the dynamic characteristics of building. The test building was a single bay, three-storey RC frame building (Fig. 17.1). The building

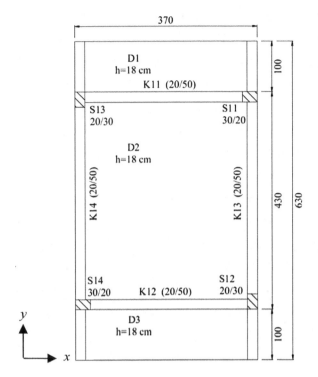

Fig. 17.2 Typical plan view of the test building (Goksu et al. 2015)

was designed and constructed to represent the common characteristics of existing seismically vulnerable building stock in Turkey (Goksu et al. 2015; Comert et al. 2016). In the study of Goksu et al. 2015, the change of the modal frequencies and the modal damping ratios of the building with increasing structural damage was identified through forced vibration tests. Different than Goksu et al. 2015, in this study, acceleration response of the building was measured under ambient vibrations. Additionally, the mode shapes of the building were introduced.

A set of quasi-static cyclic lateral loading was applied in x direction of the building in terms of incrementally increasing reversed displacement cycles. Figure 17.2 displays the x and y directions of the building in plan view. The building experienced structural damages similar to ones commonly observed in existing RC structures after earthquakes. Ambient vibration tests were carried out before and after quasi-static lateral loading cycles as well as at certain damage levels. During these dynamic tests, an output only system identification algorithm was followed to obtain a relationship between the modal parameters and the level of structural damage based on the response of the structure to ambient vibrations. The system identification process of the building showed that the modal frequencies decreased while damping ratios for different modes increased with the increasing levels of damage. Since the test building

17.2 Description and Instrumentation of the Test Building

The test building was a full-scale, three-storey RC moment resisting bare frame building with a single bay in x and y directions. It was designed to be representative of a large number of substandard RC buildings. It has insufficient construction characteristics, such as low compressive strength of concrete, inadequate thickness of concrete cover, plain reinforcing bars and poor reinforcement detailing. Moreover, the columns were weaker than beams. The height of each story was 3 m. The concrete compressive strength (fc') was 10 MPa and the yield stress of the plain longitudinal and transverse reinforcing bars was 350 MPa. The typical plan view of the building is presented in Fig. 17.2. More detailed information about the test building can be found in Goksu et al. (2015).

Ambient vibration measurements were performed through 6 uniaxial piezoelectric voltage accelerometers and a 24-bit delta-sigma dual core data acquisition system with anti-aliasing filter. Each accelerometer was mounted at the center of each story and mat foundation (A2, A4, A5 and A6) while, 2 additional accelerometers were mounted on 2 opposite corners of the third story (A1 and A3) (Fig. 17.3). The configuration given in Fig. 17.3 illustrates the arrangement of accelerometers for capturing the vibration response of the building in x direction. For capturing the response of the building in y direction, the accelerometers at the centers of the slabs (i.e. A2, A4, A5 and A6) were rotated 90°.

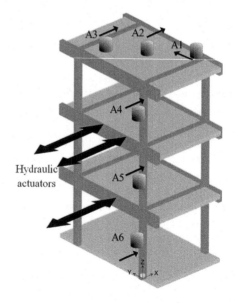

Fig. 17.3 Instrumentation of the buildings

17.3 Ambient Vibration Test Survey

17.3.1 Ambient Vibration Tests

Ambient vibration tests were performed before and after quasi-static reversed cyclic loading. Additionally, for observing the rate of changes of the dynamic characteristics with gradually increasing damage, it was also carried out at a certain damage level. The quasi-static reversed cyclic loading was applied to the building only in the x direction using one hydraulic actuator at the lower first and two other hydraulic actuators at the second story slab levels of the building (Fig. 17.3). Actuators enforced the loading in terms of incrementally increasing reversed lateral displacement cycles. Hence, targeted damage levels were achieved by pushing and pulling the building to specific ground story drift ratios (d.r.), which were calculated as the ratios of the lateral displacement of the top floor to the height of the building. It should be noted that the building was unloaded and the hydraulic actuators were detached from the buildings before dynamic measurements. The dynamic tests were carried out for x and subsequently y direction before and after quasi-static reversed cyclic loading (undamaged state and 3% d.r., respectively) while for only x direction at the damage level corresponding to 0.5% d.r..

17.3.2 Damage Pattern of the Building at the Time of Ambient Vibration Tests

Damage pattern of the one of the columns of the building at the time of dynamic measurements is given in Fig. 17.4. The damage started with bending cracks at upper and lower ends of the columns at around 0.5% d.r.. Although, the crack widths increased at around 3% d.r., they did not cause crushing of concrete and spalling of concrete cover. However, the base shear-drift ratio relationship showed a major decrease in lateral load bearing capacity (at around 20% at 3% d.r.) (Fig. 17.5).

17.3.3 Modal Identification

System identification of the building for the undamaged and damaged states was carried out using the enhanced frequency domain decomposition (EFDD) method introduced by Brincker et al. (2001a, b). The EFDD follows an output-only modal identification algorithm, which gives natural frequencies, modal damping ratios and mode shapes. The EFDD derives power spectral density (PSD) functions of the many single degree of freedom (SDOF) systems whose modal properties are equivalent to the inspected original system. This algorithm basically computes singular value decomposition (SVD) of the PSD function matrix of output channels, which

Fig. 17.4 Damage propagation of one of the columns the building (S11 column)

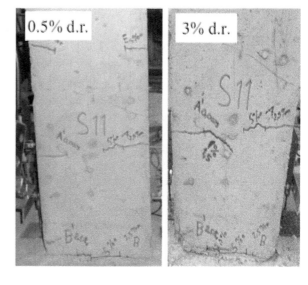

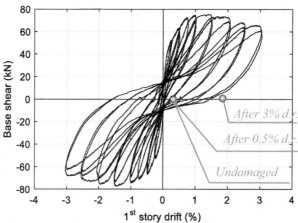

Fig. 17.5 Base shear-drift ratio relationship of the building and the vibration testing stages

are the vibration responses measured by the accelerometers from A1 to A6 in this study. The PSD function is taken back to the time domain using inverse fast fourier transform (IFFT) in order to obtain logarithmic free-decay function of the SDOF so that the modal frequency and the corresponding damping ratio can be determined.

In this study, the sampling frequency of the data was selected to be 100 Hz. For obtaining the PSD relationship, hanning window was utilized to minimize the leakage effects and the number of the fast fourier transform points was taken as 1024. Figures 17.6 and 17.7 display the singular values of the PSD matrix of the response and the mode shapes in x and y directions for the undamaged and damaged states of the building, respectively.

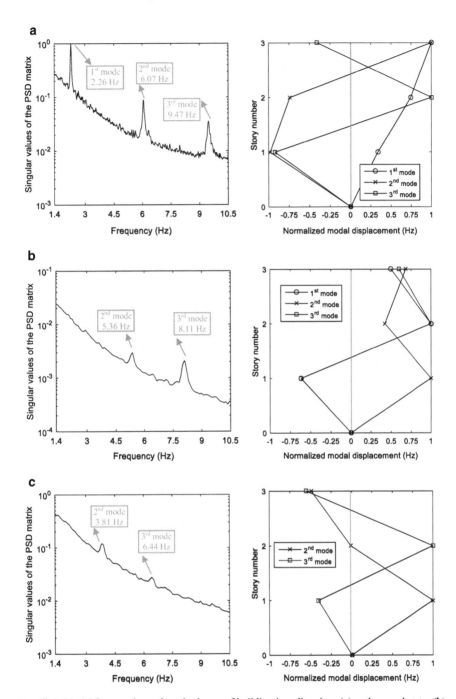

Fig. 17.6 Modal frequencies and mode shapes of building in x direction, (**a**) undamaged state, (**b**) after %0.5 d.r. and (**c**) after 3% d.r

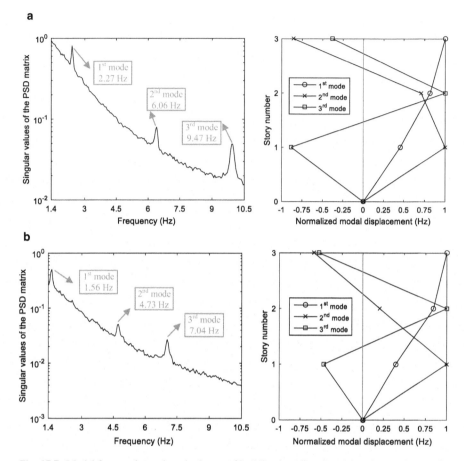

Fig. 17.7 Modal frequencies and mode shapes of building in y direction, (**a**) undamaged state, (**b**) after 3% d.r

Frequencies corresponding to the modal frequencies of the building are shown on these figures. However, the magnitudes of the peaks of the singular values of the PSD matrices for the damage states are significantly weak (Figs. 17.6b–c and 17.7b). This is probably caused by the increasing damping forces of the building due to the increasing damage, for instance, increase in friction between the crack interfaces (Fig. 17.4). Therefore, this complication in the peak picking process was discarded by inspecting the coherence function of the responses (Fig. 17.8). As seen in this figure, the frequency peaks can be captured with ease. Additionally, they are well-matched with the ones determined through EFDD algorithm. It should be noted that the first frequency peak for the damaged state of the building could not be captured due to technical limitations.

The shifting of the peaks toward lower frequencies with the increasing damage can be clearly seen in Figs. 17.6, 17.7 and 17.8. This shifting of the peaks was

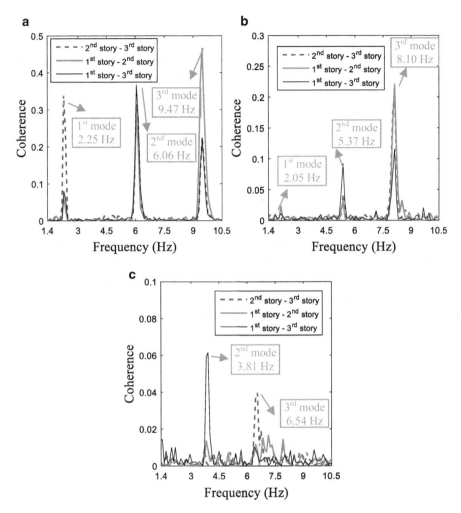

Fig. 17.8 Coherence functions in x direction, (**a**) undamaged state, (**b**) after %0.5 d.r. and (**c**) after 3% d.r

expected since the building was enforced to exhibit lateral deformations beyond its elastic limit. As aforementioned, the building was subjected to quasi-static cyclic loading only in x direction. Thus, as expected, the reduction of vibration frequencies due to structural damage observed in y direction was smaller than those observed in x direction (Table 17.1). For example, the third frequency peak as 9.47 Hz for x direction at undamaged state (Fig. 17.6a) shifted to the frequency peak as 6.44 Hz after the building was subjected to displacement cycles at 3% d.r. at the ground story (Fig. 17.6c). However, for the y direction, the third frequency peak as

Table 17.1 Identified modal frequencies and corresponding modal damping ratios

Damage states	Modal frequencies (Hz)						Modal damping ratios					
	x direction			y direction			x direction			y direction		
	1st mode	2nd mode	3rd mode	1st mode	2nd mode	3rd mode	1st mode	2nd mode	3rd mode	1st mode	2nd mode	3rd mode
Undamaged	2.26	6.07	9.47	2.27	6.06	9.47	0.018	0.028	0.033	0.048	0.047	0.045
0.5% d.r.	–	5.36	8.11	No test			–	0.048	0.039	No test		
3% d.r.	–	3.81	6.44	1.56	4.73	7.04	–	0.050	0.047	0.049	0.048	0.046

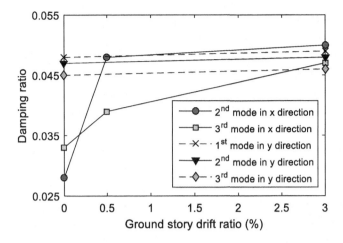

Fig. 17.9 Change in modal damping ratio as a function of achieved maximum drift ratio (damage state)

9.47 Hz at undamaged state (Fig. 17.7a) shifted to 7.04 Hz at the damaged state after the building was subjected to displacement cycles of 3% d.r. at the ground story (Fig. 17.7b). As also seen in Table 17.1, the modal damping ratios increased with increasing damage. Furthermore, the change in modal damping ratio as a function of achieved maximum lateral drift (damage state) is given in Fig. 17.9. As seen in this figure, in contrary to the modal damping ratios for y direction, the ones in x direction yielded significant jump with increasing in damage (approximately 78% and 42% increase in damping ration for the second and third mode, respectively). This can be explained with the application of quasi-static loading only in x direction of the building causing more significant damage in this direction. It should be noted that the finite element model (FEM) of the building was established to perform a modal analysis with the purpose of obtaining modal frequencies of the building by Goksu et al. (2015). The modal frequencies predicted through FEM and identified from the EFDD method are in good agreement for the first two modes in x and y directions of the undamaged state of the building.

It is worth to note that, the lateral load bearing capacity of the building was degraded by approximately 20% as it was subjected to the target ground story drift level of 3% (Fig. 17.5). However, the visual inspection made after the building was unloaded from the 3% d.r. did not show any indication of severe structural damage, which could have caused such a major decrease in lateral load bearing capacity (at around 20%). This observation indicates that the vibration based inspection may be able to provide a more realistic condition assessment for the damaged structures.

17.4 Conclusion

The dynamic characteristics of a full-scale substandard RC building were determined through ambient vibration tests before and after subjecting the building to quasi-static reversed cyclic loading. The modal characteristics of the building was obtained using the EFDD method. The results of the study are listed below;

- The modal frequencies decreased, while the modal damping ratios increased with increasing structural damage.
- Increase in the modal damping ratios of the building in x direction was more significant compared to the ones in y direction. This is due to the lateral loading, which was applied only in x direction and caused more significant damage this direction.
- Ambient vibration tests clearly showed that the building had significant structural damage while very limited information could be gathered by visual inspection. This results emphasize that the vibration based inspection may be able to give a more reliable condition assessment for the damaged structures comparing to the visual damage inspection. It should be noted that these results are limited by the specimen considered in this study.

Acknowledgements The authors are thankful to financial supports of Istanbul Development Agency (Project No: TR10/12/AFT/0050) and ITU Scientific Research Fund Department. The contributions of Prof. Dr. Z. Celep, Prof. Dr. Z. Polat, Prof. Dr. T. Kabeyasawa, Assoc. Prof. Dr. K. Kusunoki, Assoc. Prof. Dr. K. Orakcal, Assoc. Prof. Dr. E. Yuksel, Assist. Prof. Dr. U. Yazgan, Dr. C. Demir, AN. Sanver, AO. Ates, M. Comert, Dr. C. Yenidogan, O. Ozeren, E. Tore, S. Khoshkholghi, A. Moshfeghi, S. Hajihosseinlou, HF. Ghatte, I. Sarıbas, Tech. A. Sahin and 2014 summer trainees are greatly appreciated. The authors are also thankful to supports of Akcansa Co., Art-Yol Co., Boler Celik Co., Hilti-Turkey Co., Tasyapi Co., Urtim Co., and the staff of Kadikoy Municipality.

References

Brincker, R, Ventura CE, Andersen P (2001a) Damping estimation by frequency domain decomposition. In Proceedings of the 19th international modal analysis conference (IMAC), Kissimmee, pp 698–703

Brincker R, Zhang L, Andersen P (2001b) Modal identification of output-only system using frequency domain decomposition. Smart Mater Struct 10(3):441–455

Brownjohn JMW, Xia PQ (2000) Dynamic assessment of a curved cable-stayed bridge by model updating. J Struct Eng ASCE 126(2):252–260

Comert M, Demir C, Ates AO, Orakcal K, Ilki A (2016) Seismic performance of three-storey full-scale sub-standard reinforced concrete buildings. Bull Earthq Eng 15:3293. https://doi.org/10.1007/s10518-016-0023-4

Foti D, Diaferio M, Giannoccaro NI, Mongelli M (2012) Ambient vibration testing, dynamic identification and model updating of a historic tower. NDT&E Int 47:88–95

Goksu C, Inci P, Demir U, Yazgan U, Ilki A (2015) Field testing of substandard RC buildings through forced vibration tests. Bull Earthq Eng. doi:https://doi.org/10.1007/s10518-015-9799

Ivanovic SS, Trifunac MD, Novikova EI, Gladkov AA, Todorovska MI (2000) Ambient vibration tests of a seven-story reinforced concrete building in Van Nuys, California, damaged by the 1994 Northridge earthquake. Soil Dyn Earthq Eng 19(6):391–411

Kusunoki K, Tasai A, Teshigawara M (2012) Development of building monitoring system to evaluate residual seismic capacity after an earthquake. In: Proceedings of the fifteenth world conference on earthquake engineering, Lisbon, Portugal

Sanayei M, Khaloo A, Gul M, Catbas NC (2015) Automated finite element model updating of a scale bridge model using measured static and modal test data. Eng Struct 102(2015):66–79

Vidal F, Navarro M, Aranda C, Enomoto T (2014) Changes in dynamic characteristics of Lorca RC buildings from pre- and post-earthquake ambient vibration data. Bull Earthq Eng 12(5):2095–2110

Chapter 18
System Identification of a Residential Building in Kathmandu Using Aftershocks of 2015 Gorkha Earthquake and Triggered Noise Data

Yoshio Sawaki, Rajesh Rupakhety, Simon Ólafsson, and Dipendra Gautam

Abstract System identification is conducted to estimate the fundamental vibration period and damping ratio of a residential building in Kathmandu. Ground motion and structural response due to aftershocks of the 2015 Gorkha Earthquake, as well as noise data triggered by ambient vibration is used to identify the dynamic properties of the structure. In total, motion due to 3 aftershocks and 362 ambient vibration is used. The identification is based on estimating the frequency response function of the structure. When using the aftershock data, this function is estimated from power spectral density functions of motion recorded at the ground floor and roof of the structure. In case of triggered noise, it is assumed that the input motion is a white noise. Fundamental vibration period is estimated from the first dominant peak of the transfer function, and damping ratio is estimated by using the half-power bandwidth. The building being studied is a 4-storey reinforced concrete frame with masonry infill walls. The fundamental period of the building estimated from aftershock data and triggered noise data was found to be similar in the range of 0.24–0.4 s. Empirical relations available on the literature predict a fundamental period of 0.25 s for the building being studied. It can thus be concluded that the fundamental period of the building can be estimated with confidence using both aftershock and ambient vibration data. The damping ratio, however, showed greater variation. This variation is, in part, due to the inherent uncertainty in spectral estimation which requires smoothing operations that directly affect the bandwidth of the dominant peak of frequency response function.

Keywords System identification · 2015 Gorkha Earthquake · Ambient vibration · Spectral estimation

---

Y. Sawaki (✉) · R. Rupakhety · S. Ólafsson
Earthquake Engineering Research Centre, University of Iceland, Selfoss, Iceland
e-mail: yoshio@hi.is; rajesh@hi.is; simon@hi.is

D. Gautam
Structural and Geodynamics Laboratory, StreGa, University of Molise, Campobasso, CB, Italy

18.1 Introduction

Structural system identification is a method to evaluate dynamic structural properties by using the input recorded response (and excitation if available). Mainly, there are two genres of the mathematical methods for system identification, non-parametric methods (Söderström and Stoica 1989) and parametric method (Jenkins and Watt 1968). Non-parametric method, such as transient analysis, correlation analysis and spectral analysis, rely on visual inspection or curve fitting of spectral contents of recorded motion. On the other hand, parametric models, such as auto regressive model (AR), moving average model (MA), auto regressive moving average model (ARMA), etc., rely on time-domain modelling of linear systems the parameters of which are calibrated using recorded excitation and response data (see, for example, Ljung 1999). A more detailed review of structural system identification is provided by Alvin et al. 2003). In this study, non-parametric methods are used as a preliminary investigation of the dynamic properties of the structure, due to their simplicity of execution and interpretation.

This paper is based on a case study of a residential building in Kathmandu. The building was shaken by the 2015 Gorkha Earthquake and its aftershocks. Aftershock motion recorded at the ground floor and the roof of the building as well as ambient vibrations recorded at the roof are used to estimate fundamental vibration frequency and damping ratio of the building. We present a brief introduction to the earthquake and aftershocks, the recorded data, the properties of the structure, and describe the methodology used in system identification. We then present and discuss the results in terms of fundamental vibration frequency and damping ratio.

18.2 Gorkha Earthquake

A strong earthquake occurred in Nepal on 25 April 2015 at 11.26 UTC (local time 11.56). The size of the earthquake was estimated to be moment magnitude Mw 7.8 by United States Geological Survey (USGS) and local magnitude Ml 7.6 by National Seismological Centre (NSC) of Department of Mines and Geology of Government of Nepal. According to NDRRIP (2015), the numbers of the casualties and injuries are 8962 and 22,302, respectively, and 776,895 private houses were completely damaged and 298,998 of them were partially damaged. The ground motion had higher intensity in the districts east of the epicentre than that in the district west of it, which can be attributed to two main causes. As the area east of the epicentre lies directly above the rupture surface, it was more strongly affected by seismic wave radiation than the area west of the epicentre. Even though the epicentre was located near Gorkha, large slips occurred in the fault plane north of Kathmandu. Rupture propagation toward the east caused forward directivity effect, which resulted in devastating damage distribution in the area east of the epicentre (Koketsu et al. 2016). More details on the earthquake, its aftershock distribution, and ground motion

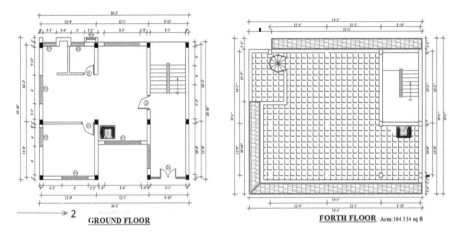

Fig. 18.1 Ground floor and roof plan of the building being studied. The small yellow pictures indicate the locations of accelerometers installed in the building

properties recorded in Kathmandu can be found in Rupakhety et al. (2017) and references therein.

18.3 Building and Instrumentation

18.3.1 Building

The building being studied is situated in Tyanglaphat, Kathmandu, Nepal. The geographical location of the building in relation to the Gorkha Earthquake and its aftershocks are presented in Rupakhety et al. (2017). The building is made of reinforced concrete frames, cast in situ. The floors are also reinforced concrete and cast in situ. Both exterior and interior walls are brick masonry with cement/sand mortar. The building is 4-storey high with a floor height of 3.15 m. The strength of concrete used in the frame and floor slab is 20 MPa, and the tensile strength of steel reinforcement is 415 MPa. Ground floor and roof plan of the building are shown in Fig. 18.1, with indications of the locations of accelerometers used to record vibration data.

18.3.2 Instrumentation

The CUSP-3CLP, a strong motion accelerometer manufactured by Canterbury Seismic Instruments (CSI) Ltd. (http://www.csi.net.nz/) were used to recorded vibrations of the building. The instruments are operated by the Earthquake

Table 18.1 List of aftershocks of 2015 Gorkha Earthquake recorded at the roof and ground floor

No.	Date	Time (UTC)	PGA (%g)	Magnitude M_w	Latitude (°, NSC)	Longitude (°, NSC)	Depth (km)
1	31/10/2015	9:37	0.3	–	28.03	85.23	–
2	28/05/2016	3.43	1.1	4.7	27.94	85.50	10
3	21/06/2016	5:48	0.4	3.6	27.93	85.21	10

Engineering Research Centre (EERC) of University of Iceland. The CUSP-3CLP contains one internal triaxial MEMS silicon accelerometer, the observable range is ±4 g and the dynamic range is 108 dB. The unit on the ground floor was installed on 1 May 2015, a few days after the 2015 Gorkha Earthquake. This unit recorded the Mw 7.3 aftershock on 12 May 2015. Ground motion data recorded by this unit during the 10 largest aftershocks are available as electronic supplement to Rupakhety et al. (2017). The range of horizontal peak ground acceleration (PGA) in these records is from 0.01 to 0.14 g. The unit on the roof was installed in the main structural system of the building at the concrete floor of the fourth level later, in October 2015. The unit on the roof has recorded some aftershocks as well as several triggered noise due to ambient vibrations.

18.4 Data

18.4.1 Aftershock Data

Three aftershocks used in this study are listed in Table 18.1 and were recorded on both the ground floor and the roof. Three component acceleration time series were recorded during each of these aftershocks. The horizontal components are called as channel-1 (see Fig. 18.1 for their orientations) and channel-2 and the vertical one is called as channel-3 hereafter. Although many aftershocks have occurred since the unit on the roof was installed, only three of them were recorded simultaneously by the unit on the roof and ground floor. On the other hand, many more aftershocks were registered by the unit on the ground floor. This is, in part due to malfunctioning of the triggering system of the unit on the roof.

18.4.2 Ambient Vibration Data

Many triggered noise data recorded from 6 October 2015 to 21 June 2016 are considered in this study. The total number of triggered noise records is 825. Many of them had peculiar drift or were of very short duration, and were not used for further analysis. It was also noticed that the noise data recorded by one of the horizontal channels was of lower quality, for example, severe drift, spiky peaks, etc., features that were not present in the other horizontal component. In total, 362 time series of triggered noise recorded by one of the horizontal components of the roof unit is used in further analysis.

18.5 Methodology

18.5.1 Using Aftershocks

The time series of ground and roof horizontal acceleration were windowed to cover 90% of cumulative Arias Intensity. The selection of the window was also verified visually, assuming that a stationary signal implies a linear build-up of Arias Intensity with time. Same time window was used for both ground and roof acceleration. The windowed time series were tapered with a Tukey window accommodating the taper over 15% of the total length of the signals. The signals were then band-pass filtered using a 4th-order zero-phase Butterworth filter in the frequency band 1–15 Hz. The power spectral density (PSD) of signals was estimated with Welch's algorithm dividing the signal into 5 segments with 50% overlap. The squared amplitude $|H(f)|^2$ of the frequency response function of the building is estimated as

$$|H(f)|^2 = \frac{S_{yy}(f)}{S_{xx}(f)} \quad (1)$$

where $S_{yy}(f)$ and $S_{xx}(f)$ are the Welch's PSD estimated from horizontal motion at the roof- and ground, respectively. An example of normalized $|H(f)|^2$ is shown in Fig. 18.2. The natural frequency ($f_n = 1/T_0$, T_0: natural period) is selected as the frequency of the first peak (red circle in Fig 18.2). The two frequencies f_1 and f_2 ($f_2 > f_1$) (black circles in Fig 18.2) correspond to the half-power bandwidth. The damping ratio ξ is estimated with the half-power bandwidth method (Papagiannopoulos and Hatzigeorgious 2011), using the following equation.

$$\xi = \frac{f_2 - f_1}{2f_n} \quad (2)$$

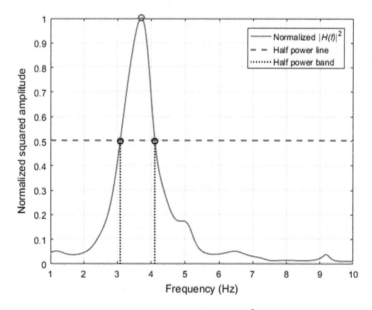

Fig. 18.2 Example of a normalized squared amplitude $|H(f)|^2$ computed from of the cahnnel-2 data of No.3 aftershock

18.5.2 Using Noise Data

18.5.2.1 Squared Amplitude of Frequency Response Function

Assuming the input motion as a white noise with a constant power spectral density P_0 at all frequencies of interest, the $|H(f)|^2$ is equal to the PSD of motion of the roof scaled by the constant P_0.

$$|H(f)|^2 = \frac{S_{yy}(f)}{P_0} \qquad (3)$$

Although the constant P_0 is unknown, the fundamental frequency and damping ratio of the structure can be estimated as was done with the aftershock data because the constant is merely a scaling factor. For simplicity, the $|H(f)|^2$ was normalized by this maximum value, and the resulting normalized spectrum is independent of P_0. This method of system identification is hereafter referred to as blind identification. For this application, 362 time series of one of the horizontal components of roof acceleration were used. The recorded noise is windowed between 5% and 95% of total Arias Intensity, and the estimation of PSD is the same as that described in the previous section.

18.5.2.2 H/V Method

Assuming that the building is very stiff in the vertical direction compared to the horizontal direction, the horizontal to vertical spectral ratio (HVSR) provides an estimate of the frequency response function of the structure. This method is frequently used for estimating the resonance site frequencies (see, for example, Nakamura 2008). It has also been found to be effective to estimate frequencies of buildings from ambient vibration measurements (see, for example, Gallipoli et al. 2009, 2010). In HVSR method, it is common to record a few tens of minutes of ambient vibrations continuously. In this application, however, we use triggered noise, each with different duration. One of the issues in the use of HVSR method is the smoothing of spectral estimates before computing the H/V ratio. Welch algorithm, as mentioned in the previous section, is effective for this purpose. As an alternative to smoothing, ensemble averaging can be used to reduce variability of PSD estimates. The assumption is then that the individual triggered signals are realizations of the same random process. With this assumption, the periodogram estimates of PSD of individual signals can be averaged to obtain a smoother average PSD function. In this work, we use 282 signals in two ways. In the first method, HVSR from each signal is estimated by using Welch's PSD of the vertical and horizontal motion. In the second method, periodogram estimates of individual signals are averaged to obtain average horizontal and vertical PSDF, and a single HVSR curve for the whole ensemble is estimated. Using the mean horizontal PSD $\overline{S_{yy}}(f)$ and mean vertical PSD $\overline{S_{zz}}(f)$ the mean H/V spectrum, which can be considered proportional to $|H(f)|^2$ is obtained as

$$|H(f)|^2 \propto \text{HVSR} = \frac{\overline{S_{yy}}(f)}{\overline{S_{zz}}(f)} \qquad (4)$$

18.6 Results

18.6.1 Results from Aftershocks Data

Figure 18.3 shows the normalized squared amplitude $|H(f)|^2$ of the RCC building estimated from the channel-1 and -2 data of each of three aftershocks, based on the method described in 5.1. Table 18.2 shows the damping ratio and natural period estimated from the normalized squared amplitude. The results show that the fundamental period in direction 1 (see Fig 18.1) are in the range 0.23–0.25 s, and that in direction 2 (orthogonal to direction 1) are in the range 0.27–0.28 s. This indicates that the building is slightly stiffer in direction 1. Since the difference is rather small and considering the uncertainties in the estimation process (for example the smoothing parameters used in the Welch algorithm), it may be considered that the building has similar fundamental period in the two horizontal directions. It is also interesting

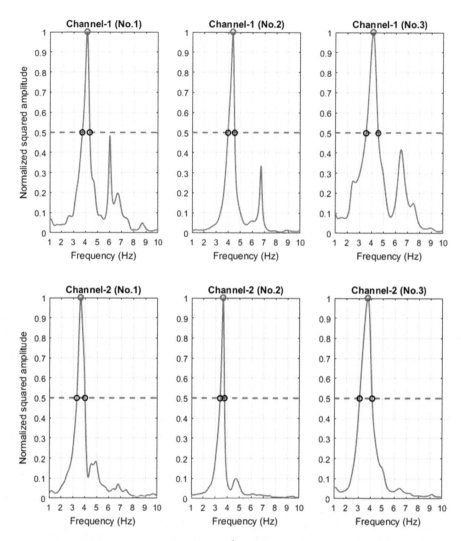

Fig. 18.3 Normalized squared amplitude $|H(f)|^2$ of the frequency response function of the RCC building computed from PSD of the cahnnel-1 and -2 data of each aftershock described in Sect. 18.4.1. On each of the figures, the red-dotted line represents the power level, the red circle denotes the natural frequency and the two black circles represent the half-power bandwidth

to note that a second peak between 6 and 7 Hz is present in the frequency response function of the building in direction 1. This indicates contribution of higher mode of vibration in direction 1, which seems to be missing in direction 2. Various empirical equations between the fundamental period of RCC frame buildings with infill walls and the building height are available in the literature (see, for example, Kocak et al. 2013). Kocak et al. (2013) performed finite element analysis of buildings with different configurations of infill walls and openings and found that for RC buildings

Table 18.2 Damping ratio and natural period estimated from channel-1 and -2 data of the three aftershocks by using the half-power bandwidth method

Data No.	Date	Time (UTC)	Channel	Damping ratio ξ	Natural period $T_0(s)$
1	31/10/2015	9:37	1	0.0751	0.2406
			2	0.0952	0.2753
2	28/05/2016	3.43	1	0.0653	0.2296
			2	0.0502	0.2779
3	21/06/2016	5:48	1	0.1237	0.2427
			2	0.1390	0.2669

having infill wall with openings, the fundamental period obtained by the finite element analysis was best predicted by the empirical equation given in Guler et al. (2009):

$$T_0 = 0.026 H^{0.9} \quad (5)$$

where H is the total height of the building measured in meters. According to this equation, the fundamental period of the building being studied here is 0.25 s, which is very close to the results obtained from system identification using aftershock data. The damping ratio estimated from the aftershock data showed much larger variation, ranging from 5% to 14%. In particular, the damping ratio estimated from the third aftershock is much larger than that estimated from the other aftershocks. Such large damping ratio is not realistic. It is important to note that the damping ratio obtained from the half-power bandwidth is very sensitive to the level of smoothing carried out in estimating the PSD. If smoothing is excessive, the peaks become wider, leading to larger damping ratios. A careful examination of the PSD of channel 2 shows that the small peak around 5 Hz observed in aftershock 1 and 2 has been smoothed out in aftershock 3 (see the bottom panel of Fig. 18.3). This in effect makes the first peak wider and therefore results in higher damping ratio. This phenomenon is, however, not observed in channel 1, and yet the estimated damping ratio is rather high.

18.6.2 Results from Ambient Vibration Data

18.6.2.1 Using the PSD of Triggered Noise

The number of triggered noise data, which had been recorded from 6 October 2015 to 21 June 2016, is 825, of which only 362 were used, the rest being of low quality. Only channel 2 data is used because many records from channel 1 were found to contain excessive drift. The estimation of fundamental period and damping ratio is based on the method as described in Sect. 18.5.2.1. Figure 18.4 shows the histograms of estimated natural period and damping ratio. The estimation was based on Welch's estimate of PSD for each triggered noise signal. The fundamental period varies from 0.27 to 0.32 s and the damping ratio varies from 2% to 10%

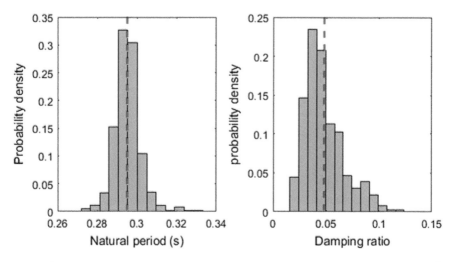

Fig. 18.4 Histogram of the natural period and damping ratio estimated from 362 normalized $|H(f)|^2$. The red-dashed line represents the mean value on each of the figures

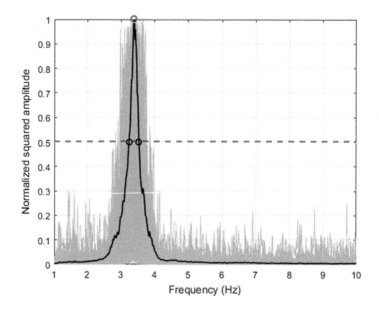

Fig. 18.5 Normalized squares amplitude $|H(f)|^2$ of the 362 triggered noise based on periodogram estimate of PSD. The solid black line represents the normalized ensemble average

approximately. As an alternative, ensemble averaging of PSD was also used. The periodogram estimates of normalized $|H(f)|^2$ of individual noise signals is shown in Fig. 18.5, where the normalized ensemble average is shown with the black curve. The period and damping ratio estimated from the normalized ensemble average

Table 18.3 Mean values and standard deviations of the natural period and damping ratio

Channel	No of signals	Mean natural period $\overline{T_0}$(s)	Standard deviation of T_0	Mean damping ratio $\overline{\xi}$	Standard deviation of ξ
2	362	0.2954	0.0186	0.0480	0.0069
2[a]	362	0.2956	–	0.0414	–

[a]The results in this row are based on ensemble averaging

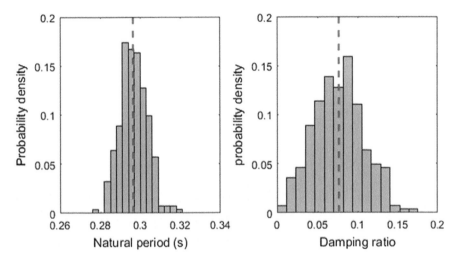

Fig. 18.6 Histogram of the natural period and damping ratio estimated from 282 mean H/V spectra. The red-dashed line represents the mean value on each of the figures

$|H(f)|^2$ are compared in Table 18.3 with the mean value of periods and damping ratios estimated from $|H(f)|^2$ based on Welch's PSD.

18.6.2.2 Mean H/V Spectrum

The horizontal PSD of channel-2 noise data and vertical PSD of channel-3 noise data (282 time series) are estimated with Welch's PSD estimate and periodogram PSD estimate. HVSR computed from Welch's PSD are used to estimate fundamental period and damping ratio from each of the 282 time series. The frequency distribution of natural period and damping ratio estimated from HVSR with Welch's estimate are shown in Fig. 18.6. The natural periods range from 0.28~0.32 s and its mean value is 0.3 s from (see Table 18.4). This value is similar to the result obtained in 6.1.1. The variation in the damping ratios is larger than that of those obtained in Sect. 6.1.1.

The periodogram estimates of horizontal and vertical PSD were used to estimate ensemble average PSD. The corresponding periodograms and their ensemble average are shown in Fig. 18.7. The HVSR obtained from the ensemble averaged PSD is

Table 18.4 Mean values and standard deviations of the natural period and damping ratio of the RCC building

Channels	PSD estimate	No of time series	Mean natural period $\overline{T_0}$(s)	Standard deviation of T_0	Mean damping ratio $\overline{\xi}$	Standard deviation of ξ
2&3	Welch's	282	0.2965	0.0067	0.0768	0.0311
2&3	Ensemble average	282	0.2994	–	0.0860	–

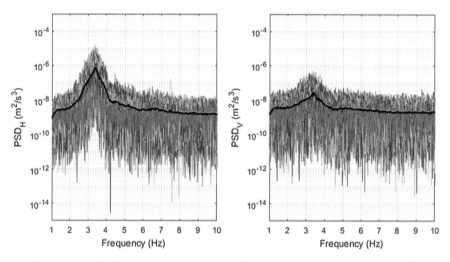

Fig. 18.7 Mean horizontal PSD (PSD_H) and mean vertical PSD (PSD_V) (black-solid line on each figure)

shown in Fig. 18.8. The fundamental period and damping ratio corresponding to the HVSR are 0.2944 s and 8.6%, respectively.

18.7 Conclusions

Preliminary results on structural system identification of a typical 4-storey RCC building with masonry infill walls in Kathmandu are presented. The results are based on three aftershocks of the 2015 Gorkha Earthquake as well as numerous time series triggered by ambient vibrations. The aftershock motion recorded at the ground floor and the roof of the 4-storey building were used to estimate frequency response function of the building, from which the half-power bandwidth method was used to estimate fundamental period and damping ratio. Fundamental periods obtained from the three aftershocks were similar, around 2.5 s, and matched well with published empirical equations applicable to similar structures. Triggered noise time series from the roof were used in system identification in two ways. In the first approach, the

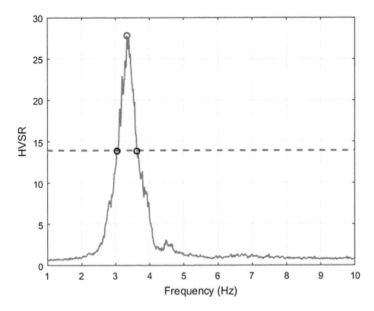

Fig. 18.8 HVSR obtained from ensemble-average PSD of 282 triggered noise time series. The red-dashed line represents half the value of the maximum of the mean H/V spectrum

building transfer function was considered directly proportional to the PSD of the recorded horizontal motion, assuming that the excitation is a white noise process. In the second approach, building transfer function was estimated from the ratio of PSD of horizontal and vertical motion. In both cases, two approaches of spectral estimation were used. The first approach is based on time averaging of signals, which is achieved through Welch's PSD estimate, and could be used to estimate fundamental period and damping ratio from each time series. The second approach was based on ensemble averaging, where the periodogram estimates of several time series were used to compute average PSD. The fundamental period of the building estimated from the noise data using these different approaches were found to be similar, around 0.3 s, which is slightly larger than that estimated from the aftershocks. The damping ratio estimated from both the aftershock and the noise data showed large variation. This is in part due to uncertainties related to smoothing of the spectral density functions, which has a direct impact on the half-power bandwidth of the estimated squared amplitudes of the building frequency response function. The results indicate that ambient vibration measurement provide estimates of building fundamental frequency similar to that obtained from stronger shaking during earthquake aftershocks. It can also be concluded that the ambient vibration measurements can be used in either the framework of HVSR method, or just the horizontal component of motion, to estimate fundamental period of buildings. In either case, it seems desirable, to make several measurements of ambient vibrations and use either time or ensemble averaging to estimate the properties of the structure. It was found that damping ratio estimated from both aftershock data and ambient vibrations were largely variable. While the mean damping ratio estimated from noise data was similar to what is generally expected for these types of buildings, there was considerable variation between different time series. This variation is, in part, due to uncertainties in estimates of actual PSD functions of recorded time series. In both time and ensemble averaging methods, the level of averaging (or smoothing) has a direct effect on the bandwidth of the dominant peak of estimated frequency response function, and therefore the damping ratio has larger variation. It would be interesting to apply parametric models of system identification, by using time series models, to estimate the damping ratio of the building and to test whether such methods result in lower uncertainty in the estimates of damping ratio.

Acknowledgements We acknowledge financial support from University of Iceland Research Fund and the national power company of Iceland, Landsvirkjun. The first author acknowledges the Government of Japan for providing him the Monbukagakusho Scholarship to support his research internship at the EERC. We thank Mr. Damodar Rupakhety for allowing us to install the accelerometers in his house and to use the collected data for the research presented herein. Mr. Rajan Dhakal prepared the plans of the building presented here, and Dr. Benedikt Halldorsson helped in configuring the accelerometers installed in the building; their contributions are gratefully acknowledged.

References

Alvin KF, Robertson AN, Reich GW, Park KC (2003) Structural system identification: from reality to models. Comput Struct 81(12):1149–1176

Gallipoli MR, Mucciarelli M, Vona M (2009) Empirical estimate of fundamental frequencies and damping for Italian buildings. Earthq Eng Struct Dyn 38:973–988

Gallipoli MR, Mucciarelli M, Šket-Motnikar B, Zupanćić P, Gosar A, Prevolnik S, Herak M, Stipčević J, Herak D, Milutinović Z, Olumćeva T (2010) Empirical estimates of dynamic parameters on a large set of European buildings. Bull Earthq Eng 8:593–607

Guler K, Yuksel E, Kocak A (2009) Estimation of the fundamental vibration period of existing RC buildings in Turkey utilizing ambient vibration records. J Earthq Eng 12(S2):140–150

Jenkins GM, Watt DG (1968) Spectral analysis and its application. Holden-Day, San Francisco

Kocak A, Kalyoncuoglu A, Zengin B (2013) Effect of infill wall and wall openings on the fundamental period of RC buildings. Earthquake Resistant Engineering Structures IX, WIT Transactions on The Built Environment 132:121–131. doi:https://doi.org/10.2495/ERES130101

Koketsu K, Miyake H, Guo Y, Kobayashi H, Masuda T, Davuluri S, Bhattarai M, Adhikari LB, Sapkota SN (2016) Widespread ground distribution caused by rupture directivity during the 2015 Gorkha Nepal earthquake. Sci Rep 6:28536. https://doi.org/10.1038/srep28536

Ljung L (1999) System identification. Wiley, New York

Nakamura Y (2008) On the H/V spectrum. In: Proceedings of the 14th world conference on earthquake Enginering, 12–17 October, Beijing, China

National Seismological Centre (NSC) (2015) http://www.seismonepal.gov.np/. Last accessed 21 June 2015

Nepal Earthquake (2015) Disaster relief and recovery information platform (NDRRIP), Nepal Disaster Risk Reduction Portal. http://drrportal.gov.np/. Last accessed 6 July 2016

Papagiannopoulos GA, Hatzigeorgious GD (2011) On the use of the half-power bandwidth method to estimate damping in building structures. Soil Dyn Earthq Eng 31:1075–1079

Rupakhety R, Olafsson S, Halldorsson B (2017). The 2015 Mw 7.8 Gorkha Earthquake in Nepal and its aftershocks: analysis of strong ground motion. Bull Earthq Eng. doi:https://doi.org/10.1007/s10518-017-0084-z

Söderström T, Stoica P (1989) System identification. Prentice Hall International, Cambridge

United States Geological Survey (USGD) (2015) http://earthquake.usgs.gov/earthquakes/eventpage/us20002926#general_summary. Last accessed 21 June 2015

Chapter 19
Damage Observations Following the M_w 7.8 2016 Kaikoura Earthquake

Dmytro Dizhur, Marta Giaretton, and Jason M. Ingham

Abstract On 14 November 2016 a magnitude M_w 7.8 earthquake struck the upper South Island of New Zealand with effects also being observed in the capital city, Wellington. The affected area has low population density but is the largest wine production region in New Zealand and also hosts the main national highway and railway routes connecting the country's three largest cities of Auckland, Wellington and Christchurch, with Marlborough Port in Picton providing connection between the South and North Islands. These transport facilities sustained substantial earthquake related damage, causing major disruptions. Thousands of landslides and multiple new faults were counted in the area. The winery facilities and a large number of commercial buildings and building components (including brick masonry veneers, historic masonry construction, and chimneys), sustained damage due to the strong vertical and horizontal acceleration. Presented herein are field observations undertaken days directly following the earthquake, with the aim to document earthquake damage and assess access to the affected area.

Keywords Kaikoura earthquake · Winery damage · Masonry damage · Road and bridge damage · Railway damage · Landslides

19.1 Introduction

On 14 November 2016 at 12:02 am local time (13 November 2016 New Zealand at 11:02:56 UTC), the moment magnitude M_W7.8 Kaikoura earthquake occurred along the east coast of the upper South Island, New Zealand (Earthquake Commission & GNS 2015). The earthquake initiated in the Waiau Plains (42°41′24.0″S 173°01′12.0″E)

D. Dizhur (✉) · M. Giaretton · J. M. Ingham
Department of Civil and Environmental Engineering, The University of Auckland, Auckland, New Zealand
e-mail: ddiz001@aucklanduni.ac.nz

Fig. 19.1 A map of NZ highlighting main areas of impact. (Source: Google Maps)

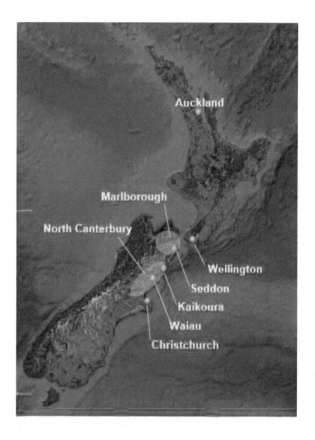

in North Canterbury, and involved multiple fault segments as the rupture generally propagated northward over 150 km to Cape Campbell in the Marlborough region.

The Marlborough Fault Zone (MFZ) is one of the most active crustal regions of New Zealand, with many mapped active faults. Several mapped fault segments participated in this earthquake, including The Humps, Hundelee, Hope, Jordan Thrust, Kekerengu, and Needles Faults. In addition, several significant rupture displacements occurred on previously unmapped fault segments, most notably the Papatea Fault near the Clarence River mouth. The most distinctive seismological aspect of the earthquake was the number of these fault segments which ruptured co-seismically in the same event (Earthquake Commission & GNS, 2015). Surface rupture along the causative faults resulted in significant localized damage to transportation infrastructure near the coast and also fault rupture-induced landslides. The strong earthquake-induced ground motions in the near-source region also resulted in substantial landslides along State Highway 1, see Fig. 19.1. Ground motions with horizontal accelerations exceeding 1.0 g were observed at four locations (two in the Waiau area, and also in Kekerengu and Ward). The ground motion recorded near the hypocentre in Waiau also exhibited 2.7 g in the vertical direction, exhibiting the so-called 'slapdown effect' with asymmetric accelerations, very similar to that at Heathcote Valley in the 2010–2011 Canterbury earthquakes.

19.2 Emergency Response and Management

In the first days to weeks following the earthquake, the multi-level governmental response to the 2016 Kaikoura earthquake focused first on tsunami-related evacuations, life safety, building and infrastructure damage assessment, access provision and control, and social welfare services including the provision of food, water and temporary accommodation to those in need. Given the relatively focused pattern of damage, there was additional resource capacity within each of the Civil Defence Emergency Management Groups (CDEM) groups to support the most heavily impacted local district and city councils.

Following the 14 November 2016 earthquake a state of emergency was only declared for the Kaikoura District Council. The Kaikoura District Council faced substantial building and infrastructure damage, coastal uplift that impacted harbour and boat access, and massive landslides blocked railroad and highway routes to the Kaikoura peninsula. With only 22 permanent staff, the emergency declaration helped the district to gain the additional staff and resources that it needed to carry out its response operations, including the substantial social welfare operations necessary to support residents as well as the large tourist population trapped on the Kaikoura peninsula.

19.3 Transportation Infrastructure

Damage to the transportation infrastructure resulted in major disruptions following the 14 November 2016 earthquake.

19.3.1 Roads

The most severe damage was experienced by the road networks along the east coast of the South Island. Landsliding, fault rupture, bridge damage (shaking and liquefaction induced) and road/rail platform instability closed a large section of State Highway 1 north and south out of Kaikoura and the main north rail line from Ward south to Cheviot. Figure 19.2 shows examples of common damage observed along this route. There was widespread slumping of bridge approaches, with main bridges north and south of Kaikoura experiencing vertical offsets between approaches and the deck, and lateral displacement of deck sections and piers. Kaikoura was inaccessible by land for 3 days, with an inland access road partially cleared initially for

Fig. 19.2 Examples of observed road damage (**a**) Upto 4 metres of vertical uplift across the main State Highway (**b**) Significant rock fall over main State Highway (−42.449696, 173.575200) (**c**) Significant damage to the road pavement and movement of retaining wall structure (**d**) Significant damage to the road pavement (**e**) Large landslide extending over main State Highway 1 (**f**) Major landslide along State Highway 1

emergency vehicles and then opened for scheduled convoys 11 days after the event. Full access to Kaikoura for two way traffic via the inland route was opened on 19 December 2016, 35 days after the event. State Highway 1 south of Kaikoura reopened to traffic during daylight hours on 21 December 2016. As of March 2017 there is little redundancy in the land transport network, with one 45 km stretch of highway serving as the only paved land transportation link between the northern and southern parts of the South Island, with all freight and normal traffic using this route. The significant increase in the level of traffic on this route will require accelerated maintenance and improvements of the surface.

19.3.2 Railway

Severe damage was experienced by the railway networks along the east coast of the South Island, see Fig. 19.3. The timeline for full reinstatement of routes as of March 2017 is not clear, with current estimates of months for road and over a year for rail. All rail transportation was halted, with freight shifted onto new coastal shipping services and additional road services.

19.3.3 Bridges

There was clear evidence that liquefaction and lateral spreading affected bridges in the areas of North Canterbury and Marlborough, and that such effects contributed to the overall damage and performance of the bridges, together with inertial loads and soil-structure interaction. There was widespread slumping of bridge approaches (see Fig. 19.4b), with main bridges north and south of Kaikoura experiencing vertical offsets between approaches and the deck, and lateral displacement of deck sections and piers. Some bridge superstructure sustained significant damage, see examples in Fig. 19.4.

19.3.4 Ports

Wellington's CentrePort suffered substantial earthquake damage, resulting in disruption to importers and the timely delivery of goods in the Wellington region and ongoing uncertainties about when freight operations will fully resume, and how long it will take to repair infrastructure to allow cruise ship tourism to fully resume operations. Marlborough Port located in Picton, also sustained structural damage resulting in major disruptions to its normal operations.

Fig. 19.3 Coastal Pacific Railway line damage (**a**) Lateral movement of the bridge abutment. No damage observed to the bridge superstructure (−42.513496, 173.506657) (**b**) Major lateral displacement of railway lines (**c**) Major horizontal and vertical movement of land adjacent to bridge approach (−41.961324, 174.054479)

Fig. 19.4 Performance of bridges (**a**) Major damage to the bridge superstructure, with all columns showing signs of plastic-hinging (−42.634382, 173.066299) (**b**) Significant spalling at the base of column (southern column). Longitudinal reinforcing bar buckling (**c**) Major settlement of abutments, Oaro overbridge #1790 (−42.512124, 173.506385) (**d**) Under bridge deck, visible hinging of the columns

19.4 Building Performance in North Canterbury and Marlborough

The upper South Island has previously experienced strong earthquakes with the 1848 Marlborough earthquake being the most destructive (McSaveney 2013) and the 2013 Cook Strait (M_w6.5) and 2013 Lake Grassmere (M_w6.6) earthquakes being the most recent (Morris et al. 2013). The M_L7.4 Marlborough earthquake in 1848 damaged (and destroyed during the aftershocks) most of the unreinforced masonry (URM) structures in Wellington and in the Marlborough region, which were then rebuilt using timber and this more earthquake resilient structural form largely contributed to reduce the level of damage and loss of life in the later earthquakes, including the M_w8.2 Wairarapa (60 km east of Wellington) in 1855 (McSaveney 2013) and subsequent earthquakes. In the area located south of Kaikoura, the large M_w7.3 North Canterbury earthquake (Amuri Hurunui District) in 1888 caused severe damage to buildings made of cob and stone masonry. As a result of all these historic strong earthquakes, most of the vulnerable structures and non-structural components have been retrofitted or unfortunately destroyed in the past, resulting in a less severe damage having occurred during the 14 November 2016 earthquake.

19.4.1 Buildings Constructed Pre-Seismic Provisions

Damage to buildings that were constructed pre-seismic code provisions was mainly observed in Waiau where the most severe horizontal (1.12 g) and vertical (3.21 g) peak ground accelerations (PGA) were recorded, resulting in significant damage to historic stone and concrete churches and to cob cottages, and out-of-plane loss of veneer from residential housing, as well as widespread damage and in some cases collapse of domestic chimneys. A number of significantly damaged registered historic buildings include: All Saints Church (1924, Waiau, Fig. 19.5c), Waiau Lodge Hotel (1910, Waiau), Cob Cottage museum (1860, Waiau, Fig. 19.5d), Watters Cob Cottage (ca. 1880, Rotherham) and St Oswald's Church (1927, Wharenui, Fig. 19.5b). Early buildings located in Havelock, Picton, and Blenheim (see Fig. 19.5a) or closer to Christchurch performed well and most were found to have a previously implemented retrofit intervention.

19.4.2 Chimneys

Limited damage to URM chimney was observed in Picton, Seddon and Rotherham, where most chimneys appeared to have been previously removed or lowered at the roofline, likely to have occurred following the 2013 Seddon and 2013 Lake

Fig. 19.5 Examples of building performance (**a**) St Joseph's Church (1917), clay brick masonry cavity-wall building (**b**) St Oswald's Church (1927), stone and concrete, collapse of parts of front parapet (**c**) All Saints Church (1924, Heritage List #3690, red tagged), showing tilting of the bell tower and significant settlement of tower foundation (**d**) Cob Cottage museum (1866, Heritage List #3682, red tagged) – earthen structure

Grassmere earthquakes (Morris et al. 2013) or earlier previous earthquakes (McSaveney 2013). The removal of the URM chimneys appeared to be the only mitigation intervention adopted in the area prior to the 2016 Kaikoura earthquake and no braced or strapped URM chimneys were identified. In Blenheim, Hanmer Springs, Clarance, and Waiau the damage level was moderate to extensive, see Fig. 19.6. Numerous examples existed of chimneys cracked at the roofline and displaced and skewed after rocking due to the intense and long shaking. In a few cases, residents were quick to remove damaged chimneys that remained standing and cover the stack, in order to eliminate risks due to aftershocks and strong winds.

19.4.3 Residential Veneer Cladding

Clay brick masonry or concrete block as an anchored wall veneer is used in New Zealand for its durability, resistance to fire and moisture, and for aesthetic reasons. A comprehensive background and description of veneers and tie types is presented in (Dizhur et al. 2013). Following the 2016 Kaikoura earthquake,

Fig. 19.6 Observations of chimney damage (**a**) Disintegration (**b**) Disintegration and rotation (**c**) Toppling above the roofline (**d**) Detachment (**e**) Detachment

moderate to extensive damage to masonry veneers was localised mainly in Seddon and Waiau areas. The damage was typically attributed to a combination of differential movement between the masonry veneer and the timber framing and the poor anchorage of ties connecting the masonry veneer to the timber wall framing, see Fig. 19.7a–b. The differential movements between the timber structure and veneers induced both in-plane cracking and the initiation of out-of-plane deformation and collapse in relation to the shaking direction. Such behaviour was also observed in hollow concrete block veneers and in more modern veneer systems Fig. 19.7c–d.

19.5 Multi-storey Building Performance in Wellington

The 2016 Kaikoura earthquake had low spectral demands in the short period range throughout the Wellington region, which resulting in most short stiff buildings experiencing shaking demands well below code-level during the earthquake. As a result URM and other building types with low periods suffered little to no damage. However, modern engineered structures with a fundamental period near 1.5 s were affected by the earthquake induced shaking. Seismic demands for such buildings often exceeded the code-level spectrum and structural components were subjected to repeated cycles of inelastic deformation.

One 5-storey ductile concrete frame building located on reclaimed land for the port lost support for precast double tee floor units during the earthquake. The frame parallel to the floor unit span showed clear evidence of beam elongation. Based on early preliminary observations made by engineers conducting post-earthquake assessments, it was apparent that damage was concentrated in 5–15 storey concrete moment frames with precast flooring systems. The following critical damage states have been observed:

(a) (b) (c) (d)

Fig. 19.7 Examples of damaged veneer dwellings (**a**) Out-of-plane collapse of masonry veneer in Seddon (**b**) Examples of damaged modern veneer dwelling, Seddon (**c**) Examples of in-plane and out-of-plane damage to hollow concrete block veneer dwelling, Waiau (**d**) Examples of damaged modern veneer dwelling, Waiau

- Reduced seating for precast floor units due to beam elongation in supporting frames.
- Widespread cracking of precast floor units. Specific concern is focused on cases of cracks transverse to the span of the unit near the support, indicating reduced capacity to gravity support loads.
- Fracture of mesh in precast diaphragm topping, specifically a concern for pre-1995 buildings where additional ties in the diaphragm are not typically provided for lateral support of columns.
- Damage to corner columns of concrete moment frames due to frame elongation and high shear demands.
- The above described buildings and damage profiles have become the focus of a targeted detailed damage assessment program developed by Wellington City Council (WCC) for approximately 80 buildings throughout the Central Business District (CBD).

19.6 Performance of The Marlborough Wine Industry

Wine production in Marlborough has experienced significant growth during the past two decades and is currently the largest wine producing region in New Zealand with 141 wineries that make up more than 75% of the country's total wine production. The wine industry in the area was strongly affected by the 2013 Cook Strait (M_w6.5) and 2013 Lake Grassmere (M_w6.6) earthquakes (Morris et al. 2013). During the 2016 Kaikoura earthquake, similar damage to that observed in 2013 was experienced by wineries facilities, see Fig. 19.8. Numerous wineries were inspected to assess damage to winemaking facilities and identify the overall impact of the 2016 Kaikoura earthquake shaking on the Marlborough wine industry. The following main observations were made:

Fig. 19.8 Observed damage to winery facilities (**a**) Infrequent and rigid connections at tank base resulting in substantial damage to hold-down brackets and pull-out from concrete substrate (0.13 g hPGA) (**b**) 60 kL tanks with no damage sustained (0.23 g hPGA) (**c**) Extreme buckling of tanks and pull out of catwalk support frames (0.23 g hPGA) (**d**) Complete collapse of slender tanks ('domino effect') (**e**) New four row stacking system showing no damage (0.20 g hPGA) (**f**) Collapse of older wooded barrel storage (0.14 g hPGA)

- Observed performance of tanks varied depending on tank typology (legged tanks and flat-bedded tanks) and tank capacity.
- Damage to legged tanks was generally concentrated at the base frame support with partial tank collapse observed in some cases.
- The extent of damage to flat-bedded tanks (larger in terms of liquid capacity) was more widely observed compared to legged tanks. Damaged was observed in various tank elements such as buckling of the stainless steel shell, creasing of the top cone, localized buckling of the tank skirt and damage of anchorage rods and bolts.

- Wine loss was observed in extreme cases, such as tank wall perforation and tank collapse.
- Damage to other winery infrastructure was observed, such as catwalks and thermal tank insulation.

19.7 Summary

On 14 November 2016 a $M_w 7.8$ earthquake struck the upper South Island of New Zealand and also Wellington, with the epicentre located approximately 4 km south of Waiau, Canterbury (100 km north of Christchurch). The most notable seismological aspect of the earthquake was the number of fault segments, part of the Marlborough Fault Zone (MFZ), which ruptured co-seismically in the same event. Surface rupture along the causative faults resulted in significant localized damage to transportation infrastructure near the coast and also fault rupture-induced landslides. The main national highway and scenic railway routes connect the country's three largest cities of Auckland, Wellington and Christchurch, with Port Marlborough in Picton providing connection between the South and North Islands, were heavily damaged and closed. Kaikoura was inaccessible by land for 3 days because of the numerous massive landslides, in addition to the coastal uplift that impacted harbour and boat access.

Because of the large number of historic strong earthquakes in the Wellington, Marlborough and Canterbury regions, most of the vulnerable structures and non-structural components had been previously retrofitted or unfortunately demolished, resulting in a less severe damage during the Kaikoura earthquake. Despite that, Waiau and Seddon represented the worst hit areas with high levels of damage observed in historic buildings (vintage concrete structures and Cob cottages) and veneer dwellings. The Marlborough region is also extremely important for the wine industry, accounting for 75% of the country's total wine production. Observed damage to wine making facilities consisted mainly of damage to storage tanks, thermal tanks isolation, supporting catwalk systems and in a few cases damage to cooling pipe systems. It is hoped that despite the very localise damage, this earthquake has provided a timely reminder for building owners and territorial authorities to take action to address earthquake related risks.

Acknowledgements Funding for this field reconnaissance undertaking was partially provided by QuakeCoRE. House owners and building occupants are greatly thanked for providing access to the inspected buildings. Great thanks are extended to all the winery owners that kindly provided access (and their time) to their facilities.

References

Dizhur D, Moon L, Ingham J (2013) Observed performance of residential masonry veneer construction in the 2010/2011 Canterbury earthquake sequence. Earthquake Spectra 29 (4):1255–1274. https://doi.org/10.1193/050912EQS185M

Earthquake Commission & GNS (2015) GeoNet project: geological hazard information for New Zealand. Retrieved from http://info.geonet.org.nz

McSaveney E (2013) Historic earthquakes. Retrieved from www.TeAra.govt.nz

Morris GJ, Bradley BA, Adam W, Matuschka T (2013) Ground motions and damage observations in the Marlborough region from the 2013 Lake Grassmere earthquake. Bull N Z Soc Earthq Eng 46(4):169–187

Chapter 20
Seismic Rehabilitation of Masonry Heritage Structures with Base-Isolation and FRP Laminates – A Case Study Comparison

Simon Petrovčič and Vojko Kilar

Abstract The cultural and historical significance of architectural heritage buildings demands intrinsic considerations regarding appropriate conservation measures that need to be undertaken in order to restore or maintain their historical values. In this paper two contemporary seismic strengthening measures with varying degrees of invasiveness and strengthening efficiency are employed in a case study numerical simulation on a typical neo-renaissance masonry heritage building. The use of fibre reinforced polymer composites and the implementation of base-isolation was considered in the study in order to achieve the desired, code-based seismic protection levels. Non-linear static analyses with incremental increases in levels of seismic intensities were conducted on mathematical models of the fixed-base, FRP-strengthened and base-isolated variants of the structure. The comparison of results based on static pushover analyses for various ranges of seismic intensity was presented, while each strengthening measure was assessed in terms of its efficiency.

Keywords Historic masonry structures · Seismic rehabilitation · Fibre reinforced polymers · Base isolation

20.1 Introduction

A series of earthquakes that took place in central Italy in August 2016 has again demonstrated that masonry heritage structures are extremely vulnerable to seismically induced loading. The typology of historic masonry construction is problematic in terms of inadequate lateral strength and in most cases structural rehabilitation is needed to achieve code-based safety requirements. Structural deficiencies, such as the fragility of masonry and foundations under tension forces, low shear strength and brittle behaviour of masonry walls, the lack of stiff floor diaphragms, irregular

---

S. Petrovčič (✉) · V. Kilar
Faculty of Architecture, University of Ljubljana, Ljubljana, Slovenia
e-mail: simon.petrovcic@fa.uni-lj.si

openings, poor and uneven construction quality all contribute to the inadequate seismic performance of masonry structures.

In general, architectural conservation guidelines (e.g. ICOMOS Charters) and guidelines for structural restoration (e.g. Eurocodes) usually lead to contradictory approaches to heritage conservation. On the one hand architectural conservation guidelines follow the minimum intervention concept by which the applied interventions should be limited, reversible and should strive toward minizing the impact into the heritage substance. On the other hand, the aim of structural restoration measures is to maximise safety and resilience of the structure by using strengthening techniques which are often very invasive and irreversible.

In this paper two contemporary seismic strengthening measures with varying degrees of invasiveness and strengthening efficiency are presented and employed in a case study numerical simulation on a typical neo-renaissance masonry heritage building.

Firstly, the initial building is strengthened with fibre reinforced polymer (FRP) composites. Compared to traditional techniques, FRP composites are lightweight, noncorrosive and exhibit high tensile strength and stiffness (Prota et al. 2008). FRP materials usually are applied in different geometric layouts on masonry panels. In the context of architectural conservation, this strengthening method is moderately invasive, since the FRP laminates are applied on the outer planes of panels (piers or spandrels).

Secondly, seismic base isolation is implemented in the initial structure. A base isolation system consists of a decupling isolation layer between the building and the ground. The isolation layer is made of devices that support the building against static actions and, in case of seismic actions, implement a decoupling effect of the dynamic response of the building compared to the ground. A base-isolated building is characterized by a smaller acceleration response, leading to a smaller amount of forces of inertia, a smaller amount of interstorey drifts and internal forces in structural elements (Kelly et al. 2007). The implementation of base isolation in existing buildings is localised to the base storey, where isolation devices are usually installed. Its installation requires significant construction works at the level of application, and given the large sizes of historic buildings, the installation works are distributed alternately on the building plan. Nevertheless, the interventions in the rest of the structure are minor and usually involve stiffening of floor slabs and interconnecting adjacent walls to achieve a global box-like behaviour. The installation of base isolation is therefore very limited: installation issues and accessibility to the equipment for inspection, maintenance, or replacement purposes are the most challenging factors to be met in the implementation of the base-isolation of existing buildings.

The paper aims to compare the two strengthening techniques in terms of the achieved level seismic resistance. In the final part of the paper comparison of results based on static pushover analyses for various ranges of seismic intensity are presented and discussed. Each selected strengthening measure is assessed in terms of its strengthening efficiency.

What is more, the authors of the paper have in their previous research (Petrovčič and Kilar 2013) proposed a refined technique for the modelling and analysis of regular unreinforced masonry (URM) structures, which is based on the equivalent frame approach, and incorporates linear beam elements and the plastic hinge concept, including the interaction between the compressive force and bending moment/ shear force in an URM panel due to lateral loading. Therefore, this study also aims in applying this technique to URM structures that have been strengthened with FRP composites, while its application to base-isolated structures has already been successfully undertaken (Petrovčič and Kilar 2016).

20.2 Analysed Masonry Building

20.2.1 Building Characteristics

In the conducted case study a masonry building typical for the neo-renaissance era in Europe was analysed. The building is a three-storey URM structure with floor plan dimensions 24.6 m × 12.9 m (directions X and Y, respectively) and a total height of 15.8 m. The examined masonry building is schematically presented in Fig. 20.1.

Detailed information regarding its geometry, mechanical parameters of masonry and modelling are presented in a previous paper by the authors (Petrovčič and Kilar 2013). The structure can be divided into four different types of planar wall assemblies, i.e. Wx-1, Wx-2, Wy-1 and Wy-2. Each planar wall of the structure has been modelled based on the equivalent frame approach, using piers, spandrels and rigid zones as indicated in Fig. 20.1. The storey masses amount to $m_{base} = 878$ tons, $m_1 = 748$ tons and $m_2 = 499$ tons, whereas the fundamental periods of the structure are $T_X = 0.39$ s (in the X-direction) and $T_Y = 0.29$ s (in the Y-direction). The masonry mechanical parameters are presented in Table 20.1.

The building is located in a moderately active seismic area, with a design ground acceleration equal to $a_g = 0.25$ g and a soft soil site that corresponds to sub-soil of class C in accordance with Eurocode 8-1. Through a preliminary pushover analysis of the fixed-base structure it has been determined that the behaviour factor of the building equals to $q = 2.5$ for the X-direction of seismic loading and $q = 2.2$ for the Y-direction (Petrovčič and Kilar 2013).

20.2.2 Equivalent Frame Modelling

In general, masonry walls often incorporate irregularities in the form of an irregular layout of openings, so that special modelling considerations have to be taken into account (Parisi and Augenti 2013). In the cases of irregular geometry a macro-element discretization of masonry elements is usually adopted, which is able to predict the seismic response with sufficient accuracy and with relatively low

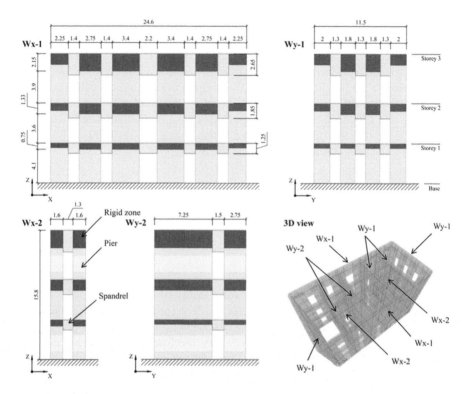

Fig. 20.1 Examined masonry building

Table 20.1 Considered masonry mechanical parameters

Parameter		Value	Parameter		Value
Masonry self-weight	(γ)	1900 kg/m^3	Young's modulus	(E)	1500 MPa
Peak compr. strength	(f_c)	2.4 MPa	Shear modulus	(G)	250 MPa
Tensile strength	(f_t)	0.3 MPa	Friction coefficient	(μ)	0.5
Cohesion	(τ_0)	0.2 MPa	Cohesion coefficient	(ζ)	0.08

computational costs compared to those involved in the case of conventional nonlinear finite element modelling approaches (Kappos and Papanikolaou 2016).

The in-plane lateral resistance of URM buildings is provided by the piers and spandrels. In addition to the vertical force in the piers due to the dead loads, these resisting elements are under horizontal seismic actions at the base of the structure subjected to shear and bending (Magenes and Calvi 1997). Depending on the width to height ratio of a masonry element and on the respective values of the normal force, bending moment and shear force, three failure mechanisms might be observed (Table 20.2).

The symbol θ_{cr} in Table 20.2 denotes the pier's chord rotation at the formation of the first crack (elastic limit). In the case of the analysed building values of θ_{cr} lie

Table 20.2 Seismic failure modes of URM (Petrovčič and Kilar 2013)

Failure mode	Damage pattern	Limit rotations for piers			Limit rotations for spandrels		
		θ_{DL}	θ_{SD}	θ_{NC}	θ_{DL}	θ_{SD}	θ_{NC}
Rocking	Tensile flexural cracking at pier corners, or vertical cracks in the more compressed corner.	θ_{cr}	$0.008\alpha_V$	$0.011\alpha_V$	0.002	0.008 l_{sp}/h_{sp}	0.015
Diagonal cracking	Diagonal cracking that typically involves both mortar joints and masonry units.	θ_{cr}	0.004	0.005	0.001	0.004	0.020
Shear sliding	In-plane sliding along a single mortar bed-joint or along bed-joints and head-joints in a stepwise fashion.	θ_{cr}	0.004	0.005			

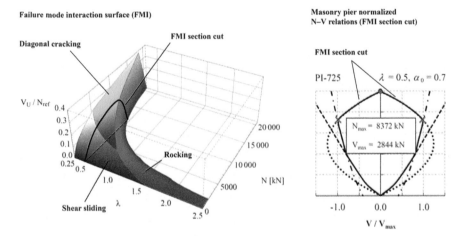

Fig. 20.2 An example of a failure mode interaction surface of a masonry pier (left) and normalized pier N-V relations (right)

between 0.003 and 0.01. The symbol α_V is the pier's shear ratio, considering the zero moment contra-flexure point and the height-to-width ratio of the pier (for details see Petrovčič and Kilar 2013). The ratio l_{sp}/h_{sp} is to length-to-height ratio of the spandrel.

The equivalent frame discretization of URM and seismic failure modes are schematically presented in Fig. 20.2. In order to accommodate different failure modes in as simple as possible and simultaneously computationally effective way, the authors proposed a new technique for the modelling and analysis of regular URM structures (Petrovčič and Kilar 2013). This technique is based on the equivalent frame approach, and incorporates linear beam elements and the plastic hinge concept. The complex seismic failure mechanism of masonry piers is expressed by a single failure mode interaction surface (an "FMI surface"), taking into account the influence of variation in the pier's vertical loading, and its bending moment distribution. The ultimate lateral strength of a masonry element is expressed as a section which cuts through the FMI surface. A single failure mode interaction plastic hinge (an "FMI hinge") for each masonry frame element is introduced by combining the specific failure modes, taking into account their minimum envelope. In the present paper a nonlinear equivalent frame 3D mathematical model of the initial (fixed-base) structure was created in the structural analysis program SAP2000 (CSI 2016) based on this technique. The model was extended to include the FRP-strengthened variant and the base-isolated variant of the building.

20.3 Seismic Stregthening

20.3.1 Fiber-Reinforced Polymers (FRP)

The primary objective of FRP strengthening is to increase the capacity of each member as well as the overall capacity of the masonry structure. For the design of the FRP strengthening of URM piers the guidelines CNR-DT 200/2004 (CNR 2004) were considered. The design approach in the context of FRP retrofitting is to strengthen the weak links in the existing structural system without drastically changing the building or its collapse mechanisms.

The shear capacity of masonry pier strengthened with FRP applied on both sides of the panel can be seen as the combination of two resisting mechanisms (CNR 2004): (i) shear forces due to friction in presence of compression loads, and (ii) for elements capable of resisting tensile stress a truss mechanism becomes active, and shear forces are carried out by equilibrium.

The CNR-DT 200/2004 defines the design shear capacity, V_{Rd}, of the FRP strengthened masonry panel in the following manner:

$$V_{Rd} = \min\left(V_{Rd,m} + V_{Rd,f}, V_{Rd,\max}\right) \quad (20.1)$$

In the above expression the $V_{Rd,m}$ denotes the masonry contribution to the shear strength of the panel:

$$V_{Rd,m} = \frac{1}{\gamma_{Rd}} \cdot f_{vd} \cdot D \cdot t, \quad (20.2)$$

while the $V_{Rd,f}$ denotes the FRP contribution, which may be evaluated as follows:

$$V_{Rd,f} = \frac{1}{\gamma_{Rd}} \cdot 0.6 \cdot \frac{A_{fw}}{p_f} \cdot f_{fd} \cdot D \quad (20.3)$$

The symbols in Eqs. (20.1), (20.2), and (20.3) have the following definitions: γ_{Rd} is the partial factor to be assumed equal to 1.20, d is the distance between the

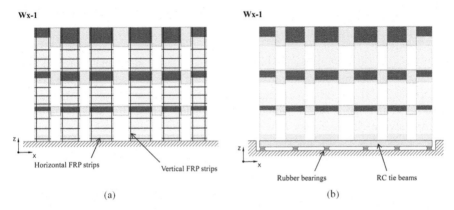

Fig. 20.3 FRP strengthened variant (**a**) and BI variant (**b**)

compression side of the masonry and the centroid of FRP flexural strengthening, t is the masonry panel thickness, f_{vd} is the design shear strength of the masonry, equal to f_{vk}/γ_M, A_{fw} is the area of FRP shear strengthening in the direction parallel to the shear force, p_f is the center-to-center spacing of FRP reinforcement measured orthogonally to the direction of the shear force, f_{fd} is the design strength of FRP reinforcement, defined as the lesser between FRP tensile failure strength and debonding strength, D and t are the width and thickness of the masonry panel, respectively.

The maximum value $V_{Rd,max}$ inducing failure of the compressed strut of the truss mechanism equals to:

$$V_{Rd,\max} = 0.3 \cdot f_{md}^h \cdot D \cdot t \quad (20.4)$$

It should be noted, that the equations above are valid only FRP strengthening placed parallel to the mortar joints, i.e. grid layout of FRP strips.

In the conducted numerical simulation, the strengthening of masonry piers with carbon FRP (CFRP) laminates has been considered. It has been considered in the study that the FRP strips have been installed on both sides of the panels and have been arranged in a grid pattern consisting of two vertical FRP strips at the edges of the pier and three horizontal FRP strips. It has been assumed that the vertical FRP strips run continuously from the bottom to the top storey (Fig. 20.3a).

The selection of FRP laminates was based on the effective axial stiffness, defined as the product $E_f \cdot \rho_{\square}$, which represents a measure of the effective axial stiffness of the external FRP reinforcement of a masonry panel, where E_f is the Young's modulus of FRP and $\rho_{\square}$ is the horizontal FRP reinforcement ratio, defined as the area of horizontal FRP strips on both sides of the panel divided by the product of the height and the thickness of the panel. Based on experimental results by Marcari and co-autohrs (Marcari et al. 2007) the effective axial stiffness equal to 60 MPa has been considered. The authors have concluded, that effective axial stiffness of the FRP strips larger than 15 MPa changes the original failure mode of the panels clearly changed from shear to a shear/flexural mode; consequently, larger gains in lateral strength were achieved. Based on this assumption CFRP strips with the width/ thickness ratio 120 mm/1,4 mm have been selected ($E_f = 21,000$ kN/cm^2, $\varepsilon_f = 0.014$).

The selected FRP characteristics have been included in the mathematical model of the initial, unstrengthened variant by using the same methodology as described in Sect. 20.2.2. Based on experimental results of other researchers (e.g. Marcari et al. 2007; Prota et al. 2008), it should also be pointed out that the installation of FRP strips appears to have very little influence on the initial stiffness of the masonry structure. This in turn implies that in FRP-strengthened masonry buildings, the distribution of seismic forces on load-bearing walls remains basically unaltered when compared to the initial, unstiffened structure. What is more, the increased strength of FRP panels does not determine significant changes to the inelastic deformation capacity of strengthened panels whose ductility is substantially very similar to that of the as-built panels (Marcari et al. 2007).

20.3.2 Base Isolation (BI)

The proposed retrofitting approach with base-isolation is based on the response control concept, whose aim is the controlling and limiting of dynamic effects on the structure by means of special isolation devices. A new isolation layer between the ground and the structure needs to be created, which consists of devices that support the building against static actions, and, in the case of an earthquake, implement decoupling of the dynamic response of the building compared to the ground.

When designing a base isolation system, the first task is to determine the desired target period of vibration of the base-isolated structure $T_{BI} = 2\pi\sqrt{m_{BI}/k_{BI}}$, which is indirectly related to the achievable decrease in the maximum stress levels (Kelly et al. 2007), where m_{BI} denotes the total mass of the structure together with the mass of the base isolation system (rubber bearings, foundation plate and RC tie beams that support the isolation devices), whereas k_{BI} denotes the total lateral stiffness of all the isolation devices by summing their individual lateral stiffnesses.

When implementing base isolation in an existing building the proper initial assessment of T_{BI} can be a demanding task. In the present study, the initial value of T_{BI} was determined by an approach developed by the authors of the paper in a previous study (Petrovčič and Kilar 2016), where the selection of T_{BI} depends on the desired level of seismic protection as set out for example in the conservation plan for the building in question. Basically, the primary task of the designer is to select a seismic return period for which the retrofitted building will undergo damage corresponding to the desired limit state, e.g. the damage limitation limit state DL (Table 20.3). In this case the T_{BI} is expressed in terms of the spectral acceleration S_a^{BI} and spectral displacement S_d^{BI} of the elastic acceleration-displacement response

Table 20.3 Performance requirements based on EC8-3, indicating the corresponding return periods and peak ground accelerations of the examined building

EC8–3 Limit state	Return period	PGA on rock
Damage limitation (DL)	95 years	0.15 g
Significant damage (SD)	475 years	0.25 g
Near collapse (NC)	2475 years	0.43 g

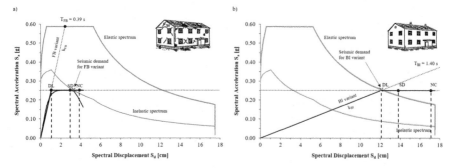

Fig. 20.4 Seismic demand for the fixed-base variant (**a**) and for the base-isolated variant (**b**)

spectrum (ADRS) at the intersection between the ADRS and the base shear capacity of the building at the DL limit state:

$$T_{BI} = 2\pi \sqrt{S_d^{BI}/S_a^{BI}} \tag{20.5}$$

In the case of the analysed building the return period $T_R = 475$ years (PGA $= 0.25$ g) was selected as the target performance level, indicating the shape of the ADRS (Fig. 20.4). The required period of vibration of the BI variant thus equals to $T_{BI} = 1.40$ s and the required lateral stiffness equals to $k_{BI} = 51272$ kN/m. The selected T_{BI} is 3.75-larger than of the FB variant for the X-direction of seismic loading, indicating a reduction of elastic spectral accelerations by a factor of 2.50.

A base isolation system consisting of natural rubber bearings (NRB) was selected, with a total of 18 NRB inserted at the foundation level and positioned in an orthogonal grid under the main loadbearing walls as presented in Fig. 20.3b (in the figure only the distribution of NRBs under wall Wx-1 is shown). In order to ensure a more uniform distribution of stresses in the NRBs, a system of RC grid beams was included in the model. The beams have a rectangular cross-section of b/h $= 80/85$ cm. The weight of the new base-isolated storey is $m_{base} = 410$ tons. The total weight of the structure with the implemented base isolation system is $m_{BI} = 2535$ tons.

Based on the calculated k_{BI} the NRBs with a diameter of 600 mm and a height of 150 mm (the height of the rubber was equal to 80 mm) were selected. The bearings are made of soft rubber and have a horizontal stiffness of 2830 kN/m per isolator, with damping equal to $\xi = 10\%$ of critical damping. The total horizontal stiffness of the selected base isolation system is $k_{BI} = 50940$ kN/m. The maximum horizontal displacement of the NRBs is equal to 15 cm, which is about 200% of the height of the rubber.

20.3.3 Eurocode 8 Requirements

Part 3 of Eurocode 8, i.e. EC8-3 (CEN 2005), which deals with the assessment and retrofitting of existing buildings, is based on recent trends regarding performance requirements and checks of compliance in terms of displacements, providing also a degree of flexibility to cover the large variety of situations arising in practice. The fundamental requirements of EC8-3 refer to the state of damage in the structure that are defined by means of three limit states presented in Table 20.3.

In order to achieve an appropriate level of protection EC8-3 takes into account a different seismic return period for each of the limit states. Longer return periods are associated with greater seismic intensities. Although it should be pointed out, that many heritage buildings have resisted earthquakes with limited damage only, even when their calculated resistance has not completely met code requirements.

This allows for the code-based ground acceleration values that are in use in new buildings to be reduced, meaning that a higher probability of exceedance is admitted. In turn, the considered return periods vary from country to country and are given in the corresponding National Annexes to EC8-3. Table 20.3 presents the return periods and corresponding peak ground acceleration (PGA) values, referring to the site of the analysed building, which is located on sub-soil class C (soft site, soil factor $S = 1.15$). The PGAs indicated in the table refer to a hard site (foundation on rock).

20.3.4 Nonlinear Static Analyses and Comparison of Results

The seismic performance was examined for the following variants: (i) the initial unstrengthened structure, i.e. fixed-base (FB) variant, (ii) FRP-Strengthened variant (FRP) and (iii) Base-Isolated variant (BI). Since it was concluded in the previous study (Petrovčič and Kilar 2013) that the X-direction seismic response is more critical than that in the Y-direction, in this paper only the seismic performance for the loading in X-direction of the building was analysed.

20.3.4.1 Capacity Curve Comparison

A nonlinear static (pushover) analysis was conducted. The demand parameter that was compared to the capacity was the top (roof) displacement. The N2 method

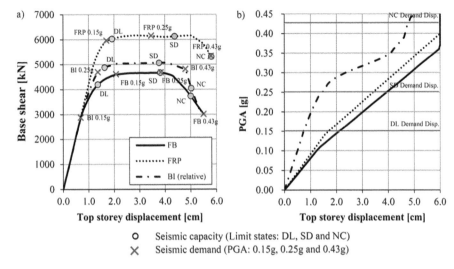

Fig. 20.5 Comparison of results – pushover curves (left) and IN2 curves (right)

(Fajfar 2000) was used to determine the demand top displacement by comparing the intersection between the capacity obtained by pushover analysis and by an inelastic response spectrum. A lateral load pattern proportional to the first mode of vibration was considered. Lateral loads (weighted, taking into account the nodal masses) were applied to the nodes corresponding to the individual storey levels.

Pushover curves of analysed variants, indicating the base shear vs. the lateral roof (top) displacement relationship, are presented in Fig. 20.5a. The protection levels which are defined in EC8-3 as the limit states DL, SD and NC are also indicated in the figure. It can be seen that the capacities of the FB and BI variants are directly comparable. It should be noted that in order to assess the effects of BI on the initial structure, the relative displacements need to be observed and compared. The relative displacements indicate the lateral top displacement with respect to the BI level, i.e. subtracting the lateral displacements due to the base isolation. In the BI case, the maximum base shear is slightly larger due to the different lateral force distribution, and due the difference in the total seismic mass of the FB and BI variants. The FRP variant, on the other hand, can withstand an approximately 30% larger maximum base shear force when compared to the initial FB variant.

20.3.4.2 Seismic Demand Displacements and Damage Patterns

The seismic demand displacements were calculated by employing the N2 method for PGA levels corresponding to the observed limit states DL, SD and NC (Table 20.3). The FB structure exhibits inadequate seismic performance for all three observed limit states, i.e. the corresponding demand exceeds the capacity requirements.

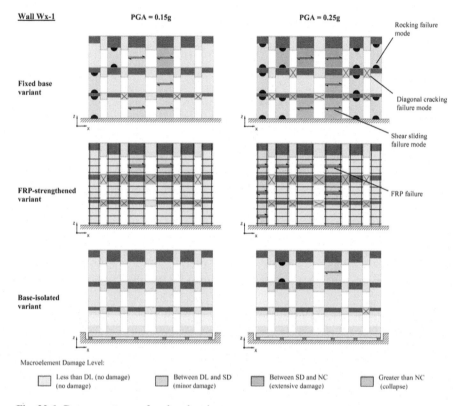

Fig. 20.6 Damage patterns of analysed variants

In the case of the FRP and BI variants for the PGA = 0.15 g and 0.25 g (the DL and SD limit states, respectively) the structure's top displacement is smaller than the displacement of the observed limit state. This means that in both cases the structure has been adequately strengthened for the analysed limit state. Moreover, the displacement of the BI variant is for both PGAs smaller than the DL limit state, indicating limited and repairable damage to the structure also in the case of PGA = 0.25 g.

When observing the demand displacement for PGA = 0.43 g, the structure's performance is inadequate for all three examined variants. In the case of the FB and also FRP strengthened variants, the structure would for this PGA value reach its maximum displacement, indicating its collapse. The BI variant would, based on the results, be heavily damaged and would be very close to collapsing.

For PGA = 0.15 g and 0.25 g the damage patterns of the structure for all three variants are presented in Fig. 20.6 for the wall Wx-1. The type of failure mechanism and the level of damage are indicated for each macroelement. The limit chord rotations are given in Table 20.2. It can be seen from Fig. 20.6 that the FB variant is heavily damaged in the case of PGA = 0.25 g. The piers of the top storey are all damaged, some of them considerably. A similar observation can be made for the piers and spandrels in the base storey. All the damage grades which are higher than the SD macroelement limit state indicate damage that is very difficult and uneconomic to repair.

The FRP strengthened variant receives significantly less damage in the case of PGA = 0.25 g. The damage mostly concentrated in the top storey, where the axials forces in the pier are smaller. It should also be noted that since the strengthening of spandrels was not considered, most of the spandrels in the model are heavily damaged.

The least amount of damage was recorded in the case of the BI variant. For the target PGA = 0.25 g it can be seen that the structure undergoes minor damage, which is likely to be repairable. Damage is localised in two piers in the top storey. This is consistent with the EC8-3 requirements for the DL limit state, which allows the occurrence of local damage, where repair measures are usually not needed or are very limited.

20.3.4.3 Top Storey Displacements for Varying PGA Levels

In the final part the relationship between the top storey displacements and the PGA was analysed by progressively increasing the PGA in a stepwise fashion. In each step the seismic demand displacement has been calculated by using the N2 method. Such an analysis, which has been given the name "incremental N2 analysis" (IN2), was first introduced and applied to infilled RC frames (Dolšek and Fajfar 2005). In the conducted IN2 analyses 45 consecutive intensity levels were considered, ranging from PGA = 0.01 g to PGA = 0.45 g, with a 0.01 g step size.

The results of the analyses are presented in Fig. 20.5b. It can be seen from the figure that the IN2 curves for the FB and FRP variants follow the same trend and that in the PGA range from approx. 0.15 g to 0.37 g the FRP variant yields about 15% smaller seismic demand displacements, leading to reduced damage in the piers and different damage pattern distributions. In the case of the BI variant, this difference is more profound. It is largest at PGA = 0.20 g where the seismic demand displacement of the BI variant equals to 34% of that of the FB variant.

20.4 Conclusions

Two seismic strengthening measures, applied to an existing unreinforced masonry structure, have been considered. In the first variant fibre reinforced polymer (FRP) composites have been used, while in the second variant seismic base isolation (BI) has been implemented. It can be concluded that both strengthening measures provide sufficient lateral strength increases to meet code-based requirements.

The BI variant of the building obtains significantly less seismically induced damage than the FRP variant. Moreover, the seismic strengthening with BI is also less invasive for the building, since all major interventions are concentrated at the base storey. Nevertheless, installation of BI in existing buildings is extremely demanding and is limited only to buildings that meet specific requirements. The strengthening with FRP is, on the other hand, appropriate for most buildings – depending on the allowed degree of intervention as set out in the conservation plan (conservation of façades with frescoes or other artistic assets). Additionally, the refined microelement modelling technique for URM structures, developed by the authors in previous studies, was successfully extended to both analysed seismically strengthened variants of the building. This means that the developed technique eases not only the implementation of seismic failure modes of URM, but also the implementation of strengthening measures in various general purpose structural analysis computer programs, not specifically designed for URM analyses.

Acknowledgements The results presented in this article are based on work that was supported by the Slovenian Research Agency (grant number P5-0068). The authors greatly acknowledge the financial support for this research.

References

CEN – European Committee for Standardisation (2005) Eurocode 8: design of structures for earthquake resistance – part 3: general rules, seismic actions and rules for buildings, design code EN 1998-3. CEN, Brussels

CNR – Advisory Committee on Technical Recommendations for Construction (2004) Guide for the design and construction of externally bonded FRP systems for strengthening existing structures. CNR, Rome

CSI – Computers and Structures (2016) SAP2000 (v18.2.0) – Structural Analysis Program. Berkeley, USA

Dolšek M, Fajfar P (2005) Simplified non-linear seismic analysis of infilled reinforced concrete frames. Earthq Eng Struct Dyn 34:49–66

Fajfar P (2000) A nonlinear analysis method for performance-based seismic design. Earthq Spec 16(3):573–592

Kappos AJ, Papanikolaou VK (2016) Nonlinear dynamic analysis of masonry buildings and definition of seismic damage states. Open Construct Build Technol J 10(1):192

Kelly JM, Robinson BWH, Skinner RI (2007) Seismic isolation for designers and structural engineers. Robinson Seismic Limited, Petone

Magenes G, Calvi GM (1997) In-plane seismic response of brick masonry walls. Earthq Eng Struct Dyn 26(11):1091–1112

Marcari G, Manfredi G, Prota A, Pecce M (2007) In-plane shear performance of masonry panels strengthened with FRP. Compos Part B 38(7):887–901

Parisi F, Augenti N (2013) Seismic capacity of irregular unreinforced masonry walls with openings. Earthq Eng Struct Dyn 42(1):101–121

Petrovčič S, Kilar V (2013) Seismic failure mode interaction for the equivalent frame modeling of unreinforced masonry structures. Eng Struct 30(54):9–22

Petrovčič S, Kilar V (2016) Seismic retrofitting of historic masonry structures with the use of base isolation—modeling and analysis aspects. Int J Archit Herit:1–8

Prota A, Manfredi G, Nardone F (2008) Assessment of design formulas for in-plane FRP strengthening of masonry walls. J Compos Constr 12(6):643–649

Chapter 21
From Seismic Input to Damage Scenario: An Example for the Pilot Area of Mt. Etna Volcano (Italy) in the KnowRISK Project

Raffaele Azzaro, Salvatore D'Amico, Horst Langer, Fabrizio Meroni, Thea Squarcina, Giuseppina Tusa, Tiziana Tuvè, and Rajesh Rupakhety

Abstract In this paper we present a multidisciplinary approach aimed at assessing seismic risk regarding non-structural damage. The study has been carried out in the framework of the European KnowRISK Project and focuses on the pilot area of Mt. Etna volcano (Italy). Both instrumental data and as well as macroseismic observations provide unique opportunities for testing innovative and classical approaches for assessing seismic risk. Starting from the seismic hazard analysis, we first identify a test site (Zafferana) affected by non-structural damage. We produce seismic scenarios based on macroseismic and ground-motion data and finally obtain the relevant risk map using the Italian census data to classify buildings into vulnerability classes and a model to predict damage distribution.

Keywords Mt. Etna · KnowRISK project · Seismic scenarios · Ground motion parameters · Risk map

21.1 Introduction

Simulating earthquake scenarios is a multidisciplinary field of investigation aimed at assessing the level of seismic risk to which an area is exposed. A variety of methods, leading to deterministic or probabilistic seismic scenarios, have been developed according to different levels of accuracy required. The most complete are generally

R. Azzaro (✉) · S. D'Amico · H. Langer · F. Meroni · T. Squarcina · G. Tusa · T. Tuvè
Istituto Nazionale di Geofisica e Vulcanologia, Rome, Italy
e-mail: raffaele.azzaro@ingv.it

R. Rupakhety
Faculty of Civil and Environmental Engineering, School of Engineering and Natural Sciences, University of Iceland, Reykjavík, Iceland

Director of Research, Earthquake Engineering Research Centre (EERC), University of Iceland, Selfoss, Iceland

based on numerical models taking into account ground-motion predictive relationships, local geologic conditions, historical macroseismic intensity data, together with building typologies and their vulnerability. The seismic scenarios are usually referred to large earthquakes causing structural damage, with the aim of estimating casualties, economic and social costs of recovery etc.

On the other hand, damage to non-structural elements of buildings (partitions, ceilings, cladding, electrical and mechanical systems, furniture etc.) is known to cause injuries, losses, business interruption, and limit the functionality of critical facilities, such as hospitals and schools, causing a significant impact on earthquake resilience.

In this paper we present the procedure and preliminary results of seismic scenarios focused on non-structural damage in the pilot area of Mt. Etna volcano, Italy. In methodological terms, we first carried out the hazard analysis addressed to map the volcano's areas that are most exposed to macroseismic intensities typically representative of non-structural damage, and to identify the earthquakes to be assumed as scenario. Then we performed ground motion scenarios instrumentally derived or calculated by means of synthetic simulations, that have been estimated according different geological site conditions (hardrock, D-type soils); in particular, we selected as significant parameters both response and drift spectra. Finally we derived risk maps that depict the extent of non-structural damage following the scenario earthquake, using the Italian census data to classify buildings into vulnerability classes and a model to predict damage distribution as a function of the intensity degree in the European Macroseismic Scale (EMS, see Grünthal 1998). In doing this, we assume that non-structural damage corresponds to grades 1, 2 and 3 on EMS.

21.2 Seismic Hazard

Several studies aimed at assessing seismic hazard at Etna have been carried out in the last years by means of a probabilistic approach based on the use of macroseismic data (Azzaro et al. 2013a, and references therein). These analyses indicate that hazard is determined by two distinct seismotectonic regimes: (i) large regional earthquakes occurring in eastern Sicily ($6.2 \leq M_W \leq 7.4$) referred to long exposure times (50 years), (ii) local events due to seismic sources of the volcano associated with short exposure times (30 years or less). Although characterized by different magnitude ranges and occurrence rates, both produce the same values of intensity (degrees VIII and IX) in the study area.

In particular volcano-tectonic earthquakes, although of low energy ($M_W \leq 5.3$), produce heavy damage (up to IX degree EMS) due to the shallowness of hypocenters (Alparone et al. 2015) but affect small areas because of the rapid attenuation of seismic intensity (Azzaro et al. 2006, 2013b). Most of them are located in the eastern flank of Etna (Fig. 21.1), crossed by a system of active faults, but it is noteworthy that all the historical destructive events (some ten with epicentral intensity $I_0 \geq$ VIII EMS) occurred in the area among the towns of Acireale, Giarre and Zafferana, the most densely populated zone of the volcano (Meroni et al. 2016). For this reason we selected this sector as study area of the KnowRISK project.

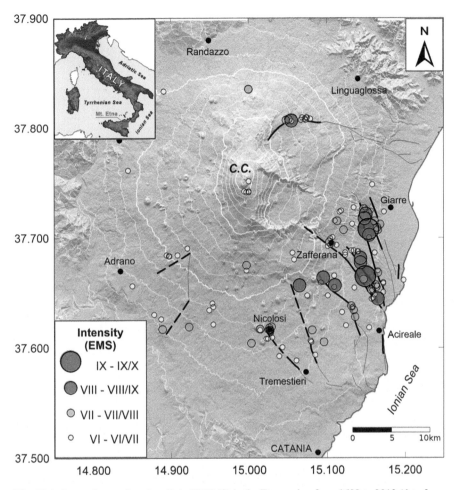

Fig. 21.1 Damaging earthquakes ($I_0 \geq$ VI EMS) in the Etna region from 1600 to 2013 (data from CMTE Working Group 2017). Black lines indicate the main faults (from Azzaro et al. 2016); C.C. central craters of Etna

21.2.1 Occurrence Probability of Non-structural Damage

The following hazard analysis has been carried out thorough the SASHA code (D'Amico and Albarello 2008). As input data we considered earthquake parameters and intensity database from the Macroseismic Catalogue of Etnean Earthquakes, spanning from 1600 to 2013 (CMTE Working Group 2017). In particular, we selected the earthquakes above the damage threshold ($I_0 \geq$ VI EMS), obtaining a dataset of 140 events with M_W ranging from 3.7 to 5.3, associated with 4432 intensity values. The probabilistic hazard (hazard curves at a site) has been calculated for an exposure time of 30 years.

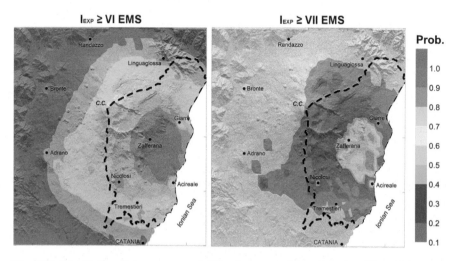

Fig. 21.2 Distribution of occurrence probability for expected intensities I_{exp} VI and VII EMS calculated for an exposure time of 30 years. Dashed line indicates the study area

In the EMS scale (Grünthal 1998), non-structural damage (grades 1, 2 and 3) is confined mainly between degrees VI and the VII, while starting from degree VIII the contribution of structural damage becomes predominant (grades 3, 4 and 5). Since the aim of the KnowRISK project is to investigate the risk due to non-structural damage, we focused on mapping occurrence probabilities relevant for expected intensities I_{exp} VI and VII.

Figure 21.2 shows that the study area has a probability greater than 50% to suffer extensive non-structural damage related with degree VII EMS, while the probability increases to 80% at least if referred to slight non-structural damage determined by degree VI EMS. Results confirm that seismic hazard in this sector of the volcano is high not only for destructive earthquakes but also for moderate events.

In particular, we selected Zafferana as test site since it has a high probability to be struck with an intensity at least of degree VII EMS in 30 years as well as there is availability of census data on the building stock.

21.2.2 Disaggregation Analysis

In this section we identified the 'design earthquake' for Zafferana, that is the event that more contributes to the hazard of this site (Albarello 2012). It means to identify the most significant magnitude/epicentral distance bins that can be associated with seismic events occurring in the past; note that it not necessarily corresponds to a single earthquake, since more events may be located at similar epicentral distances and have comparable magnitudes.

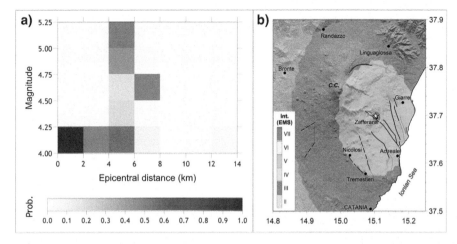

Fig. 21.3 (a) Disaggregation of M_W (bin 0.25) versus epicentral distance for the town of Zafferana. (b) Probabilistic seismic scenario referred to the 1984 earthquake parameters: I_0 VII EMS; M_W 4.2. Expected intensity is represented on a grid with intermodal distance of 1 km; star represents the epicentre of the earthquake

Using a tool of the SASHA code, the probability values of I_{exp} VII EMS obtained for the test site were summed for classes of epicentral distances (2 km) and magnitude (0.25 unit), and then normalized to provide the disaggregation of data.

The result is shown in Fig. 21.3a where hazard related with I_{exp} VII at Zafferana is mainly due to small size earthquakes – M_W ranging from 4.00 to 4.25 – having epicentre very close to this site, up to 6 km away. Conversely, stronger earthquakes occurring in the eastern flank contribute less to the hazard. This conclusion suggested us to consider as seismic scenario for non-structural damage a moderate event located nearby our test site.

21.2.3 Macroseismic Intensity Scenario

As suggested by the disaggregation analysis, the first seismic scenario was calculated using the parameters of the October 19th, 1984 earthquake, which are derived only through macroseismic data (CMTE Working Group 2017). This event indeed had epicentre very close to Zafferana and M_W 4.2, calculated by means of the intensity-magnitude relationship in Azzaro et al. (2011). The intensity estimated for the site of Zafferana is VII EMS, corresponding to the maximum intensity value of the macroseismic field; nearby localities experienced lower intensity degrees. For this reason we assume an epicentral intensity I_0 equal to degree VII.

The scenario was calculated in terms of macroseismic intensity according to the method reported in Azzaro et al. (2013b), Meroni et al. (2016) and Rotondi et al. (2016). The software used for the analysis is PROSCEN. In practise, by adopting a probabilistic approach based on the Bayesian statistics, we reproduce the decay of the intensity from the source according to an isotropic attenuation models, and then calculate the probability distribution of the intensity at a given site (I_s) conditioned on the epicentral intensity (I_0) of the earthquake and the epicentre-site distance through a binomial-beta model. The mode of the smoothed binomial distribution is taken as the intensity at site.

Figure 21.3b shows the intensity scenario for the 1984 earthquake. The pattern of the expected intensities well reproduces the observed intensity data points, especially for higher EMS values, highlighting a strong attenuation of intensity in short distances. Non-structural damage related with degrees VII and VI potentially affects a significantly urbanised zone around Zafferana. Some misfit with observed data concerns the extension of degree IV, that can be explained taking into account that diagnostics collected during the macroseismic survey in order to assess lower degrees are more subjective; however, it should be noted that degree IV does not imply the presence of damage.

The 1984 scenario represents the input for generating the risk map described in Sect. 21.4.4. In addition, we also considered a second seismic scenario referred to a stronger event located ca. 4 km far from Zafferana, to be used for simulating ground motion scenarios (Sect. 21.3). In such a case we selected the October 29th, 2002 earthquake, M_W 4.8.

21.3 Ground Motion Scenarios

Numerical ground motion simulation methods for the construction of earthquake scenarios are now currently being applied for estimating the seismic hazard. In this study, in order to develop the scenarios of earthquake ground motion, we consider, besides peak ground amplitudes and the response of a damped single degree-of-freedom (SDOF) system – the so called drift spectra, which are particularly useful in estimating drift demands in buildings subjected to pulse-like ground motions (Iwan 1997) and can provide a better understanding of non-structural damage.

Being strong events rare in the area, and consequently instrumental data are scanty, we include weak motion data (see Tusa and Langer 2016) to calibrate the input of synthetic simulation of stronger events.

21.3.1 Synthetic Simulation

Synthetic simulations of ground motion scenarios are based on the code EXSIM (Boore 2009) with slight modifications (see Langer et al. 2015). The scenarios have been generated for the two events before discussed, i.e. the 1984 and 2002 earthquakes representing sources close and far, respectively, to the test site of Zafferana.

21 From Seismic Input to Damage Scenario: An Example for the Pilot Area...

The input parameters (in terms of seismic velocities, attenuation factor, F_{max}, radiation pattern, partitioning factor) used for simulation of ground motion of these events, were chosen following Langer et al. (2015) (see Table 1 in Langer et al. 2016). From the point of view of site response, two types of soils were considered: (i) hardrock site (Site H), for which no specific amplification factors were used; (ii) D-type soil (Site D), for which the site amplification functions given in Scarfì et al. (2016) were applied.

The obtained simulated ground motion parameters for both scenarios in the site of Zafferena are listed in Table 21.1, while the estimated response spectra are reported in Fig. 21.4. It is evident that the response spectra are characterized by a maximum seismic response at short periods for both earthquakes and, as expected, the introduction of site effects (Site D) causes significant amplifications of ground motion at same periods.

From the Housner Intensity (HI) we estimate an "equivalent" macroseismic intensity exploiting the empirically derived formula by Chiauzzi et al. (2012):

$$I_{EMS} = I_{eq} = \max\,[1.41\,\ln\,(HI) + 7.98, 0.27\,\ln\,(HI) + 6.02]$$

We use this relation as a proxy for comparing numerical results to the observed macroseismic intensities when strong motion data are unavailable.

Table 21.1 Ground motion parameters for the two scenarios using a factor of 0.9 for the partition between the two horizontal components. PGA, peak ground acceleration; PGV, peak ground velocity; HI, Housner Intensity derived on the base of response spectra

1984 eq., M_W 4.2, Ep. distance = 0 km			
	PGA (gal)	PGV (cm/s)	HI (cm) (I_{eq})
Site H	19.6	1.7	11 (5.5)
Site D	130	13	67 (7.6)
2002 eq., M_W 4.8, Ep. distance = 7 km			
	PGA (gal)	PGV (cm/s)	HI (cm) (I_{eq})
Site H	14	1.4	11 (5.5)
Site D	77	11	73 (7.7)

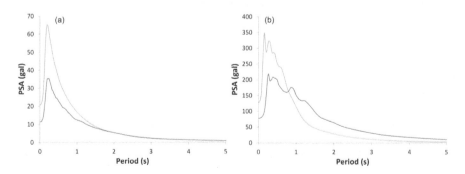

Fig. 21.4 Response spectra calculated for the test site of Zafferana referred to: blue, 1984 M_W 4.2 eq. (epicentre); orange, 2002 M_W 4.8 eq. (Ep. dist. = 7 km). (**a**) Hardrock, (**b**) Soft Soil (D-type)

Moving on from the earthquake scenarios relevant for the site Zafferana, we carried out first tests on the so called drift spectra. The interstory drift ratio, defined as the difference in lateral displacements in between two consecutive floors normalized by the interstory height, is the response parameter that is best correlated with damage in buildings (Miranda and Akkar 2006). The drift spectrum is based on a relatively simple linear model. Contrary to classical response spectra, the method is based on a model that consists of a combination of a flexural beam and a shear beam.

The lateral stiffness ratio, α, is a dimensionless parameter that controls the degree of participation of overall flexural and overall shear deformations in the continuous model, thus controlling the lateral deflected shape of the model. A value of α equal to zero represents a pure flexural model, while α equal to ∞ corresponds to a pure shear model.

Critical parameters in the model are the parameter α, the interstory height (H in meters) and its relation to the natural period of the building (T in s), and the damping (only in the damped model). In order to be consistent with commonly used value in response spectra, we have been using a constant damping of 5%.

From a seismological point of view, the geological condition of the sites is among the most critical issues. Setting here the parameter α equal to 20, and using the following empirical relation (Crowley and Pinho 2004):

$$T = 0.1\,H$$

we obtain the drift spectra for the test site shown in Fig. 21.5.

The peak ground motion parameters, response spectra and drift spectra underscore the relevance of geological site conditions. Indeed, in the case of soft soil conditions (D-type), we could observe critical values for both the simulated earthquakes. In the case of the smaller event, that is the M_W 4.2 eq. having epicentre very close to Zafferana, we note higher frequencies (>1 Hz), which stand for smaller buildings (H < 10 m according to the formula height / natural period).

In the case of the larger event, the M_W 4.8 eq. having epicentre 7 km far from the test site, the spectral values (both in response spectra and drift spectra) are somewhat lower

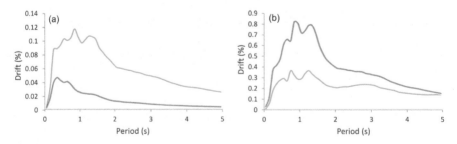

Fig. 21.5 Drift spectra (average over 12 azimuth directions) calculated for the test site of Zafferana referred to: blue, 1984 M_W 4.2 eq. (epicentre); orange, 2002 M_W 4.8 eq. (Ep. dist. = 7 km). (**a**) Hardrock, (**b**) Soft Soil (D-type)

in the high frequency range. However values of the response spectra at long periods (i.e. larger buildings having $T_0 > 1$ s) are found to be lower for the smaller event.

Moreover, the average drift spectrum is below the critical value of 0.5% reported in the EC8. However, note that standard deviations of the values shown in Fig. 21.5 amount to 20–30% for the M_W 4.8 scenario, and up to 60% in the M_W 4.2 case. That means that there is a fair possibility of drift reaching critical values (in the sense of EC8) for both scenarios assuming soft soil conditions.

21.4 Seismic Risk Maps

The Mt. Etna area is highly urbanized, with many towns and villages located all around the volcano at different altitudes up to 700 m a.s.l. In particular, the southern and eastern flanks are the most populated areas, with inhabited centres being very close to each other. Moreover, a dense network of roads, power lines and methane pipelines covers the territory. The study area considers part of the south-eastern flank of the volcano, over an area of approximately 510 square kilometers including 28 municipalities with a population of ca. 400,000 inhabitants.

21.4.1 Vulnerability of Residential Buildings

To carry out a vulnerability analysis on a regional scale, the size of the building stock can be inferred from data collected during the Italian census, when correctly adapted for the purpose of the vulnerability evaluation (Meroni et al. 1999, 2000). The Italian National Institute of Statistics (ISTAT) census data on residential buildings, disaggregated by census sections, has been used as a proxy of the elements exposed at seismic risk. They provide a uniform cover of the whole country; however, the information available only makes it possible to estimate the total number of buildings and their total volume, bringing a poor classification in terms of age and a few typological parameters. Frequencies for groups of homogenous structures can be deduced from the ISTAT data on residential buildings, with respect to a number of typological parameters: vertical structures, age of construction, number of storeys, state of maintenance and state of aggregation with adjacent buildings (Table 21.2). The 1991 census (ISTAT 1991) was used in this study, as the latest data from 2001 and 2011 are not directly usable. In fact, strict legal rules on confidentiality of information, in force from 1996, impose data providing in an aggregated form only, with no chance to intersect multiple independent variables at least at a municipality level. This limitation does not allow a data crossing on the typological characteristics, critical in the procedure of seismic vulnerability evaluation. The

Table 21.2 Typological classes of buildings identified from ISTAT data

Structural typology	Building age	Number of floors	Structural context	Level of maintenance
Masonry buildings	Age < 1919	1 or 2 floors	Isolated buildings	Good
Reinforced Concrete (RC) buildings	$1919 \leq$ age ≤ 1945	3, 4 or 5 floors	Block of buildings	Low
Soft storey buildings	$1946 \leq$ age ≤ 1960	6 or more floors		
Other typologies	$1961 \leq$ age ≤ 1971			
	$1972 \leq$ age ≤ 1981			
	Age > 1981			

1991 census data have been updated to 2011, by adding all the necessary information for vulnerability assessments coming from the comparison of the same census variables reported in the following surveys at a municipal level, the smallest area in which cross-check data are allowed.

21.4.2 A Model for EMS Vulnerability Classifications of ISTAT Census Data

The building vulnerability can be assessed from the ISTAT data and classified into six classes (A to F), according to the EMS scale, by assigning a score of vulnerability I_v. The classification procedure is consistent with a vulnerability assessment at a national scale (Meroni et al. 2000) calibrated on more than 28,000 detailed GNDT vulnerability forms (Benedetti and Petrini 1984). In that work, referring to the municipalities in which the GNDT I^{st} and IInd level forms were available, the average vulnerability indices have been evaluated for homogenous groups of buildings based on the census variables of the ISTAT data.

The adopted method proposed by Bernardini et al. (2008) describes a deterministic classification of groups of buildings defined on the 1991 ISTAT data; this proposal uses an additional parameter, i.e. the possible date of seismic classification of the territory. This parameter is consistent with the criteria suggested by the EMS scale which introduced vulnerability classes for buildings constructed with antiseismic protection criteria, becoming stricter and stricter (D, E, F). This is primarily important for RC buildings, but certainly not negligible even for masonry buildings. This method is defined on five parameters, specified in Table 21.3 for each of the five types of vertical structures provided by the ISTAT data.

Table 21.3 Parameters for classifications of the ISTAT data (from Bernardini et al. 2008)

k (type)	1 Soft storey	2 R.C.	3 Masonry	4 Other	5 Unknown
$I_v{}^1{}_1$ (k)	50	45	60	55	52
Delta_i (k)	−20	−20	−25	−20	−22
Delta_j (k)	−10	−15	−15	−15	−15
Manut (k)	−10	−10	−10	−10	−10
Classif (k)	−10	−20	−10	−10	−15

The indices *Delta_i* and *Delta_j* of the second and third rows refer, respectively, to the ranges of the construction age (or total retrofitting) of the buildings and to the typological factors on buildings aggregation and number of storeys.

The factors *Manut* and *Classif* can reduce the I_v, and they can be applied respectively if:

- the group of buildings is declared in a good state of maintenance (in the year 1991);
- the group of buildings was built after the date of the seismic classification of the territory.

Therefore, the mean of I_v index for each group of buildings is defined by the relation:

$$I_v(i, j, k) = I_v{}^1{}_1(k) + Delta\_i(k) * (i - 1)/5 + Delta\_j(k) * (j - 1)/5 + Manut(k) + Classif(k)$$

Therefore, the classification into vulnerability classes of the EMS scale is evaluated according to the score ranges specified in Bernardini et al. (2008).

21.4.3 Updated Assessment of the Exposed Elements

The final result of our analysis is the classification of residential buildings in six vulnerability classes of the EMS scale (A–F) (Grünthal 1998). Afterwards, this distribution expressed at municipality level, was updated to 2011, as already mentioned above.

To bypass the restrictions imposed by new aggregated data in the two last censuses, the evolution of the settlements on the territory has been studied through a census variable considered in the last three ISTAT censuses at a municipal level that is "*the surface of the housing area occupied by at least one resident person*". Applying these percentages of increase or reduction on the overall values should allow to redistributing such variations on the vulnerability classes derived from the 1991 ISTAT data only and obtaining their projection to the year 2011.

Nevertheless, the increasing trend of residential buildings has probably affected the distribution in vulnerable classes, as the newer buildings should have a better construction quality and a lower seismic vulnerability. It is assumed that the increase or decrease of buildings detected by ISTAT on the territory is ruled by the principle of conservation of buildings having better conditions from the point of view of vulnerability, with the phasing out of buildings in worse conditions in terms of seismic performance.

We proceeded with the assignment of the vulnerability classes according to the methodology described by Bernardini et al. (2008). Faced with a tiny increase in the amount of built-up environment (about 12% over the entire study area), we have obtained a different distribution in the vulnerability classes; as expected, an increase in the lower vulnerability classes (D, E) and a smaller decrease in the most vulnerable classes (from A to C), were noted. This overall pattern describes a generalized decrease in the building's vulnerability in the last 20 years in the studied area.

Finally, an average vulnerability index was calculated for each census sections of the municipality in the zone of analysis (0–1); the elaboration consists of a weighted sum of the volumes in each of the vulnerability classes multiplied by the average score of each vulnerability class. The adopted numerical scores of the index are related to the central values of the vulnerability classes ranges (from A to F) deducted from Bernardini et al. (2007). The geographical distribution of the mean vulnerability index for residential buildings evaluated in each census section, is shown in Fig. 21.6.

21.4.4 Seismic Risk Evaluation

Seismic risk in an urban region follows two main steps: (i) exposure geo-referenced inventory and vulnerability classification of assets at risk; and (ii) vulnerability characterization according to damage models. In this work, damage models are selected according to the macroseismic evaluation of the seismic scenario provided in previous sections, so a macroseismic method for the vulnerability assessment of buildings has been adopted. The damage model proposed by Lagomarsino and Giovinazzi (2006) and revised in Bernardini et al. (2007) has been successfully applied in previous Portuguese and Italian seismic risk studies (Sousa 2006, 2008; D'Amico et al. 2016). This model classifies the building stock following the vulnerability table of the EMS scale and predicts damage distributions, conditioned by an intensity level, for each damage grade of the scale.

According to this model, the seismic building vulnerability is described by a vulnerability index varying between 0 and 1, independently from the hazard severity level. The authors estimated an expected damage degree, μ_D, for a building typology according to the following equations (Bernardini et al. 2007):

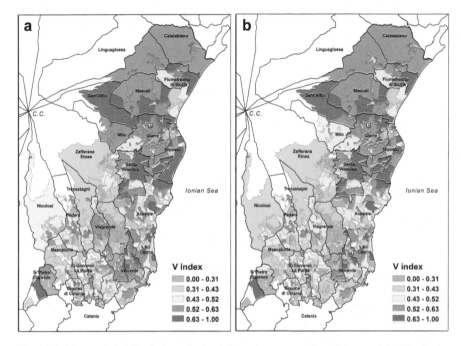

Fig. 21.6 Mean vulnerability index calculated for each census section of the municipalities in the study area, referred to the year 1991 (**a**), and 2011 (**b**)

$$\mu_D = 2.5 + 3\tanh\left(\frac{I + 6.25 V_I - 12.7}{3}\right) \times f(V_I, I) \quad (21.1)$$

with $f(V_I, I)$ defined as:

$$f(V, I) = \left\{ \begin{array}{l} e^{\left(\frac{V}{2} \times (I-7)\right)}, I \leq 7 \\ 1, I > 7 \end{array} \right\} \quad (21.2)$$

where μ_D is the mean damage grade (grade 1, slight; grade 2, moderate; grade 3, heavy; grade 4, very heavy; and grade 5, collapse) of D, the random variable damage, I is the intensity and V_I is the vulnerability index.

Figure 21.7 shows the risk map referred to the 1984 M_W 4.2 earthquake with epicenter in Zafferana. Physical structures exposed to the earthquake impact are evaluated in 5 levels of damage severity expressed in terms of damaged volume of building stock. Given the moderate energy of the shaking scenario, the best description of the results in terms of non-structural damage is to map slight damage as a combination of the low damage degrees of the EMS scale (D1 + D2 + 60%D3) calculated for each census sections of the municipalities in the study area. In this way, it is possible to define the area where non-structural damage is concentrated.

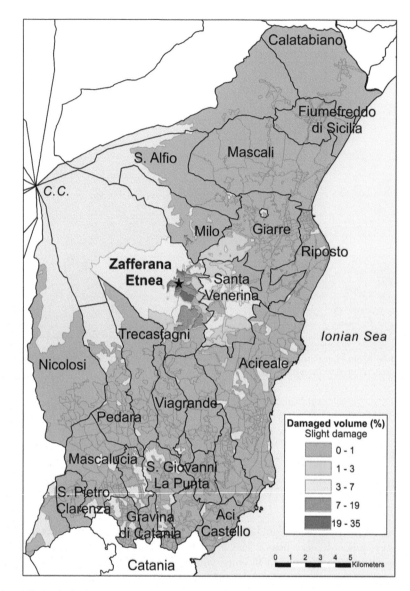

Fig. 21.7 Slight damage calculated for each census section of the municipalities in the eastern flank of Etna area due to the October 19th, 1984 Zafferana earthquake ($M_w = 4.2$, $I_0 = $ VII EMS)

Acknowledgements *Know your city, Reduce seIsmic risK through non-structural elements*, the KnowRISK project was co-financed by the EU-Civil Protection Financial Instrument with the Grant Agreement N. ECHO/SUB/2015/718655/PREV28. It is a project that involved four different European research centers and universites under the coordination of the Instituto Superior Tecnico (Portugal). The partners are the Istituto Nazionale di Geofisica e Vulcanologia (Italy), the Laboratorio Nacional de Engenharia Civil (Portugal) and the Earthquake Engineering Research Centre (University of Iceland). We acknowledge Carlos Sousa Oliveira, Mario Lopes, and all the KnowRISK group for the fruitful discussions and support.

References

Albarello D (2012) Design earthquake from site-oriented macroseismic hazard estimates. Boll Geof Teor Appl 53(1):7–18

Alparone S, Maiolino V, Mostaccio A, Scaltrito A, Ursino A, Barberi G, D'Amico S, Di Grazia G, Giampiccolo E, Musumeci C, Scarfì L, Zuccarello L (2015) Instrumental seismic catalogue of Mt. Etna earthquakes (Sicily, Italy): ten years (2000–2010) of instrumental recordings. Ann Geophys 58(4):S0435

Azzaro R, Barbano MS, D'Amico S, Tuvè T (2006) The attenuation of seismic intensity in the Etna region and comparison with other Italian volcanic districts. Ann Geophys 49(4/5):1003–1020

Azzaro R, D'Amico S, Tuvè T (2011) Estimating the magnitude of historic earthquakes from macroseismic intensity data: new relationships for the volcanic region of Mount Etna (Italy). Seismol Res Lett 82(4):533–544

Azzaro R, D'Amico S, Peruzza L, Tuvè T (2013a) Probabilistic seismic hazard at Mt. Etna (Italy): the contribution of local fault activity in mid-term assessment. J Volcanol Geotherm Res 251:158–169

Azzaro R, D'Amico S, Rotondi R, Tuvè T, Zonno G (2013b) Forecasting seismic scenarios on Etna volcano (Italy) through probabilistic intensity attenuation models: a Bayesian approach. J Volcanol Geotherm Res 251:149–157

Azzaro R, D'Amico S, Tuvè T (2016) Seismic hazard assessment in the volcanic region of Mt. Etna (Italy): a probabilistic approach based on macroseismic data applied to volcano-tectonic seismicity. Bull Earth Eng 17(7):1813–1825

Benedetti D, Petrini V (1984) On seismic vulnerability of masonry buildings: proposal of an evaluation procedure. L'industria delle Costruzioni 18:66–78

Bernardini A, Giovinazzi S, Lagomarsino S, Parodi S (2007) Matrici di probabilità di danno implicite nella scala EMS98. XII Convegno ANIDIS "L'ingegneria sismica in Italia", Pisa, CD-ROM

Bernardini A, Salmaso L, Solari A (2008) Statistical evaluation of vulnerability and expected seismic damage of residential buildings in the Veneto-Friuli area (NE Italy). Boll Geof Teor Appl 49(3–4):427–446

Boore DM (2009) Comparing stochastic point-source and finite-source ground-motion simulations: SMSIM and EXSIM. Bull Seism Soc Am 99:3202–3216

Chiauzzi L, Masi A, Mucciarelli M, Vona M, Pacor F, Cultrera G, Gallovic F, Emolo A (2012) Building damage scenarios based on exploitation of Housner intensity derived from finite faults ground motion simulations. Bull Earth Eng 10:517–545. https://doi.org/10.1007/s10518-011-9309-8

CMTE Working Group (2017) Catalogo Macrosismico dei Terremoti Etnei, 1600–2013. INGV Catania. http://www.ct.ingv.it/macro/

Crowley H, Pinho R (2004) Period height relationship for existing European reinforced concrete buildings. J Earth Eng 8:93–119

D'Amico V, Albarello D (2008) SASHA: a computer program to assess seismic hazard from intensity data. Seismol Res Lett 79(5):663–671

D'Amico S, Meroni F, Sousa ML, Zonno G (2016) Building vulnerability and seismic risk analysis in the urban area of Mt. Etna volcano (Italy). Bull Earthq Eng 14(7):2031–2045

Grünthal G (ed) (1998) European macroseismic scale 1998 (EMS-98). European Seismological Commission, subcommission on Engineering Seismology, working Group Macroseismic Scales. Conseil de l'Europe, Cahiers du Centre Européen de Géodynamique et de Séismologie 15, Luxembourg, p 99. http://www.ecgs.lu/cahiers-bleus/

ISTAT (1991) 13° censimento generale della popolazione, 1991. Dati sulle caratteristiche strutturale della popolazione e delle abitazioni, Roma

Iwan WD (1997) Drift spectrum: measure of demand for earthquake ground motions. J Struct Eng 123:397–404

Lagomarsino S, Giovinazzi S (2006) Macroseismic and mechanical models for the vulnerability and damage assessment of current buildings. Bull Earthq Eng 4:415–443

Langer H, Tusa G, Scarfì L, Azzaro R (2015) Ground-motion scenarios on Mt. Etna inferred from empirical relations and synthetic simulations. Bull Earthq Eng 14(7):1917–1943. https://doi.org/10.1007/s10518-015-9823-1

Meroni F, Petrini V, Zonno G (1999) Valutazione della vulnerabilità di edifici su aree estese tramite dati ISTAT, Atti 9° Convegno Nazionale ANIDIS: L'ingegneria Sismica in Italia, Torino (CD-ROM)

Meroni F, Petrini V, Zonno G (2000) Distribuzione nazionale della vulnerabilità media comunale. In: Bernardini A (ed) La vulnerabilità degli edifici. CNR-GNDT, Roma, pp 105–131

Meroni F, Zonno G, Azzaro R, D'Amico S, Tuvè T, Oliveira CS, Ferreira MA, Mota de Sá F, Brambilla C, Rotondi R, Varini E (2016) The role of the urban system dysfunction in the assessment of seismic risk in the Mt. Etna area (Italy). Bull Earthq Eng 14(7):1979–2008. https://doi.org/10.1007/s10518-015-9780-8

Miranda E, Akkar S (2006) Generalized interstory drift spectrum. J Struct Eng 132:840–852. https://doi.org/10.1061/ASCE0733-94452006132:6840

Rotondi R, Varini E, Brambilla C (2016) Probabilistic modelling of macroseismic attenuation and forecast of damage scenarios. Bull Earthq Eng 14(7):1777–1796. https://doi.org/10.1007/s10518-015-9781-7

Scarfì L, Langer H, Garcia-Fernandez M, Jimenez JM (2016) Path effects and local elastic site amplification: two case studies on Mt Etna (Italy) and Vega Baja (SE Spain). Bull Earthq Eng 14:2117. https://doi.org/10.1007/s10518-016-9883-x

Sousa ML (2006) Risco Sísmico em Portugal Continental. PhD thesis, IST, UTL, Lisbon, Portugal

Sousa ML (2008) Annualized economic and human earthquake losses for the Portuguese mainland. In: Proceedings of the 14EWEE, China

Tusa G, Langer H (2016) Prediction of ground-motion parameters for the volcanic area of Mount Etna. J Seismol 20:1–42. https://doi.org/10.1007/s10950-015-9508-x

Chapter 22
Seismic Performance of Non-structural Elements Assessed Through Shake Table Tests: The KnowRISK Room Set-Up

Paulo Candeias, Marta Vicente, Rajesh Rupakhety, Mário Lopes, Mónica Amaral Ferreira, and Carlos Sousa Oliveira

Abstract In the scope of the KnowRISK research project a set of tests were carried out on the LNEC-3D shake table in order to assess the seismic performance of several non-structural elements and building contents. The aim was to create a room that was as realistic as possible, not only in terms of spatial arrangement but also in terms of furniture and decorative objects. This would allow the presence of daily-life objects and furniture that can represent hazard inside regular homes during an earthquake. Damages were observed with increasing intensity seismic motions. In some of the tests different non-structural protective measures were implemented in order to observe how they mitigate the damage. Videos of the entire set-up were recorded during the tests, for the KnowRISK interventions, and accelerations were measured in a wardrobe and a bookcase. This allowed to obtain qualitative as well as quantitative data about the objects' seismic performance.

Keywords Non-structural elements · Shake table tests · Seismic performance · Protective measures

---

P. Candeias (✉) · M. Vicente
Laboratório Nacional de Engenharia Civil, Lisbon, Portugal
e-mail: pcandeias@lnec.pt

R. Rupakhety
Faculty of Civil and Environmental Engineering, School of Engineering and Natural Sciences, University of Iceland, Reykjavík, Iceland

Director of Research, Earthquake Engineering Research Centre (EERC), University of Iceland, Selfoss, Iceland

M. Lopes · M. A. Ferreira · C. S. Oliveira
Instituto Superior Técnico, Lisbon, Portugal

Department of Civil Engineering, Architecture and Georesources, CEris, Lisbon, Portugal

22.1 Introduction

In the event of an earthquake, losses, both human and material, are expected to a certain degree. The mitigation of the risks associated to non-structural elements and building contents presents a challenge as limited information is available about their seismic behaviour and implementation of protective measures to the public in general.

In the scope of the KnowRISK project (Know your city, Reduce seISmic risK through non-structural elements), a series of shake table tests were carried out in order to gain insight into the seismic performance of non-structural elements and building contents. A set-up was assembled on the LNEC-3D shake table to simulate a bedroom, with furniture and several decorative elements easily identifiable by the selected target-group, to collect images and produce videos to use later in the KnowRISK intervention with the target-groups.

Real earthquake motions, selected from the Iceland earthquake database, were used in the tests. Two signals, from two different seismic events and recorded at a floor level in two different buildings, a fourteen storey RC office building located in Reykjavik and the two storey RC Selfoss Town Hall building, were selected to represent different response spectra. Increasing intensity seismic motions, scaled up from the original ones, were imposed to assess the seismic performance of the objects placed in the KnowRISK room set-up.

22.2 The KnowRISK Room Set-Up

In order to meet the objectives that were set by the KnowRISK project, it was decided to simulate a real scale bedroom. The assembled model would represent a teenagers' room since that met the age range of the target-group of the KnowRISK intervention. Testing a teenagers' room would make it possible to produce support material to be used during the risk communication actions. Even if teenagers are not the main decision-makers in their household it is important to point out that they can influence their parents and, if not, they can hold to the non-structural protective measures tested and adopt them later on their adulthood.

The aim was to create a room that was as realistic as possible, not only in terms of spatial arrangement but also regarding furniture and chosen decorative objects. This would allow the presence of daily-life objects and furniture that can represent hazard inside regular homes during an earthquake.

The LNEC-3D shake table has a platform with dimensions of 4.60 m by 5.60 m, with an approximate height clearance of 8 m. These characteristics allowed to assemble a real scale model for the shake table tests. In order to allow better image capture only two walls were built, forming a corner on the south-west side of the shake table, a floor and an upper beam element to support a lamp (Fig. 22.1). The motions were imposed only in the East-West direction of the shake table.

The workable room area of approximately 16 square meters allowed to create several recreational spaces. Inside this room the following spaces were created:

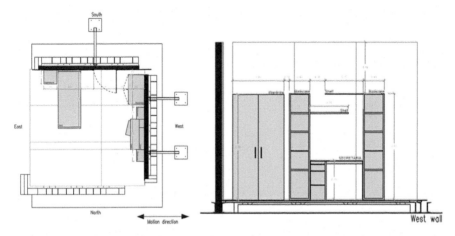

Fig. 22.1 The KnowRISK room set-up: Plan view (left) and West wall view (right)

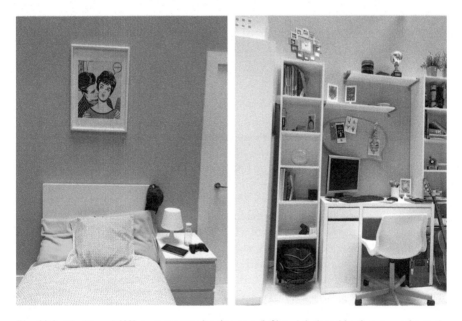

Fig. 22.2 The KnowRISK room set-up: sleeping area (left) and desk and bookcases on the study area (right)

(i) sleeping and dressing, formed by a single sized bed, a small nightstand and a wardrobe, standing across the bed; (ii) study area, where a desk, chair, two tall bookcases and some shelves, fixed to the wall, were placed; and (iii) play and relax area, occupying the central space with a carpet, a small bench and pillows.

For the decoration of the space several elements were used (Fig. 22.2). The focus was to have objects that one can quickly associate to a twenty-first century teenager. The shelves were filled with books, magazines, pictures, notebooks and some old

Fig. 22.3 The KnowRISK room set-up completed, before the shake table tests

toys wondering around. Movie posters and pictures were placed on the wall in front of the desk. On the desk, a computer and some pens and pencils. Higher up, a sports related trophy was placed on the top book shelf.

During the design process of the KnowRISK room, efforts were made to recreate critical situations in terms of hazard. The idea was to create a space that, due to its layout, could have tall and heavy furniture falling down and blocking an exit or even tall bookcases that could fall on top of a bed, in case an earthquake occurs. In order to strengthen this idea, a doorway was added to the South wall with the tall and heavy wardrobe close by on the West side (Fig. 22.3).

Everything regarding the KnowRISK room set-up was arranged to be as close as possible to a real room, as portrayed in Fig. 22.3, mainly due to the need of having all the furniture and objects that can cause injuries and material losses during an earthquake. Ultimately, this real looking space would allow the film crew to capture images of a type of room that the KnowRISK target-group could relate to.

22.3 Selection of Seismic Motions

Given the purpose of simulating a bedroom with furniture inside a housing building, the shake table input motions were recorded at a floor level and not on the ground level. Two particular locations were selected in Iceland, one in Reykjavik, a fourteen

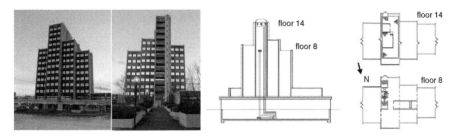

Fig. 22.4 View of the Reykjavik fourteen storey RC office building (left) and instrumentation plan (right)

Fig. 22.5 View of the Selfoss Town Hall two storey RC building (left) and instrumentation plan (right) on the ground floor (in blue) and 2nd floor (in red)

storey RC office building (Fig. 22.4), and another one in Selfoss, the Town Hall two storey RC building (Fig. 22.5). The site condition of the Reykjavik building is rock. The geology at the Selfoss building site is more complicated, basically lava rock on the surface, potentially with some soft layers underneath the lava. Without more detailed information, it can be assumed, for the current purpose, as a rock site.

Table 22.1 shows some information about the earthquake records from the Iceland database (Ambraseys et al. 2001) that were analysed. They were selected based on various aspects such as the magnitude of the event, the signal response spectrum and the knowledge about the existence of non-structural damage caused by them.

The records from the 2000-06-17 and 2000-06-21 events were used to perform a simplified dynamic identification of both buildings and obtain their Frequency Response Functions (FRFs) from the input-output relations. The results, summarised in Tables 22.2 and 22.3, show three distinct vibration frequencies for the Reykjavik fourteen storey RC office building between 1.5 Hz and 3.9 Hz and six vibration frequencies for the Selfoss Town Hall two storey RC building in the range from 6.3 Hz to 10.65 Hz. Furthermore, they show that the Reykjavik building is considerably more flexible than the Selfoss building, thus affecting the frequency content of the floor motions.

Table 22.1 List of records analysed from the Iceland earthquake database

Earthquake event	Record duration (s)	Building location	Building level	Sensor location	Sensor direction
2000-06-17	119.04	Reykjavik	8th floor	Middle floor	EW, NS
2000-06-17	119.04	Reykjavik	14th floor	Middle floor	NS
2000-06-17	119.04	Reykjavik	14th floor	North and South sides	EW
2000-06-21	101.12	Reykjavik	8th floor	Middle floor	EW, NS
2000-06-21	101.12	Reykjavik	14th floor	Middle floor	NS
2000-06-21	101.12	Reykjavik	14th floor	North and South sides	EW
2000-06-17	70.00	Selfoss	2nd floor	East and West sides	NS
2000-06-17	70.00	Selfoss	2nd floor	Middle floor	EW
2000-06-21	58.00	Selfoss	2nd floor	East and West sides	NS
2000-06-21	58.00	Selfoss	2nd floor	Middle floor	EW
2008-05-29	217.00	Selfoss	2nd floor	East and West sides	NS
2008-05-29	217.00	Selfoss	2nd floor	Middle floor	EW

Table 22.2 Reykjavik fourteen storey RC office building main vibration frequencies

Mode number	Frequency (Hz)	Direction
1	1.56	EW
2	2.20	NS
3	3.86	EW

Table 22.3 Selfoss Town Hall two storey RC building main vibration frequencies

Mode number	Frequency (Hz)	Direction
1	6.30	NS
2	7.76	Torsion?
3	7.86	EW
4	8.15	EW
5	9.77	NS
6	10.64	EW

22.4 Shake Table Tests Sequence

For the shake table tests the following two signals were selected from the ones presented previously: (i) the Mw6.5 2000-06-17 event, Reykjavik building (located 78 km from the epicentre), 8th floor, EW direction; and (ii) the Mw6.3 2008-05-29 event, Selfoss building (located 8 km from the epicentre), 2nd floor East, NS direction. The actual records were pre-processed prior to tuning them for the LNEC-3D shake table in order to obtain compatible displacement and acceleration time histories with a duration of 60 s while retaining the original response spectra, as

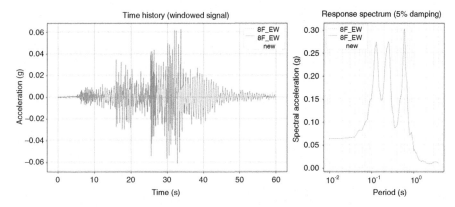

Fig. 22.6 Original signal (60 s window) from the Mw6.5 2000-06-17 event, Reykjavik building (located 8 km from the epicentre), 8th floor, EW direction (8F_EW), and signal pre-processed for the shake table (8F_EW new)

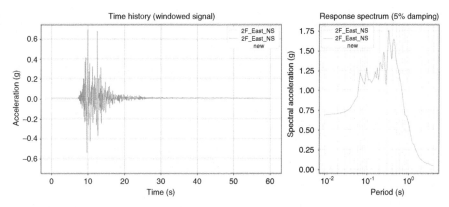

Fig. 22.7 Original signal (60 s window) from the Mw6.3 2008-05-29 event, Selfoss building (located 8 km from the epicentre), 2nd floor East, NS direction (2F_East_NS), and signal pre-processed for the shake table (2F_East_NS new)

shown in Figs. 22.6 and 22.7. These signals were then imposed to the LNEC-3D shake table only in the East-West direction (see Fig. 22.1) with increasing scale factors.

A total of ten tests were carried out (Table 22.4), the first four without any protective measures, to assess the seismic performance of the objects placed in the KnowRISK room set-up for increasing motion intensities, and the next four with several protective measures. The last two tests were carried out without protective measures in order to assess again the seismic performance of the furniture elements, now to a different input motion on the shake table.

The main characteristics of the selected earthquake records (tests number 1 and 9) are presented in Table 22.5 together with the corresponding ground motion parameters processed in the same way as described above.

Table 22.4 Shake table tests sequence

Test number	Signal	Scale factor	Protective measures
1	2000-06-17 event, Reykjavik building, 8th floor, EW direction	1	No
2	2000-06-17 event, Reykjavik building, 8th floor, EW direction	2	No
3	2000-06-17 event, Reykjavik building, 8th floor, EW direction	3	No
4	2000-06-17 event, Reykjavik building, 8th floor, EW direction	10	No
5	2000-06-17 event, Reykjavik building, 8th floor, EW direction	10	Yes
6	2000-06-17 event, Reykjavik building, 8th floor, EW direction	10	Yes
7	2000-06-17 event, Reykjavik building, 8th floor, EW direction	10	Yes
8	2000-06-17 event, Reykjavik building, 8th floor, EW direction	10	Yes
9	2008-05-29 event, Selfoss building, 2nd floor East, NS direction	1.0	No
10	2008-05-29 event, Selfoss building, 2nd floor East, NS direction	1.4	No

Table 22.5 Main characteristics of the selected earthquake records (floor and ground parameters)

Test number	Peak floor acceleration	Peak floor velocity	Peak floor displacement	Peak ground acceleration	Peak ground velocity	Peak ground displacement
1	0.06 g	5.0 cm/s	18 mm	0.04 g	2.8 cm/s	17 mm
9	0.68 g	60.0 cm/s	91 mm	0.54 g	50.8 cm/s	88 mm

22.5 Shake Table Tests Results

22.5.1 Observed Seismic Behaviour and Non-structural Protective Measures

The seismic behaviour of the various non-structural elements and building contents changed qualitatively between different tests, as the protective measures were being applied. A brief description of the observed seismic behaviour of the KnowRISK room set-up is presented next.

In Test 3, with the earthquake record already scaled up by a factor of 3 and no protective measures, it was seen that: (i) Bookcases start to rock impacting against the wall, especially bookcase A (the one standing between the wardrobe and the desk); (ii) Some objects toppled, especially those on top shelves of the bookcases (i.e. watch standing on top of bookcase A).

Fig. 22.8 Damages observed in the KnowRISK room set-up at the end of Test 4

In Test 4 (Fig. 22.8), when the earthquake record was scaled up by a factor of 10, and still no protective measures in place, the following damages were seen: (i) Objects fall from higher shelves; (ii) Bookcases start to rock impacting against the wall; (iii) Drawers under the bed and on the desk start to move, opening and closing repeatedly; (iv) Bed shifts slightly from one side to the other (in the end of the test the bed had shifted approximately 10 cm to one side); (v) Computer screen falls down; (vi) Bookcases fall, the first to topple was bookcase B and shortly after bookcase A; and (vii) Wardrobe doors open and close repeatedly.

Based on the damages observed in this last test, non-structural protective measures were implemented in order to prevent human and material losses as a consequence of the falling of non-structural elements. A few examples of the protective measures implemented include the anchoring of the tallest furniture elements to the wall, to prevent them from toppling, and various other measures like the use of non-slip mats or drawers with child-protective latches (Figure 22.9) (O'Neill et al. 2017).

The experiment was used to test certain protective solutions, however not all of those solutions worked as intended. Examples of the protective solutions adopted and their performance are described below:

(i) For bed wheel-drawers, as protective measure hook and loop tape was used. Additional content was added to provide the necessary mass of a typical drawer

(approx. 2 kg). Bed drawers start to move, opening and closing, despite the adoption of protective measures, which confirm that this type of tape should be used only with lightweight items. Drawers should have latches and if possible the wheels removed.
(ii) Desk drawers start to open and close repeatedly (no protection measures were implemented in this case).
(iii) To prevent desktop computer from falling, child-protective latches (similar to Fig. 22.9) were used to attach the equipment to the desk. The desktop computer passed the first shake table test, but fell in the following tests due to the presence of some water in the desk (the fish bowl falls) and the provided self-adhesive stopped being sticky after several attempts. Using chains or adjustable nylon straps attached to the back of the desktop would be the best solution to secure desktop computers. It is extremely important to routinely check any protective measures you have already taken to see if they are still effective.
(iv) Books placed higher in the bookcases fall (no protective measures were implemented in this case). A protective solution can be to install a thin metal or plastic wire, a wood dowel, or an elastic guardrail across the front of each shelf (FEMA E-74 2012).

The above discussion shows that further research is needed to produce solutions aiming at preventing furniture and appliances from falling. However, an important qualitative reduction in the overall damage was observed, since the toppling of the tall bookcases was prevented, even for a scale factor of 10 of the original earthquake record.

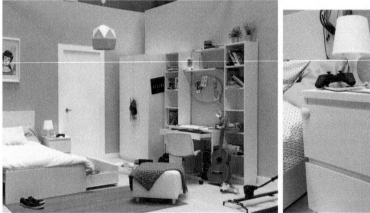

Fig. 22.9 Example of damages despite the implementation of non-structural protective measures like anchoring of tall furniture elements to the walls (left) and fastening of drawers (right), at the end of Test 5

22.5.2 Seismic Performance of the Non-structural Elements

In what concerns the data processing of these tests, we will be looking at the seismic performance of the furniture elements that were instrumented with accelerometers, the wardrobe (identified as ROUPEIRO in the figures below) and the bookcase B (identified as ESTANTE in the figures below). Comparisons will be performed with the motions measured in the shake table and, assuming that some interaction might occur, also in the wall against which they were placed. In the former case the sensors placed on the shake table will be used as reference (identified in the figures below as ST_POS1T for displacement and ST_ACC1T for acceleration) whereas in the latter case the accelerations measured in the sensors closer to each of the furniture elements will be used (SW_T for wardrobe and NW_T for bookcase B).

The focus will be on the results of Test 4, when the furniture was still unanchored, and Test 5, the first with the furniture anchored. As mentioned before, the use of the protective measures changes the seismic behaviour of the non-structural elements and building contents, both in qualitative and quantitative terms.

The response of the wardrobe in tests 4 and 5 are shown in Figs. 22.10 and 22.11, respectively, together with the shake table input. The data acquisition system trigger was not synchronised with the shake table control trigger, reason why the signals start at different time instants. The accelerations measured in the wardrobe are fairly symmetric in Test 4, with a peak value of around 2.3 g, and indicate a series of medium intensity shocks. In Test 5, the response of the wardrobe is again almost symmetric in terms of acceleration but now the peak value is only around 1.1 g, almost half of the previous peak value. This is a clear reduction that is obtained by just anchoring the wardrobe to the wall.

Moreover, looking at the acceleration response spectrum (Fig. 22.12), it is visible that the response of the wardrobe in Test 4 is closer to the shake table input and wall spectrum, for periods above 0.5 s. On the contrary, in Test 5 the wardrobe response

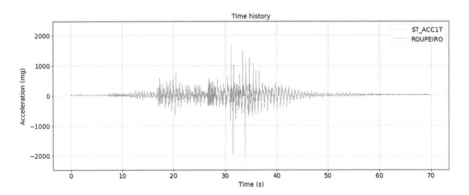

Fig. 22.10 Shake table input (ST_ACC1T) and wardrobe response (ROUPEIRO) in Test 4

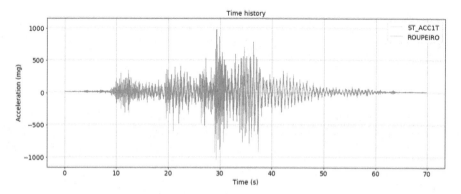

Fig. 22.11 Shake table input (ST_ACC1T) and wardrobe response (ROUPEIRO) in Test 5

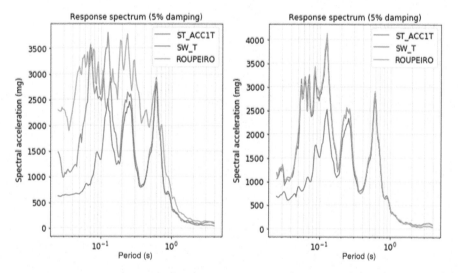

Fig. 22.12 Response spectra in the shake table, West wall and wardrobe in Test 4 (left) and Test 5 (right)

spectrum is very similar to the wall spectrum throughout the entire spectrum, including in the low period range, indicating that the anchors used were effective in connecting the wardrobe to the wall.

The response of bookcase B is clearly marked by a series of intense shocks until it topples in Test 4, for a scale factor of 10 relative to the original earthquake time history (Fig. 22.13). A joint analysis of the bookcase response with the shake table displacement, velocity and acceleration time histories, not illustrated here, shows that the toppling does not occur simultaneously with the peak of neither of these quantities. After being anchored to the wall, in Test 5, the accelerations measured in bookcase B are nearly symmetric and much lower than in the unanchored case, with a peak value around 1.5 g (Fig. 22.14).

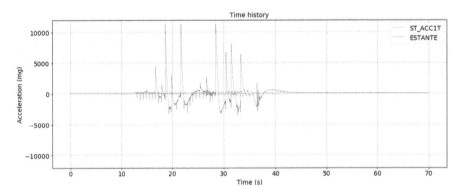

Fig. 22.13 Shake table input (ST_ACC1T) and bookcase B response (ESTANTE) in Test 4

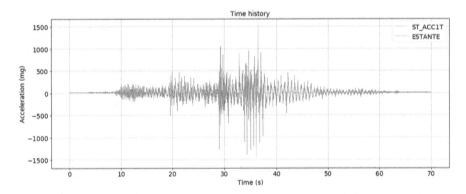

Fig. 22.14 Shake table input (ST_ACC1T) and bookcase B response (ESTANTE) in Test 5

In what regards the acceleration response spectra (Fig. 22.15), it is not possible to elaborate on the response of the bookcase B in Test 4 because it is largely affected by the toppling that occurred in that test. In relation to Test 5, it is visible that the bookcase B presents a response spectrum that is close to both the spectra from the shake table and the wall for periods above 0.2 s, where these spectra are similar. However, below this period value, where the shake table and wall spectra show relevant differences, the spectrum from the bookcase is much closer to the wall spectrum. This is naturally a consequence of anchoring the bookcase B to the wall.

22.6 Conclusions

In the scope of the KnowRISK project a series of shake table tests were carried out with a real scale room set-up, fulfilling two main objectives. First, to produce material for the KnowRISK awareness interventions related with hazard and

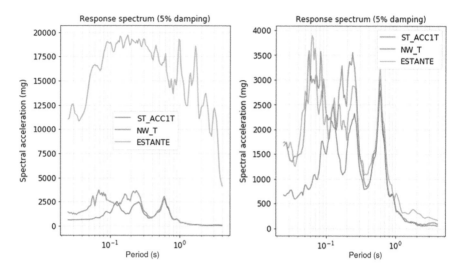

Fig. 22.15 Response spectra in the shake table, West wall and bookcase B in Test 4 (left) and Test 5 (right)

non-structural protective measures. Second, to assess the seismic performance of a set of non-structural elements and building components.

The room set-up, designed to simulate a typical teenager bedroom, included various easily recognisable elements. Moreover, the layout was designed in order to recreate various hazardous situations such as tall furniture and stocking of heavy objects in the top shelves.

The tests showed that the use of anchoring systems to connect tall furniture elements to the walls is effective in reducing losses, by preventing toppling of the furniture. Further studies of this kind are, however, required to keep on improving the protective measures for other types of elements.

Additionally, the results obtained from the accelerometers installed in the wardrobe and in a bookcase showed the differences in their seismic response before and after the implementation of protective measures, namely the anchoring to the walls. The large shocks associated to a free standing object near a wall are eliminated, and the response becomes associated with that of the wall used for support, with smaller peak accelerations.

Acknowledgements This project was financed by the European Commission's Humanitarian Aid and Civil Protection (Grant agreement ECHO.A.5 (2015) 3916812) – KnowRISK (Know your city, Reduce seISmic risK through non-structural elements). More information about the KnowRISK project is available at the internet address https://knowriskproject.com/.

References

Ambraseys N, Sigbjornsson R, Suhadolc P, Margaris B (2001). The European strong motion data base. http://www.isesd.hi.is/ESD_Local/frameset.htm

FEMA E-74 (2012). Reducing the risks of nonstructural earthquake damage: a practical guide, Fourth Edition. Federal Emergency Management Agency

O'Neill H, Ferreira MA, Oliveira CS, Lopes M, Solarino S, Musacchio G, Candeias P, Vicente M, Silva DS (2017). KnowRISK practical guide for non-structural earthquake risk reduction. In: Proceedings of international conference on earthquake engineering and structural dynamics, Reykjavik, Iceland

Chapter 23
KnowRISK Practical Guide for Mitigation of Seismic Risk Due to Non-structural Components

Hugo O'Neill, Mónica Amaral Ferreira, Carlos Sousa Oliveira, Mário Lopes, Stefano Solarino, Gemma Musacchio, Paulo Candeias, Marta Vicente, and Delta Sousa Silva

Abstract Good performance of non-structural elements can be decisive in saving lives and costs when an earthquake strikes. The European project KnowRISK aims to educate and encourage households to take the necessary precautionary measures to protect people, houses, and contents. Preparedness and prevention act on community resilience. Within the KnowRISK project, the idea of a Practical Guide has been conceived suggesting seismic mitigation solutions for non-structural components to non-experts stakeholders. It is intended to guide people into the first steps of prevention in a straightforward manner, minimizing or avoiding injuries, damage, and long-term financial consequences. The novelty of the Guide belongs to his philosophy: a path through increasing challenges corresponds to a growing level of safety. The idea is that anyone can mitigate seismic risk in its own environment by adopting simple and low cost measures. The Practical Guide may contribute to increase risk awareness. This kind of initiatives if undertaken at larger scales may also enhance social resilience.

H. O'Neill (✉)
Instituto Superior Técnico, Lisbon, Portugal

M. A. Ferreira · C. S. Oliveira · M. Lopes
Instituto Superior Técnico, Department of Civil Engineering, Architecture and Georesources, CERis, Lisbon, Portugal
e-mail: monicaf@civil.ist.utl.pt

S. Solarino
Istituto Nazionale di Geofisica e Vulcanologia, Centro Nazionale Terremoti, Roma, Italy
e-mail: stefano.solarino@ingv.it

G. Musacchio
Istituto Nazionale di Geofisica e Vulcanologia, Roma, Italy

P. Candeias · M. Vicente · D. S. Silva
Laboratório Nacional de Engenharia Civil, Lisbon, Portugal
e-mail: pcandeias@lnec.pt

Keywords Non-structural elements · Preparedness · Community resilience · Practical guide

23.1 Introduction

Traditionally, non-structural components are the architectural, mechanical and electrical components found in buildings, as well as exterior and interior elements that are not part of the structural system, which means they are not designed to transfer loads. Loss or failure of these elements can occur even when no structural failure is observed. This kind of damages can affect the safety of the occupants of the building and safety of others who are in its vicinity. Non-structural building elements may be easily retrofitted and, in most cases, at a low cost. Mitigation techniques can also be applied to reduce the damage.

However, the public is not adequately informed about the danger of non-structural failures and especially on how to fix non-structural components. To fill this knowledge gap, we designed a Practical Guide to be used as a reference by the public to locate, fix and avoid non-structural damages. This Practical Guide is being prepared based on the outputs of Task C (Non-structural seismic risk reduction) of the KnowRISK (2016) project, and takes into account the local culture and needs of each of the participating countries: Portugal, Italy and Iceland. Previous works developed by FEMA (FEMA 2012), Petal (2003) and Quake Safe Guide (Earthquake Commission 2012) were used in the first versions of this Practical Guide.

The Practical Guide addresses essentially non-structural issues found in our homes, whose solutions can be executed by the occupants themselves, with minimal skills at almost no cost. This is expected to encourage immediate public engagement into preparedness, by providing mitigation measures that can be carried out by anyone at anytime. However, more demanding retrofitting are also proposed: for these, seeking the help of a professional is recommended referring to a Portfolio, prepared in Task C.4 of the KnowRISK project, where suggestions that are more technical are collected and published.

The Practical Guide is intended to be a KnowRISK product for laypersons, standard for the three participant countries, and it uses images and symbols extensively, rather than text that would need to be translated into the different languages.

Because the public might non be favourable in undertaking the proactive actions suggested in the Guide, care has been devoted to find effective and persuasive message that are a good compromise between what they want to hear and what we want them to do. A greater safety of the living place is the main goal of the KnowRISK project, but this is achieved only if people are acquainted with the potential dangers of their living place. The learning is expected to be in steps, where the reader explores the guide several times before acting: first to realize how unsafe the own environment is, and then to follow the suggestions contained in the guide. The Practical Guide must then be handy and appealing for frequent use.

23.2 Layout and Contents

The main concern of the Practical Guide is to portrait common non-structural safety hazards in the home, as well as providing their solutions (Fig. 23.1). The chosen media format is a foldable brochure, an always-at-hand guide that can be distributed in strategic communication actions to reach the widest audience.

The solutions have been grouped into categories representing a four steps path, from passive interest to active engagement, and accomplish to a wide range of needs the public might have. These needs have been a matter of study by Task D (Going into target-communities) of the KnowRISK (2016) project and are in part discussed by Musacchio et al. (this volume). Each step has an associated colour that highlights to the level of risk and urgency: Red, Orange, Yellow and Green, in order of decreasing risk and urgency. These are the colours often used for weather warnings and resemble those of the traffic light; the public is already familiar with them and can intuitively estimate what level of risk is associated.

The display of the practical solutions occupies the central focus of the foldable guide. It is adaptable to different degrees of motivation, from passive interest to active engagement. A polyptych unfolds in the length of eight panels featuring a detailed description of the problems and solutions using symbols and images for various specifications. The four main categories, described in the next chapter, of the guide can be summarized in:

- Eminent danger,
- Material losses,
- Secure anchoring,
- Investment enhancements.

Fig. 23.1 Polyptych of the Practical Guide. The reader can follow a process to increase own safety from very simple actions, that may not increase significantly security, to more demanding measures. The "status" of safety is marked by the colors of the traffic light

The main target of the guide is to help the reader to locate vulnerable non-structural elements, and to find information on how to mitigate the potential risk of damage. However also additional support and institutional information are provided. In particular, the reader can find

- Initial front page,
- A brief introduction to the Practical Guide,
- An explanation of non-structural elements,
- A map of Hazard for each participating country,
- Space for complementary information,
- Acknowledgement and 'disclaimer'.

This part of the Practical Guide is intended to give a general view of the hazard of the place where the user lives, a general description of what to look for, useful information on how to know more about non-structural damages.

23.3 Practical Solutions and Categories

The Practical Guide helps in identifying viable non-structural loss mitigation techniques. It suggests measures that can be taken, not only at home but also in classrooms, offices, and public spaces. These suggestions are a collection of protective measures identified during the project or taken from guides and reports about earthquake safety worldwide. It also lists solutions specifically suitable for the local culture and needs of each of the participating countries.

Most of these solutions have been designed to make use of materials available at local hardware stores (Fig. 23.2), and tested in the shaking table at LNEC – Laboratório Nacional de Engenharia Civil, in Lisbon (Fig. 23.3). Details of the testing can be found in Candeias et al. (2017).

DIY (do it yourself) solutions with low cost are given preference, for example, moving heavy or large items to the floor or lower levels of shelves, properly hanging mirrors and pictures, installing latches on kitchen cabinets, etc. A detailed description of the problems and solutions is depicted by symbols and images.

One section of the instructions – What you need – is dedicated to materials and tools required, with a classification for the expected price and expertise.

Starting with immediate survival concerns, the solutions evolve from quick-fixing with zero cost (such as rearrange furniture and contents) and progressively increase in investment of time and financial expense, as well as responsibility, proactivity and knowledge. The solutions at the end of the categories path should reach the highest expertise and cost required, at which point an alert is given to consult an accredited professional and look for further detailed guidance in the Portfolio of solutions, from the KnowRISK project.

Fig. 23.2 Example of protective solutions tested on the shake table at the LNEC

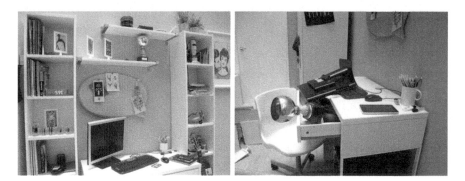

Fig. 23.3 Left: prior the shaking table test with unsecured objects. Right: after the shaking table test

23.3.1 Category 1: Eminent Danger

This section contains the most basic and immediate issues that might interest even those less inclined to this learn about non-structural risk and mitigation possibilities. Simple solutions without spending money, such as moving objects to a better position to avoid injuries or block exits, are proposed (Fig. 23.4).

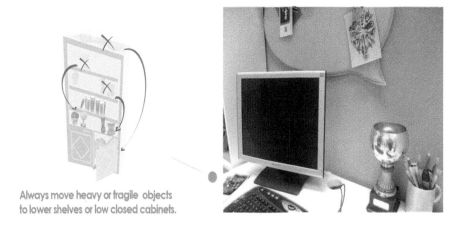

Fig. 23.4 Move heavy objects (like trophies) from higher to lower levels of shelves

23.3.2 Category 2: Material Losses

The next recommendation is to secure things that can cause injury, things that are of monetary and/or emotional value, and things which when damaged can impair continuity of educational/business activity. The goal is to prevent objects from falling or flying and enhance safety. Measures such as placing adhesive pads on valuable china, using non-slip mats, strapping screens are good and low cost solutions, for which the required material are available in several hardware stores (Figs. 23.5, 23.6 and 23.7).

23.3.3 Category 3: Secure Anchoring

This section includes measures that can increase the level of safety by taking additional measures, which require more investment and time. Fastening fans, chandeliers, flowerpots, frames, mirrors; securing tall furniture to the wall to avoid toppling (Figs. 23.8 and 23.9); protecting shelves with wires; using glazing film or close drapes on windows are some of the measures recommended.

23.3.4 Category 4: Interventions that Require a Professional

This category includes long-term investment solutions, with high cost and requires technical intervention. Replacing windows with safety glass (tempered), reinforcing

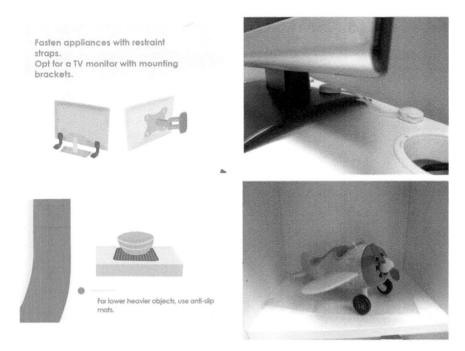

Fig. 23.5 Top: Example of flat screen safety strap. During the shake table tests the provided self-adhesive stopped being sticky after several attempts. Bottom: non-slip rubber mat to prevent objects from falling or flying

Fig. 23.6 Non-slip rubber mats can prevent movement of objects, but these should be used in dry surfaces. During the shake table tests, the fish bowl fell down due to the presence of water. Use of double-sided adhesive take is advised to secure fish bowls

chimneys and balconies, securing satellite dishes and solar power plates, improving gas pipelines and wiring, securing bunkbeds together (top & bottom) and anchoring them to a wall are some examples.

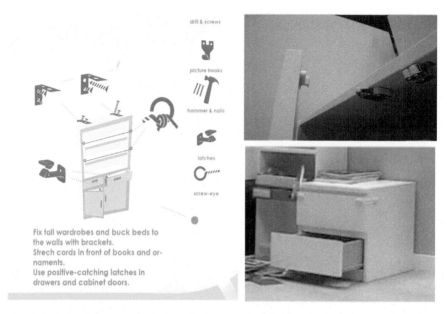

Fig. 23.7 Installation of latches on cabinet doors and drawers to keep them closed during an earthquake. The bottom picture shows the condition of drawers after the shaking table test

Fig. 23.8 Top-heavy furniture should be secured to a wall

Fig. 23.9 Condition of unsecured (left) and secured (right) after shaking table tests

23.4 Additional Information

The following information is also given:

23.4.1 Initial Front Page

It has a cover image that raises awareness to the connection between earthquakes and the contents of houses. (Fig. 23.10).

23.4.2 Introduction

This section includes a hazard map for each country (Fig. 23.11), designed specifically for the public, to help people become more aware of earthquake hazards across the country and the need for protection measures It also contains a general overview and explanation of the guide, as sequence of progressive solutions.

Fig. 23.10 Front page image

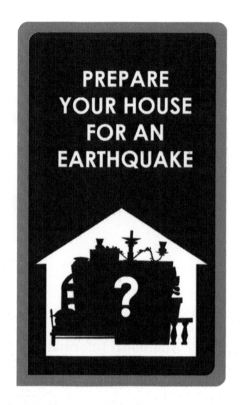

23.4.3 Acknowledgement and 'Disclaimer'

It has the logos of the institutions and the list of collaborators and sources.

The disclaimer warns the reader that the suggested solutions do not ensure full safety:

> This is just a Practical Guide. Individual situations may vary. Whenever necessary consult a specialist. You can find more information in the Portfolio.

Given the cultural differences of each country, that in turn reflect in different living styles and habitudes, the disclaimer is a necessary reminder.

Fig. 23.11 Non-structural elements and hazard map

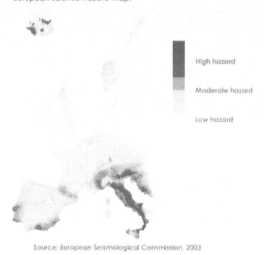

23.5 Discussion and Conclusions

Education is always a big challenge in that it demands to establish a relationship between the researcher and the public, to gain trust before conveying the message, to be convincing and exhaustive but not authoritative.

KnowRISK Practical Guide is supposed not only to increase the level of education and to raise awareness of the readers, like in a standard educative process, but also to encourage the reader to undertake safety actions. It has some pros and cons in these regards. The con is that the guide warns that we may be in danger. Depending on where we live and how our house is set, we may be exposed to non-structural damages. Knowing that we are at risk makes us feels uncomfortable and a sort of defence gate must be overpassed to gain the trust from the reader. The "human" solution in such situations is often to neglect the problem: for this reason the guide has to be a "friendly" list of suggestions more than frightening warnings.

The pros are that the readers do not have a "time frame" to adhere to. They can devote as much time as they like and can go through the guide several times without any stress to act immediately; can decide what level of safety is acceptable: and accordingly decide on the different levels of intervention.

Given this, it must be remarked that the KnowRISK Practical Guide is a unique attempt not only to encourage a culture of safety, but also provide suggestions on how to increase it. To our knowledge, it is the first time that such an approach, where guide suitable for the layperson, is carefully designed and prepared in Europe. Most of the suggestions contained in the guide have been tested on the shaking table tests carried out during this project, which increases our confidence in their usefulness. It is not expected to be applicable in all situations, but will certainly be useful in raising awareness of seismic risk due to non-structural components and making the public realize that mitigation measures are not as difficult and costly as perceived by many. This Practical Guide will contribute to risk awareness if not to safety, and we strongly believe that this kind of initiatives should be undertaken in larger scales to get enhance social resilience. The Practical Guide will be complemented by the Portfolio, which provides solutions that are economically feasible and easy to implement, as well as more complex solutions related with architectural or electrical/mechanical components.

Acknowledgements This study was financed by the European Commission's Humanitarian Aid and Civil Protection (Grant agreement ECHO.A.5 (2015) 3916812) – KnowRISK (Know your city, Reduce seISmic risK through non-structural elements).

References

Candeias P, Vicente M, Rupakhety R, Lopes M, Ferreira MA (2017) Testing non-structural elements in LNEC shaking table. In: Proceedings of the international conference on structural dynamics and earthquake engineering (ICESD) in honour of late Pprof. Ragnar Sigbjörnsson, 12–14 June 2017, Reykjavik, Iceland

Earthquake Commission (2012) Easy ways to quake safe your home. New Zealand. http://www.eqc.govt.nz/fixfasten/guide

FEMA (Federal Emergency Management Agency) (2012) Reducing the risks of nonstructural earthquake damage – a practical guide, 4th ed. FEMA-E74, Washington DC

KnowRISK (Know your city, Reduce seISmic risK through non-structural elements) (2016–2017) project. Co-financed by European Commission's Humanitarian Aid and Civil Protection Grant agreement ECHO/SUB/2015/718655/PREV28

Musacchio G, Ferreira MA, Falsaperla S, Piangiamore GL, Solarino S, Crescimbene M, Pino NA, Lopes M, Oliveira CS, Silva DS, Rupakhety R, the KnowRISK Team (this volume) Set up of a communication strategy on seismic risk: the Italian case of the KnowRISK project

Petal M (2003) NSM non-structural risk mitigation – handbook. Disaster Preparedness Education Project, Istanbul

Chapter 24
A Study of Rigid Blocks Rocking Against Rigid Walls

Gudmundur Örn Sigurdsson, Rajesh Rupakhety, and Símon Ólafsson

Abstract The safety of typical bookshelf is investigated using an inverted pendulum approach. A general bookcase is modelled as a rocking block and its response to damped harmonic excitation is simulated. The rocking of the block is considered as one sided since these household items generally stand parallel to a wall. This approach adds restraints to the rocking behaviour. The overturning spectra for the bookcase is presented for one-sided and two-sided rocking for a range of damped harmonic excitations, demonstrating the chaotic behaviour of the inverted pendulum model. Furthermore, results indicate one-sided rocking to be more prone to overturning, making it a more unstable system than its two-sided counterpart.

Keywords Rocking · Rigid block · Overturning · Inverted pendulum

24.1 Literature Review

Housner formulated the free rocking of rigid block as an inverted pendulum problem, and derived an analytical solution of the associated non-linear differential equation (Housner 1963) for some special cases. The solution was only valid when the slenderness of the block is small (slenderness ratio, $\alpha < 20°$). Solutions for three different types of base motion, namely a constant acceleration, a single sine pulse, and earthquake excitation ground motion were studied by Housner (1963). It was shown that increasing the slenderness of the block increases its probability of overturning by earthquake motions. Furthermore, it was shown that taller blocks were more stable to overturning, due to an unexpected scaling effect proportional to R (distance between the centre of mass and one of the edges), than shorter blocks with the same aspect ratio. In other words, the larger the distance R, the more stable the block becomes. This observation, although seemingly counter-intuitive, was

G. Ö. Sigurdsson (✉) · R. Rupakhety · S. Ólafsson
Earthquake Engineering Research Centre, University of Iceland, Reykjavík, Iceland
e-mail: gos12@hi.is; rajesh@hi.is; simon@hi.is

observed during the Chilean Earthquake of May 1960, when several elevated water tanks with an unstable appearance survived the event while similar reinforced concrete towers suffered severe damage.

Zhang and Makris (2001) studied the transient two-sided rocking response of a rigid block subjected to trigonometric pulses and near-source ground motions in detail. Toppling of smaller blocks were found to be more sensitive to Peak Ground Accleration (PGA) while that of larger blocks was more sensitive to incremental ground velocity (Makris and Black 2001). In other words, smaller blocks with a high frequency parameter ($p \approx 2$) were likely to topple due to short pulses, whereas larger blocks ($p \approx 1$) would topple due to long duration pulses. The overturning potential is therefore mostly dependant on duration and acceleration of a pulse. The overturning response was found to be ordered and predictable under these types of excitation.

A very thorough formulation of the analytical solution was presented and demonstrated by Zhang and Makris (2001). Analytical, as well as linear and non-linear numerical solutions were used and compared to investigate the overturning scenarios of rigid blocks subjected to cycloidal pulses. They found that the non-linear solution was very close to the linear one when computing an overturning spectrum for a rocking block, hence the linear solution would be a potentially faster algorithm providing the same results. It was further demonstrated that the friction needed to sustain rocking increased linearly with the acceleration of the pulse.

Makris and Konstantinidis (2001) examined the differences of the oscillatory response between a linearly elastic SDOF system and an inverted pendulum system (slender rigid block). They concluded that these were two fundamentally different dynamic systems and one should not be used to draw conclusions regarding the behaviour of the other.

Petrone et al. (2016) investigated the stability of a two-sided rocking of a medicine cabinet. They evaluated different intensity parameters as predictors of overturning. The results indicated that the overturning of small blocks ($R < 1m$) was *PGA* dominated, and that of larger blocks ($R > 2m$) were mostly dominated by Peak Ground Velocity (*PGV*). These results are consistent with the findings of Zhang and Makris (2001) regarding frequency parameters, which is expected since these parameters are directly related. The effect of rigid walls, against which medicine cabinets are generally placed, was not modelled by Petrone et al. (2016) but was suggested for future modelling.

24.1.1 One-Sided Rocking of Rigid Blocks

Unlike free-standing blocks, which can rock about both edges at their base, there are some situations where the rocking occurs about only one edge. One such scenario is when one of the corners at the base is anchored to the floor. Hogan (1992), building on previous work of Shaw and Rand (1989), investigated the stability of one-sided rocking where one corner was permanently anchored to the floor. This model is equivalent to an inverted pendulum hitting a rigid sidewall, and only considers positive rotations. In Hogan's model, both viscous damping and a coefficient of restitution provide energy dissipation.

Very few studies have addressed one-sided rocking problem. Historically, rocking of rigid blocks was studied in order to infer intensity of large earthquakes based on performance of free-standing blocks such as monuments and tombstones. A problem similar to the one studied by Hogan (1992) is the rocking of household objects, such as slender furniture placed against walls, during earthquakes. Other examples of such problems are medicine cabinets, data storage disk cabinets, etc., placed against walls. Unlike in the case of two-sided rocking, where the change of centre of rotation results in energy loss, modelled as restitution coefficient, one-sided rocking occurs about a single axis. The main source of energy dissipation in this case is due to impact of the rocking block against the wall. When the impact is not completely elastic and/or the wall is not perfectly rigid, some loss of energy at impact is expected. The loss of energy at impact is most likely related to impact velocity (see, for example, Jankowski 2007).

This study addresses the problem of one-sided rocking of household furniture placed against rigid walls, and the motivation derives from the activities of the EU-funded project KnowRISK (Know your city, Reduce seISmic risK through non-structural elements). Full-scale tests on rocking of furniture were performed on the shaking table operated by Laboratório Nacional De Engenharia Civil (LNEC), Portugal. This study presents numerical modelling of the same furniture, and presents some preliminary results.

24.2 Problem Formulation and Solution

The main subject of interest in this study is the common bookshelf, this household item usually holds numerous valuables that are costly to replace as well as pose a risk to the inhabitants of the living space. The goal is to estimate the risk of overturning given a specific type of shaking, namely far-field and near-field ground motions. In recent earthquakes in South Iceland, no serious harm was inflicted to the people in the area. However, considerable damage occurred due to sliding and overturning of objects such as cabinets, bookshelves, dressers, electric appliances and even store shelves.

The numerical model used here is the same as the inverted pendulum, except for a rigid barrier on one side of the block, ensuring that the rocking is strictly one-sided (Hogan 1992). The object being studied is the IKEA BRIMNESS bookshelf (see Table 24.1 for details), which was also tested on the shake table operated by LNEC.

Table 24.1 Geometry of BRIMNES bookcase

Dimension	Length (cm)
Height	190
Width	60
Depth	35

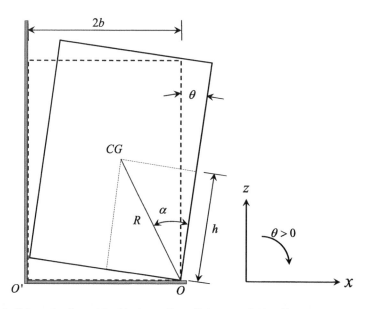

Fig. 24.1 Schematic of a rocking block standing adjacent to a rigid wall

The geometry of the block is as shown in Fig. 24.1. It is considered to oscillate about the centre of rotation O. It is assumed that friction is sufficiently large enough to prevent sliding motion. The height to the centre of gravity is h and half-width of the block is b. The distance from the centre of rotation to the centre of gravity is $R = \sqrt{h^2 + b^2}$, and the angle of slenderness is $\alpha = \arctan(b/h)$.

The second moment of inertia of the block is $I_O = (4/3)mR^2$ where m is the total mass of the block, which is assumed to be uniformly distributed and therefore with a centre of gravity at the geometric centre of the block. This model can then be simulated as a SDOF system where the only degree of freedom is the rotation. There are two forces/moments acting on the model, a restoring force, which is a gravity force acting at the centre of gravity, and a horizontal overturning force. The overturning force in this case is caused by the base motion due to earthquake excitation.

The rocking motion is described by the following equation:

$$I_O\ddot{\theta} + mgR\sin(\alpha - \theta) = -m\ddot{u}_{\ddot{E}_g}R\cos(\alpha - \theta), \quad \theta > 0 \quad (24.1)$$

where $\ddot{u}_{\ddot{E}_g}$ is the ground acceleration, g is acceleration due to gravity, and m is the total mass of the block. If the rocking were to be in the opposite direction, the following would represent that case:

$$I_O\ddot{\theta} + mgR\sin(-\alpha - \theta) = -m\ddot{u}_{\ddot{E}_g}R\cos(-\alpha - \theta), \quad \theta < 0 \quad (24.2)$$

$10# Combining Eqs. 24.1 and 24.2 yields

$$\ddot{\theta}(t) = -p^2 \left\{ \sin\left(\alpha\mathrm{sgn}[\theta(t)] - \theta(t)\right) + \frac{\ddot{u}_{Eg}}{g} \cos\left(\alpha\mathrm{sgn}[\theta(t)] - \theta(t)\right) \right\} \quad (24.3)$$

Where p is the frequency parameter (for all rectangular blocks) and is defined as:

$$p = \sqrt{\frac{3g}{4R}} \quad (24.4)$$

It is clear from this equation that only the length of the inverted pendulum plays a part here, whereas the slenderness α controls the force directions as well as point of impact.

The excitation is modelled as a damped sinewave where the amplitude and frequency can be varied and the damping ratio is taken to be 5% of critical.

$$\ddot{u}_{Eg}(t) = a_p \sin\left(\omega_{Ep} t + \psi\right) e^{-\xi\left(\omega_{Ep} t + \psi\right)} \quad (24.5)$$

where a_p is the wave amplitude, circular forcing frequency is ω_p, and damping ratio is ξ. The phase of the sine wave is fixed as

$$\psi = \sin^{-1}\left(\alpha g / a_p\right) \quad (24.6)$$

to ensure initiation of rocking (Housner 1963).

The impact is assumed to be perfectly elastic meaning that there is no energy loss. This implies that at the moment of impact, the kinetic energy is conserved and the rotation is reversed. In reality, however, some dissipation can be expected. This approach differs from the free-standing case, were the kinetic energy loss is a function of the block slenderness.

$$r = \left[1 - (3/2)\sin^2\alpha\right]^2 \quad (24.7)$$

This equation only applies for two-sided rocking. The energy loss described by this equation is caused by the centre of rotation being shifted instantly from one corner to the next, upon impact.

24.3 Modeling and Simulation

Since the equation of motion is a non-linear one, a numerical integration is required to solve the problem for an arbitrary excitation. An analytical solution can be used for a pulse type excitation as presented by Housner (1963), by assuming a small slenderness α. The differential equation is solved numerically in MATLAB by using the Newmark-beta algorithm (Newmark and Rosenbluth 1971). The algorithm is set up to solve Eq. 24.3 and has a stopping condition which detects when the block hits the ground/wall and reverses the velocity at the next time step. The performance and

Table 24.2 Dynamic parameters of the BRIMNES shelf

Parameter	Value
Slenderness (α)	0.1822 (−)
Radius (R)	96.6 (cm)
Frequency (p)	2.7598 (rad/s)
r	−1

accuracy was successfully verified against the standard ODE solver available in MATLAB which uses a Runge-Kutta scheme. For the model being studied, the dynamic parameters are presented in Table 24.2. These are descriptive for this specific shelf adjacent to a rigid wall. It is to be noted that $R < 1$ and p is slightly larger than 2, and the system might therefore be sensitive to short pulses as has been demonstrated in previous studies of two-sided rocking.

When the dynamic properties have been defined, overturning spectra can be evaluated as a function of the excitation amplitude and angular frequency. The requirement for the block to start overturning is:

$$a/g > \alpha \quad (24.8)$$

The range of values used for amplitude is therefore between the initiation acceleration $g\alpha$ and extreme earthquake acceleration, which in our case is chosen as $2g$. The frequency range is taken as the probable values of the natural frequency of typical building, which generally have a natural period between $T_p = 0.05$ and 1 s. The angular frequency is computed as $w_p = 2\pi/T_p$. The overturning spectra can furthermore be normalized by the slenderness and frequency parameters to obtain a more general solution. The normalized quantity is then derived as

$$\frac{a_p}{g\alpha} > 1 \quad (24.9)$$

When the normalized frequency ω_p/p equals 1, resonance like phenomenon occurs, making the system highly unstable.

24.4 Simulation Results

The numerical simulation is run using the above-mentioned setup, and the resulting overturning spectra for one-sided rocking is presented in Fig. 24.2. The yellow colour indicates simulations where the block overturns and the blue colour indicates scenarios where overturning does not occur.

The results indicate, as expected, that overturning is more probable for higher amplitude shaking. The required amplitude for overturning seems to decrease with increase in the pulse period. At a period close to 1, an amplitude of greater than 0.3 g is required to overturn the shelf. The results indicate multiple bifurcations, typical of

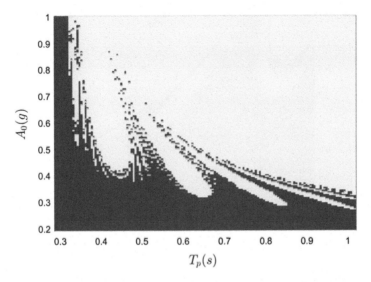

Fig. 24.2 The overturning spectra of the Brimnes bookcase for single-sided rocking of rigid block

chaotic systems. For example, at a given period, a lower amplitude excitation might overturn the shelf, which may not be overturn by a slight higher amplitude excitation of the same frequency. As the frequency ratio increases, the bifurcation leaves get wider, indicating more orderly behaviour.

The overturning spectrum for two-sided rocking of the same shelf is presented in Fig. 24.3. A similar trend is observed, where overturning becomes more likely for higher amplitudes and higher periods. However, the two-sided rocking shelves appear to be more stable, as the overturning starts at a relatively higher amplitude as well as higher periods.

This could be explained by the fact that the two-sided rocking dissipates energy due to the coefficient of restitution, which in this case is less than 1 as described by Eq. 24.7. Another contributing factor could be that two-sided rocking has a wider range of rotation giving the block a larger range to rock about without overturning.

24.5 Rocking Response to Earthquake Ground Motions

This section provides some example simulation of the rocking of bookshelf to two base excitations. The first one was recorded on the second floor of the Selfoss Town Hall during the 29 May 2008 Ölfus Earthquake in South Iceland. The moment magnitude of the earthquake was 6.3 and the epicentral distance was approximately 5 km (the peak floor acceleration, PFA is as indicated in the Fig. 24.4). The simulation shows that considerable rocking takes place, however the bookshelf does not overturn.

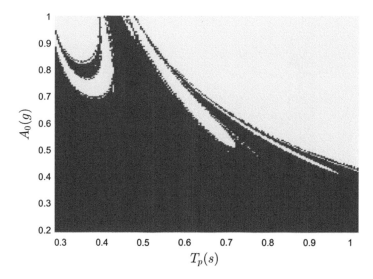

Fig. 24.3 The overturning spectra of the Brimnes bookcase for two-sided rocking of rigid block

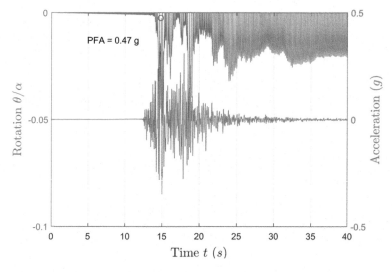

Fig. 24.4 Numerical simulation of a one-sided rocking block on the second floor of a building close to a moderate size earthquake

The second motion was recorded during the Montenegro Earthquake of 15 April 1979. The epicentral distance in this case is 65 km and the moment magnitude is 7. The results are presented in Fig. 24.5. Despite the lower peak acceleration, this motion overturns the block. Both events were strike-slip and were recorded on rock sites.

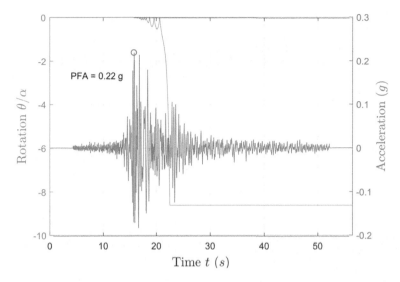

Fig. 24.5 Numerical simulation of a one-sided rocking block excited by far-field ground motion

This test shows that weaker motion (in terms of peak acceleration) overturns the block whereas the stronger one does not. In Fig. 24.6, the frequency content of the two ground motions are compared in terms of their power spectral density. The two time series have similar energy content in the frequency range 1–5 Hz. The one from the Ölfus Earthquake has higher power at low frequencies, and some peaks around 8–10 Hz, which is most likely due to the response of the building where the motion was recorded. The frequency parameter of the Brimnes shelf is also in the figure. It is interesting to note that despite higher energy content at the frequency parameter of the Brimnes shelf, the motion from Ölfus earthquake does not overturn the shelf. On the other hand, the motion with much lower power at the frequency parameter of the shelf overturns it. This result seems counter-intuitive as the former motion would be expected to resonate with the natural oscillation of the shelf. This apparent contradiction can be explained by the fact that the frequency parameter is relevant only for free rocking of the shelf. For forced rocking (for example due to base motion), the frequency of oscillation of the shelf depends on the excitation, and varies during rocking. This assertion is visible in the rocking response of the shelf plotted in Figs. 24.4 and 24.5, where the rocking frequency is much higher in the first case than in the second case. This implies that, irrespective of the free rocking frequency of the shelf, the motion from the Ölfus Earthquake induces high frequency rocking of the shelf, and despite higher root mean square acceleration (and peak ground acceleration), overturning does not occur. The motion from Montenegro Earthquake, however causes low frequency rocking (the rocking period before overturning is approximately 1 s) of the shelf, resulting in its overturning. The results indicate that the overturning safety of objects such as these is highly dependent on the peculiarities of base motion. The non-linear nature of the problem results in

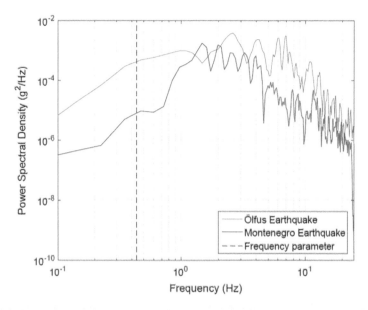

Fig. 24.6 Comparison of the power spectral density (PSD) of the ground motions used in this study. The PSD was estimated from the strong-motion part of the time series using the Welch's method

excitation dependent frequency of rocking, and the response of the system is not orderly and easily predictable. It is noted that these conclusions are based on the two example motions, and a more detailed analysis using different types of base motion is needed before the these observations and conclusions can be generalized.

24.6 Discussion and Conclusions

One-sided rocking of a rigid block, representing a bookshelf, is numerically investigated using the inverted pendulum model. A solution method for objects placed next to a rigid wall is derived assuming that loss of energy during impact with the wall is negligible. The solution method is a slightly modified version of the Newmark Beta algorithm, which is commonly used to solve differential equation of motion of single degree of freedom oscillators subjected to arbitrary forces. Numerical simulations using harmonic base motion with exponentially decreasing amplitude were performed to compute overturning spectra for one-sided and two-sided rocking. The results indicate that one-sided rocking is less-stable than two-sided rocking, an observation valid for the type of excitation used in this study. One of the reasons behind this observation might be that while energy dissipation is ignored in the one-sided rocking problem, change in rotation axis in the two-sided rocking results in loss of energy, which was modelled by using a coefficient of restitution. In real situation, some loss of energy during impact with the wall is expected during one-sided rocking. Consideration of such phenomenon is expected to result in more stable behaviour of rigid blocks rocking in one direction.

A case study using two base excitations is presented: one is the floor acceleration recorded during an earthquake in South Iceland, and the other is the ground acceleration recorded during an earthquake in Montenegro. The results indicate that the intensity (generally measured in terms of peak acceleration) is a poor indicator of stability against overturning. Comparison of the frequency content of the two ground motions did not reveal any obvious clue about the role of frequency content of the two ground motions in causing overturning of the bookshelf. It appears that whether the shelf overturns or not is dependent on the peculiarities of the base motion, although it is not clear from this case study what those peculiarities are. Base motion that cause high-frequency rocking seem to result in more stable response of the shelf than those that cause low-frequency rocking. It is however not obvious what features of base motion used in this case study are responsible for the different frequencies of rocking of the shelf. It is evident that the frequency parameter, which is related to free rocking frequency of rigid blocks, is not relevant for forced rocking. The inherent non-linear nature of the problem results in excitation dependent rocking frequency and the response seems to be more chaotic than orderly.

Acknowledgements This study was partially financed by the European Commission's Humanitarian Aid and Civil Protection (Grant agreement ECHO.A.5 (2015) 3916812) – KnowRISK (Know your city, Reduce seISmic risK through non-structural elements). We also acknowledge the support from University of Iceland Research Fund.

References

Hogan S (1992) On the motion of a rigid block, tethered at one corner, under harmonic forcing. Proc Royal Soc Lond A 439:35–45. The Royal Society
Housner GW (1963) The behavior of inverted pendulum structures during earthquakes. Bull Seismol Soc Am 53(2):403–417
Jankowski R (2007) Theoretical and Experimental assessment of parameters for the non-linear viscoelastic model of structural pounding. J Theor Appl Mech 45(4):931–942
Makris N, Black CJ (2001) Rocking response of equipment anchored to a base foundation. Pacific Earthquake Engineering Research Center, Berkeley
Makris N, Konstantinidis D (2001) The rocking spectrum and the shortcomings of design guidelines. Pacific Earthquake Engineering Research Center, Berkeley
Newmark NM, Rosenbluth E (1971) Fundamentals of earthquake engineering. Prentice Hall, Englewood Cliffs
Petrone C, Di Sarno L, Magliulo G, Cosenza E (2016) Numerical modelling and fragility assessment of typical freestanding building contents. Bull Earthq Eng:1–25

Shaw SW, Rand RH (1989) The transition to chaos in a simple mechanical system. Int J Non-Linear Mech 24(1):41–56

Zhang J, Makris N (2001) Rocking response of free-standing blocks under cycloidal pulses. J Eng Mech 127(5):473–483

Chapter 25
Finite Element Model Updating of a Long Span Suspension Bridge

Øyvind Wiig Petersen and Ole Øiseth

Abstract Errors and uncertainties in numerical models of structures affects the ability of these models to accurately predict the dynamic behaviour. However, model updating techniques can be used to calibrate the models based on experimental data. This paper presents a case study of sensitivity-based model updating applied to the Hardanger Bridge, a long span suspension bridge. Thirteen stiffness and mass parameters are chosen to represent the system uncertainties in a finite element (FE) model. Thirty vibration modes from system identification based on acceleration data is used to calibrate the FE model, using identified natural frequencies and mode shapes as objectives. In the updated model the average error in natural frequencies is reduced from 3.65% to 1.28%. The MAC numbers for the updated modes range from 0.678 to 0.999. The study indicates FE models of large suspension bridges can be significantly improved, but many uncertainties related to modelling simplifications are still present.

Keywords Suspension bridge · System identification · Model updating

25.1 Introduction

Numerical modelling of engineering structures is an important tool in the prediction of the dynamic behaviour of a real structure. In civil engineering, a finite element (FE) model is typically used to assess a structure's response to dynamic loads such as earthquake excitation, wind or traffic. However, the FE models include simplifications and assumptions on uncertain system parameters, which inherently leads to uncertainties in the predicted response.

Model updating techniques can be used to reduce the uncertainties of FE models given that measurement data of the relevant structure is available, see e.g. the survey

---

Ø. W. Petersen (✉) · O. Øiseth
Norwegian University of Science and Technology, NTNU, Trondheim, Norway
e-mail: oyvind.w.petersen@ntnu.no; ole.oiseth@ntnu.no

Fig. 25.1 The Hardanger Bridge

by Mottershead and Friswell (1993). The fields of structural health monitoring also make of use of model updating methodology to assess health and detect damage, see Carden and Fanning (2004) or Doebling and Farrar (1998). The most popular approach is to estimate unknown system parameters by calibrating the predicted outputs of the numerical model to the measured output data. For civil engineering structures, vibration data is often the most convenient type of data to acquire in a state of operation, where accelerometers or strain gauges often are cost-effective and easy to operate. For large civil engineering structures such as bridges, ambient excitation (wind or traffic) and subsequent system identification is usually preferred for modal data acquisition.

Many engineering challenges are still faced in modelling of large bridges. Practical case studies of very large bridges found in literature indicate that errors of 3–15% in predicted natural frequencies prior to updating are typical, see e.g. Zhang et al. (2000, 2001), Benedettini and Gentile (2011), Zhong et al. (2016), Hong et al. (2010) and Schlune et al. (2009). However, the previous studies also demonstrate that a significant improvement is attainable by model updating.

In this article we demonstrate an application of the sensitivity method in model updating to a case study of the Hardanger Bridge (Fig. 25.1). The bridge is located in the Hardanger fjord, Norway, and is distinguished for its slender design despite its long span length of 1310 m. A good understanding of the dynamic behaviour is imperative for long-span bridges, which commonly experience large-amplitude excitation. The bridge is also subject to other research studies, whereby it is desired to have a validated structural model. In this contribution, a parametrized FE model of the bridge is calibrated to match the natural frequencies and mode shapes which are obtained from a system identification.

25.2 System Equations

Consider a suspension bridge modelled with n_{DOF} degrees of freedom (DOFs). The equations of motion to wind loading can be formulated as follows:

$$M_0\ddot{u}(t) + (C_0 - C_{ae})\dot{u}(t) + (K_0 - K_{ae})u(t) = f(t) \tag{25.1}$$

where $u(t) \in \mathbf{R}^{n_{DOF}}$ is the response vector. M_0, C_0 and $K_0 \in \mathbf{R}^{n_{DOF} \times n_{DOF}}$ represents the mass, damping and stiffness matrices related to the structure only. C_{ae} and $K_{ae} \in \mathbf{R}^{n_{DOF} \times n_{DOF}}$ are the aerodynamic added damping and stiffness from the fluid-structure interaction. Equation 25.1 can be rewritten on state-space form as follows:

$$\dot{x}(t) = A_c x(t) + B_c f(t) \tag{25.2}$$

where $\mathrm{x}(t) = [u(t) \ \dot{u}(t)]^T \in \mathbf{R}^{2 \cdot n_{DOF}}$ is the state variable and system matrices in continuous time now are defined as:

$$A_c = \begin{bmatrix} 0 & I \\ -M_0^{-1}(K_0 - K_{ae}) & -M_0^{-1}(C_0 - C_{ae}) \end{bmatrix}, \quad B_c = \begin{bmatrix} 0 \\ M_0^{-1} \end{bmatrix} \tag{25.3}$$

To obtain the modes of the system in Eq. 25.2 the complex eigenvalue problem can be solved:

$$(A_c - \lambda I)\psi = 0 \tag{25.4}$$

The solution to Eq. 25.4 give modes that occur in complex conjugate pairs (ψ_r, $\psi_r^* \in \mathbf{C}^{2 \cdot n_{DOF}}$):

$$\psi_r = \begin{bmatrix} \phi_r \\ \phi_r \lambda_r \end{bmatrix}, \quad \psi_r^* = \begin{bmatrix} \phi_r^* \\ (\phi_r \lambda_r)^* \end{bmatrix} \tag{25.5}$$

The complex roots associated with the modal pairs are given by:

$$\lambda_r, \lambda_r^* = -\xi_r \omega_r \pm i \omega_r \sqrt{1 - \xi_r^2} \tag{25.6}$$

where ω_r is the undamped natural frequency and ξ_r is the damping ratio of mode r. In the absence of wind and when the structural damping is small, the terms K_{ae}, C_{ae} and C_0 vanishes, and the problem in Eq. 25.4 reduces to:

$$\left(M_0^{-1} K_0 - \omega_r^2 I\right) \phi_r = 0 \tag{25.7}$$

The modes shapes ϕ_r now becomes real valued and are commonly referred to as the still-air modes.

Fig. 25.2 Left: FE model; right: detail of girder-pylon connection

25.3 Finite Element Model

A model of the Hardanger Bridge is created in the FE element software Abaqus, see Fig. 25.2. The steel box girder, stiffeners and internal diaphragms are modelled with shell elements (S4R) using the geometry adopted from the construction blueprints. Shell elements are chosen over beam elements since the use of beam sections require assumptions to be made on e.g. the equivalent torsional stiffness of the girder. However, this modelling choice comes at the cost of increased processing time; the model has approximately $5*10^5$ DOFs. In this case, however, the model accuracy is highly valued, thus the computational cost is deemed as an acceptable price. The concrete pylons have a hollow core structure and are also modelled using shell elements. An equivalent elastic modulus is used for the concrete-rebar composite material, i.e. the material of the pylons is assumed to behave homogeneously.

Timoshenko beam elements (B31) are used for the hangers and cables, taking into account the effective stiffness from the cable tension. The DOFs at the base of the pylons are fixed, while the cables are pinned at the two end points where they are anchored to bedrock. The far ends of the girder is connected to the pylons using springs, see Fig. 25.2. The springs are meant to represent the real-life bearings which hinders pendulum motion of the girder, but still allows for temperature expansion. The value of the assigned springs stiffness's are highly uncertain and remains to be updated. An evenly distributed mass is added to the top of the girder to account for asphalt and other non-structural elements.

25.4 System Identification

The system identification is performed using data from a monitoring system installed at the Hardanger Bridge. The system consists of 20 accelerometers and 9 anemometers and continuously monitors the dynamic activity and wind velocities. For more information on the workings of the monitoring system, see Fenerci et al. (2016). A 120 min long time series recorded 13 January 2016 was selected for the system identification. In the duration of the recording, the measured wind velocity ranged between 3 and 6 m/s. Low wind velocities are desired to minimize the prominence of the aerodynamic terms C_{ae} and K_{ae} in Eq. 25.1, so that the identified system are

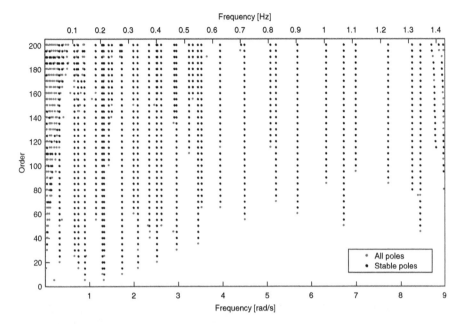

Fig. 25.3 Stabilization diagram for Cov-SSI

dominated by the structural properties only. The installed accelerometers measure in all three directions, meaning $n_d = 60$ acceleration output signals are available. The acceleration data was low-pass filtered (Chebyshev type I, cut-off 2 Hz) and resampled from 200 Hz to 5 Hz. Natural frequencies and mode shapes was identified by means of the popular tool covariance-driven stochastic subspace identification (Cov-SSI) (Peeters and De Roeck 2001). It is found that a maximum time lag of $\tau = 20$ s (corresponding to 100 time steps in the block-Hankel matrix of the Cov-SSI algorithm) produces a stabilization diagram where the poles are stable and well distinguished, see Fig. 25.3. In total, $n_m = 30$ modes were identified, listed in Table 25.1. The natural frequencies range from 0.05–1.31 Hz. The lowest horizontal, vertical and torsional mode among the identified are pictured in Fig. 25.4.

25.5 Model Updating Scheme

A set of parameters $\theta \in \mathbf{R}^{n_p}$ is used to represent the uncertain variables of the model, typically elastic moduli, spring stiffness's, densities or geometry. From a parameter sensitivity study it is found that the $n_p = 13$ mass and stiffness parameters listed in Table 25.3. have a significant influence on the natural frequencies and mode shapes. The elastic modulus and density of the girder and main cables are found to have the largest impact on the natural frequencies. A slightly less influence is observed for the hanger tension force and the non-structural mass on the girder. The elastic modulus

Table 25.1 Identified modes from the system identification

Mode no.	Mode type	$\bar{f}_r$ [Hz]	Mode no.	Mode type	$\bar{f}_r$ [Hz]
1	Horizontal	0.052	16	Vertical	0.471
2	Horizontal	0.105	17	Pylon	0.516
3	Vertical	0.119	18	Pylon	0.529
4	Vertical	0.142	19	Vertical	0.547
5	Horizontal	0.183	20	Torsion	0.560
6	Vertical	0.206	21	Vertical	0.628
7	Vertical	0.212	22	Vertical	0.715
8	Cable/horizontal	0.230	23	Vertical	0.808
9	Vertical	0.276	24	Torsion	0.828
10	Horizontal	0.318	25	Vertical	0.905
11	Vertical	0.333	26	Vertical	1.007
12	Torsion	0.374	27	Torsion	1.069
13	Vertical	0.401	28	Vertical	1.112
14	Cable/horizontal	0.418	29	Vertical	1.227
15	Horizontal	0.464	30	Torsion	1.314

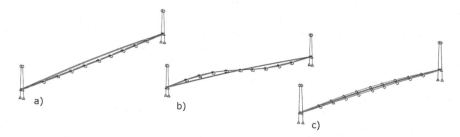

Fig. 25.4 Three of the identified modes: (**a**) horizontal mode (0.052 Hz); (**b**) vertical mode (0.119 Hz); (**c**) torsion mode (0.374 Hz)

and density of the concrete primarily dictates the two modes with large pylons motion. The spring stiffness controlling the connection between the girder and pylons are seen to primarily influence the mode shapes.

The values of these parameters are calibrated in the model updating, which is formulated as an optimization problem where the goal is to minimize the difference between the measured outputs and the corresponding model predictions. In this case the identified natural frequencies ($\bar{f} \in \mathbf{R}^{n_m}$) and MAC values from the identified mode shapes ($\bar{\phi} \in \mathbf{R}^{n_d \times n_m}$) are used. The MAC value between two modes is defined by:

$$MAC(\phi_i, \bar{\phi}_j) = \frac{|(L\phi_i)^T \bar{\phi}_j|}{\|L\phi_i\|_2^2 \|\bar{\phi}_j\|_2^2} \qquad (25.8)$$

where the binary matrix $L \in \mathbf{R}^{n_d \times n_{DOF}}$ selects the observed DOFs from all the DOFs of the model. Here, the imaginary part of $\bar{\phi}$ is neglected since it is assumed contributions from aerodynamic terms is very small. A common problem in model updating of bridges is that many of the natural frequencies are closely spaced. The mode matching is therefore facilitated by the following expression (Simoen et al. 2015):

$$1 - MAC(\phi_i, \bar{\phi}_j) + | 1 - \frac{f_i}{\bar{f}_j} | \tag{25.9}$$

Here, a model mode (i) is compared with an identified mode (j). A matching pair is obtained when the quantity in Eq. 25.9 reaches a minimum value. The sensitivity method is applied for optimization, a gradient-based method which linearizes the output residuals at the current parameter values θ_0:

$$\begin{aligned} \bar{f} - f(\theta) &\approx \bar{f} - \left(f(\theta_0) + G^{(f)}|_{\theta=\theta_0} \Delta\theta\right) \\ 1 - MAC(\theta) &\approx 1 - \left(MAC(\theta_0) + G^{(MAC)}|_{\theta=\theta_0} \Delta\theta\right) \end{aligned} \tag{25.10}$$

Here, $G^{(f)}$ and $G^{(MAC)} \in \mathbf{R}^{n_m \times n_p}$ are the Jacobians, and $\Delta\theta \in \mathbf{R}^{n_p}$ is the parameter step. It is necessary to perturb the model n_p times to compute the first order sensitivity of the model predictions with respect to the parameters. The following normalized least squares objective function is used (Simoen et al. 2015):

$$F(\theta) = \frac{1}{2} \sum_{r=1}^{n_m} a_r \frac{(\bar{f}_r - f_r(\theta))^2}{\bar{f}^2_r} + \frac{1}{2} \sum_{r=1}^{n_m} b_r (1 - MAC_r)^2 \tag{25.11}$$

where a_r and b_r assigns the error penalty weighting to the individual mode. The objective function is minimized with respect to the parameter step $\Delta\theta$. The parameters are then updated to new values ($\theta := \theta_0 + \Delta\theta$) and the process is repeated until the objective function converges.

25.6 Model Updating Results and Discussion

The focus in the model updating is put on the lowest half of the modes, since these are more important for the predicted response to wind loads. In addition, the identified frequencies are deemed as more certain than the identified mode shapes. Thus the following weighting coefficients in Eq. 25.11 are chosen: $a_{1-15} = 128$, $a_{16-30} = 32$, $b_{1-15} = 32$ and $b_{16-30} = 8$. Ten iterations were performed until a convergence in the objective function was obtained. The value of the objective function was reduced from 6.25 to 3.24 in the updating procedure.

Table 25.2 shows the values of the modal parameters prior and post update. The average error in the natural frequency is reduced from 3.65% to 1.28%. Five of the ten modes with the largest initial errors has less than 1% error in the updated model. The largest updated frequency error is 6.57% (mode 3), which can be considered fairly low taking into account the number of examined modes.

Table 25.2 Initial and updated values for the modes

Mode no. (r)	$\bar{f}_r$ [Hz]	Initial f_r [Hz] (error)	Updated f_r [Hz] (error)	Initial MAC	Updated MAC
1	0.052	0.051 (−1.67%)	0.051 (−1.48%)	0.999	0.999
2	0.105	0.101 (−3.82%)	0.105 (−0.86%)	0.993	0.992
3	0.119	0.110 (−7.44%)	0.111 (−6.57%)	0.988	0.988
4	0.142	0.141 (−0.91%)	0.141 (−0.58%)	0.994	0.993
5	0.183	0.177 (−3.26%)	0.186 (1.28%)	0.993	0.994
6	0.206	0.200 (−2.97%)	0.201 (−2.55%)	0.990	0.990
7	0.212	0.210 (−0.93%)	0.212 (−0.35%)	0.996	0.996
8	0.230	0.234 (1.70%)	0.236 (2.84%)	0.962	0.975
9	0.276	0.274 (−0.61%)	0.277 (0.29%)	0.998	0.998
10	0.318	0.306 (−3.71%)	0.320 (0.64%)	0.991	0.990
11	0.333	0.328 (−1.49%)	0.332 (−0.44%)	0.993	0.993
12	0.374	0.358 (−4.36%)	0.374 (−0.05%)	0.964	0.963
13	0.401	0.396 (−1.30%)	0.401 (0.10%)	0.998	0.998
14	0.418	0.387 (−7.33%)	0.395 (−5.42%)	0.773	0.903
15	0.464	0.449 (−3.24%)	0.464 (0.04%)	0.691	0.700
16	0.471	0.459 (−2.68%)	0.466 (−1.09%)	0.983	0.985
17	0.516	0.501 (−2.91%)	0.512 (−0.74%)	0.606	0.866
18	0.529	0.508 (−4.10%)	0.519 (−2.01%)	0.740	0.804
19	0.547	0.535 (−2.26%)	0.546 (−0.35%)	0.998	0.998
20	0.560	0.512 (−8.57%)	0.558 (−0.35%)	0.974	0.975
21	0.628	0.610 (−2.75%)	0.627 (−0.17%)	0.959	0.851
22	0.715	0.695 (−2.88%)	0.712 (−0.37%)	0.997	0.995
23	0.808	0.768 (−4.86%)	0.790 (−2.17%)	0.944	0.963
24	0.828	0.769 (−7.07%)	0.837 (1.10%)	0.986	0.978
25	0.905	0.870 (−3.81%)	0.896 (−0.97%)	0.995	0.995
26	1.007	1.015 (0.83%)	0.967 (−3.91%)	0.436	0.678
27	1.069	0.991 (−7.29%)	1.068 (−0.12%)	0.992	0.993
28	1.112	1.066 (−4.15%)	1.101 (−0.96%)	0.990	0.988
29	1.227	1.184 (−3.53%)	1.235 (0.60%)	0.912	0.764
30	1.314	1.221 (−7.09%)	1.311 (−0.18%)	0.979	0.989

The MAC values are seen to both increase and decrease, partly due to the low error penalty assigned. Some modes have no significant change in MAC numbers, as not all the mode shapes are sensitive to the chosen parameters. The modes experiencing a significant change in MAC numbers are typically higher vertical modes, which tend to be sensitive to the end boundary conditions of the girder. This also indicates that the model is also sensitive to how the girder-pylon connection is modelled.

The updated parameter values are listed in Table 25.3. A clear trend of underestimating the natural frequencies was seen in the initial model. Consequently the elastic moduli of the girder increases in the updating, while the girder density and non-structural mass decreases. The changes in the elastic moduli and densities are

Table 25.3 Parameters in the model updating

Parameter	Unit	Initial value	Updated value	Change
E-modulus girder	MPa	2.100e + 05	2.205e + 05	5.0%
Density girder	kg/m^3	7850	7585	−3.4%
Poisson ratio girder	–	0.300	0.251	−16.2%
E-modulus main cable	MPa	2.000e + 05	1.890e + 05	−5.5%
Density main cable	kg/m^3	7850	7535	−4.0%
E-modulus pylon	MPa	4.000e + 04	3.876e + 04	−3.1%
Density main pylon	kg/m^3	2500	2269	−9.2%
Non-structural mass girder	kg/m^2	220	188	−14.5%
Spring side, vertical	N/m	1.000e + 08	7.787e + 09	7686.7%
Spring Centre, longitudinal	N/m	1.000e + 09	2.319e + 09	131.9%
Spring Centre, vertical	N/m	1.000e + 08	5.576e + 09	5475.8%
Spring Centre, rotation	Nm/rad	1.000e + 09	1.080e + 07	−98.9%
Hanger tension	kN	9.039 + 02	9.314 + 02	3.0%

within a range which can be considered realistic. The Poisson's ratio for the girder, largely influencing the frequencies of the torsional modes, is reduced from 0.3 to 0.25. This updated value is below material values commonly given for steel. A looser interpretation of what a parameter represents might be necessary; the Poisson's ratio controls the shear properties of the material. The stiffness of the boundary springs experiences the largest changes, but also have the highest level of uncertainty to begin with. Very large changes in these parameters are therefore acceptable.

25.7 Conclusion

FE models of structures can be calibrated using model updating techniques. This paper presented a case study of the sensitivity method in FE model updating with application to the Hardanger Bridge, a long span suspension bridge. A system identification of the bridge was performed using data from 20 triaxial accelerometers. Thirty modes with natural frequencies in the range 0.05–1.31 Hz were identified by Cov-SSI. The identified natural frequencies and MAC numbers were used as objectives for calibration. After 10 iterations with the sensitivity method, the average error in natural frequencies was reduced from 3.65% to 1.28%. The updated MAC numbers range from 0.67 to 0.99. The changes in model parameters are within acceptable limits. This case study shows that numerical models of suspension bridges can be improved using FE updating techniques. However, accurate modelling of large bridges remains a challenge since uncertainties on the parameters and modelling techniques are still present.

References

Benedettini F, Gentile C (2011) Operational modal testing and FE model tuning of a cable-stayed bridge. Eng Struct 33(6):2063–2073

Carden EP, Fanning P (2004) Vibration based condition monitoring: a review. Struct Health Monit 3(4):355–377

Doebling SW, Farrar CR (1998) A summary review of vibration-based damage identification methods. Shock Vib Dig 30(2):91–105

Fenerci A, Øiseth O, Rönnquist A (2016) Long-term monitoring of wind field characteristics and dynamic response of a long-span suspension bridge in complex terrain. Eng Struct (submitted)

Hong AL, Ubertini F, Betti R (2010) Wind analysis of a suspension bridge: identification and finite-element model simulation. J Struct Eng 137(1):133–142

Mottershead JE, Friswell MI (1993) Model updating in structural dynamics: a survey. J Sound Vib 167(2):347–375

Peeters B, De Roeck G (2001) Stochastic system identification for operational modal analysis: a review. J Dyn Syst Meas Control 123(4):659–667

Schlune H, Plos M, Gylltoft K (2009) Improved bridge evaluation through finite element model updating using static and dynamic measurements. Eng Struct 31(7):1477–1485

Simoen E, De Roeck G, Lombaert G (2015) Dealing with uncertainty in model updating for damage assessment: a review. Mech Syst Signal Process 56:123–149

Zhang QW, Chang CC, Chang TYP (2000) Finite element model updating for structures with parametric constraints. Earthq Eng Struct Dyn 29(7):927–944

Zhang QW, Chang TYP, Chang CC (2001) Finite-element model updating for the Kap Shui Mun cable-stayed bridge. J Bridg Eng 6(4):285–293

Zhong R, Zong Z, Niu J, Liu Q, Zheng P (2016) A multiscale finite element model validation method of composite cable-stayed bridge based on Probability Box theory. J Sound Vib 370:111–131

Chapter 26
Characterization of the Wave Field Around an Existing End-Supported Pontoon Bridge from Simulated Data

Knut Andreas Kvåle and Ole Øiseth

Abstract The environmental excitation and the dynamic response are currently being monitored on the Bergsøysund Bridge, an existing end-supported pontoon bridge. Wave radars are monitoring the one-point sea surface elevation at six different locations. As the wave excitation is considered the main concern for vibration-based design of similar bridges, an appropriate description of the sea state characterizing the wave excitation is crucial. Furthermore, it is considered a necessity for an assessment of the quality of response predictions by comparison with measurements. In the current paper, time simulations of wave elevation are used to identify the already-known sea states. The Fourier Expansion Method (FEM) and Extended Maximum Entropy Principle (EMEP) are applied for this purpose. The results provide valuable insights about both the identification methods and the sensor layout.

Keywords Wave modelling · Floating bridge · Wave field · Wave spreading

26.1 Introduction

Pontoon bridges may help to overcome the ever-increasing challenges of modern bridge engineering, by utilizing the buoyancy of the water. The subject of water waves is well-described and thoroughly discussed in the scientific world, but the vast knowledge is not necessarily accessible to the bridge engineer society. The simulation techniques available are mainly verified on structures of limited physical reach, and studies attempting to verify simulated floating bridge behaviour are very scarce. The waves will in most cases represent the main excitation source for low pontoon bridges, even though traffic and wind excitation also will play significant roles for

---

K. A. Kvåle (✉) · O. Øiseth
Faculty of Engineering Science and Technology, Department of Structural Engineering, NTNU, Norwegian University of Science and Technology, Trondheim, Norway
e-mail: knut.a.kvale@ntnu.no; ole.oiseth@ntnu.no

certain structures. To be able to verify the simulated behaviour of a floating bridge, the identification of the wave excitation is a crucial sub-task to solve. The wave excitation is commonly assumed directly dependent on the sea surface elevation, such that a successful characterization of the wave field surrounding the investigated structure will be sufficient for this purpose.

The Bergsøysund Bridge is an existing end-supported pontoon bridge, and is the structure under consideration in the current paper. The bridge is currently extensively instrumented to capture the environmental excitation sources acting on it and its dynamic response. In the current paper, no data from the monitoring system will be analysed. However, the geometry and positions of the wave radars will be used in conjunction with simulated realistic sea states, to assess how well the methods can capture the parameters of the simulated sea state with the current sensor set-up. Should they fail to identify the benchmark cases, a revision of the layout would have to be considered. The experience drawn from this will be highly valuable when later comparing response predictions with monitored response. The two-dimensional directional wave spectral density is the quantity that is attempted identified in the current paper. Aspects regarding the simulation of wave data important for a realistic benchmark situation are discussed.

26.2 The Bergsøysund Bridge

The Bergsøysund Bridge is a pontoon bridge that links the islands Aspøya and Bergsøya on the west-coast of Norway ($62°59'12.8''$N, $7°52'26.5''$E). The 931 meters long bridge (Fig. 26.1), constructed by a steel truss which is supported on seven lightweight concrete pontoons, has been in operation for nearly 25 years. Lack of side support makes the bridge particularly interesting. The bridge is extensively

Fig. 26.1 The Bergsøysund Bridge. (Photograph by NTNU/K.A. Kvåle)

26 Characterization of the Wave Field Around an Existing End-Supported...

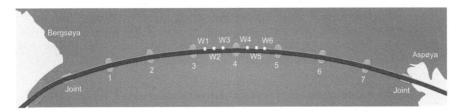

Fig. 26.2 Wave radar layout

Table 26.1 Coordinates of wave radars

Sensor	x [m]	y [m]	z [m]
W1	−73.0	−1.1	−6.2
W2	−53.0	−0.1	−6.2
W3	−33.0	0.6	−6.2
W4	33.0	0.6	−6.2
W5	53.0	−0.1	−6.2
W6	73.0	−1.1	−6.2

The origin is located on top of the bridge deck, midspan, with positive x pointing tangential to the bridge and towards Aspøya

monitored by means of accelerometers, anemometers, wave radars, and a single global navigation satellite system (GNSS) displacement sensor. Wave radars are monitoring the single-point wave elevation at six locations, as indicated in Fig. 26.2 and Table 26.1. The monitoring system is extensively described in Kvåle and Øiseth (2017). Other work on the Bergsøysund Bridge include simulation studies (Kvåle et al. 2016) and identification of the modal parameters (Kvåle et al. 2015, 2017a, b).

26.3 Describing Irregular Sea Surface Elevation

26.3.1 Stochastic Description of Irregular Sea Surface Elevation

A thorough review of relevant theory of stochastic modelling of irregular sea surfaces is given by Hauser et al. (2005), but the details required to illustrate how the methods work are repeated in the following. More details can be found in, e.g., Longuet-Higgins et al. (1963) and Sigbjörnsson (1979).

The random sea surface elevation is modelled as a function of location in space $\{r\}$ and time t, and can be expressed mathematically as follows:

$$\eta(\{r\}, t) = \int e^{i\{\kappa\}\cdot\{r\} - i\omega t} dZ_\eta(\{\kappa\}, \omega) \qquad (26.1)$$

where the wave number vector $\{\kappa\}$, the circular frequency ω, and the spectral process corresponding to the sea surface elevation Z_η, are introduced. For stationary and homogeneous random wave fields, this spectral process is related to the wave spectral density in the following manner:

$$E\left(dZ_{\eta_p}(\{\kappa\},\omega)dZ_{\eta_q}(\{\kappa\},\omega)^H\right) = S_{pq}(\{\kappa\},\omega)d\kappa_x d\kappa_y d\omega \qquad (26.2)$$

The cross-spectral density between the wave elevations at the probing locations p and q, separated by the distance vector $\{\Delta r\}$, as depicted in Fig. 26.3, is written as follows:

$$S_{pq}(\omega) = \int_{-\pi}^{\pi} S_\eta(\omega,\theta)\gamma_{pq}(\omega,\theta)d\theta = \int_{-\pi}^{\pi} S_\eta(\omega,\theta)e^{i\{\kappa\}\cdot\{\Delta r\}}d\theta \qquad (26.3)$$

where $\gamma_{pq}(\omega,\theta)$ is introduced as a 2D coherence function distributed over directions, and the following definitions are introduced:

$$\{\Delta r\} = \begin{Bmatrix} \Delta x \\ \Delta y \end{Bmatrix}, \quad \{\kappa\} = \kappa\begin{Bmatrix} \cos\theta \\ \sin\theta \end{Bmatrix} \qquad (26.4)$$

Here, $\{\Delta r\} = \{r_q\} - \{r_p\}$, such that Δx and Δy give the distances in x- and y-direction between the points under investigation. Under the assumption of the dispersion relation and deep-water waves, the wave number κ can be written as follows for $\omega > 0$:

$$\kappa = \frac{\omega^2}{g} \qquad (26.5)$$

Equation 26.2 can be rewritten using angle and frequency (2D) rather than frequency and wave number vector (3D), as follows:

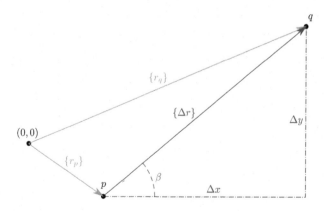

Fig. 26.3 Points p and q on the sea surface

$$E\left(dZ_{\eta_p}(\omega,\theta)dZ_{\eta_q}(\omega,\theta)^H\right) = S_{pq}(\omega,\theta)d\theta d\omega \qquad (26.6)$$

26.3.2 Time Simulation of Sea Surface Elevations from a Stochastic Description

The starting point for the simulation is the mathematical expression of the sea surface elevation given in Eq. 26.1. From Eq. 26.6, the amplitude corresponding to the contribution to the total wave elevation, from a regular wave with frequency ω and direction of propagation θ, within the region defined by $\Delta\omega$ and $\Delta\theta$, is found as follows:

$$a(\omega,\theta) = \sqrt{2S_\eta(\omega,\theta)\Delta\omega\Delta\theta} \qquad (26.7)$$

This is illustrated in Fig. 26.4. Relying on this, a realization of the sea state can be simulated as follows:

$$\eta(\{x\},t) = \sum_{k=1}^{N}\sum_{r=1}^{R}\sqrt{2S_\eta(\omega,\theta)\Delta\omega\Delta\theta}\, e^{i\kappa\{\sin\theta,\cos\theta\}\cdot\{x\}+i\alpha}e^{-i\omega t} \qquad (26.8)$$

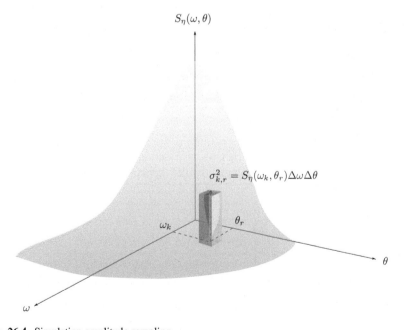

Fig. 26.4 Simulation amplitude sampling

where the following decomposition is usually applied: $S\eta(\omega, \theta) = S\eta(\omega)D(\omega, \theta)$. Furthermore, the expression inside the sum can be discretized, as follows:

$$\eta(\{x\}, t) = \sum_{k=1}^{N} \left[\sum_{r=1}^{R} \sqrt{2S_\eta(\omega, \theta) \Delta\omega \Delta\theta} \exp(i\kappa\{\sin\theta_r, \cos\theta_r\} \cdot \{x\} + i\alpha_r) \right] e^{-i2\pi(k-1)n/N} \quad (26.9)$$

This is simply the FFT of the following expression:

$$B_k = \left[\sum_{r=1}^{R} \sqrt{2S_\eta(\omega, \theta) \Delta\omega \Delta\theta} \exp(i\kappa\{\sin\theta_r, \cos\theta_r\} \cdot \{x\} + i\alpha_r) \right] \quad (26.10)$$

The approach described above is commonly referred to as the double summation method. The method results in a non-ergodic simulated wave elevation (Jefferys 1987; Nwogu 1989). The method is, however, very efficient due to the utilization of the FFT algorithm, and is considered sufficient for the current case study. The time domain simulations in the current study have durations of 30 min, and are simulated with a sampling rate of 2 Hz.

26.4 Identification of Wave Parameters from Single-Point Elevations

Both applied methods rely on the estimation of the cross-spectral densities between pairs of wave elevations, a task considered as relatively uncertain for finite length time series. To estimate the spectral density, Welch's method, with 60 sub-divisions combined with a zero-padding factor of 8, was used.

26.4.1 Modelled Sea State

The directional wave spectral density is assumed to be characterized by the one-parameter Pierson-Moskowitz wave spectral density (Pierson and Moskowitz 1964) and the cos2s directional distribution (Longuet-Higgins et al. 1963), as follows:

$$S_\eta(\omega) = \frac{A}{\omega^5} e^{-B/\omega^4} \quad (26.11)$$

$$D(\theta) = C \cos^{2s}\left(\frac{\theta - \theta_0}{2}\right) \quad (26.12)$$

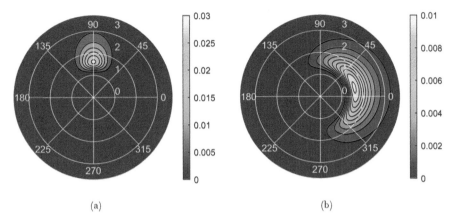

Fig. 26.5 Modelled two-dimensional wave spectral density. (**a**) Case 1: $H_s = 0.6m$, $s = 30$, $\theta_0 = 90°$. (**b**) Case 2: $H_s = 0.6m$, $s = 3$, $\theta_0 = 15°$

where $A = \alpha g^2$; $B = 3.11/H_s^2$; $\alpha = 0.0081$; H_s is the significant wave height (SWH), the mean wave height of the highest third of the waves; C is a constant ensuring that the integral of the distribution is 1; s is the spreading parameter, describing the spreading of the waves; and θ_0 is the mean wave direction. Two different cases were modelled, characterized by the following parameters:

1. $H_s = 0.6m$, $s = 30$, $\theta_0 = 90°$
2. $H_s = 0.6m$, $s = 3$, $\theta_0 = 15°$

The resulting directional wave spectral densities are illustrated in Fig. 26.5.

26.4.2 Initial Assessment of the Current Sensor Layout

According to Goda (1981) and Massel and Brinkman (1998), the layout of a spatial array can be optimized by cohering to the following guidelines:

1. No wave radar pair should have the same distance vector $\{\Delta r\}$.
2. The distance vectors should be distributed uniformly in the widest possible range.
3. At least one wave radar distance should be below ¼ of the shortest wave component to consider, i.e., $\lambda_{min} = \dfrac{2\pi}{\kappa_{max}} = \dfrac{2\pi g}{\omega_{max}^2}$. Here, λ_{min} is the minimum relevant wave length, and κ_{max} (via. Airy wave theory) and ω_{max} (via. The dispersion relation for deep-water waves) the corresponding maximum peak wave number and maximum peak frequency, respectively.

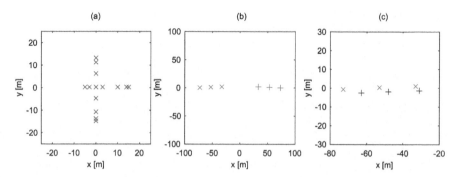

Fig. 26.6 Sensor layout. (**a**) Donelan et al. (1985). (**b**) Present layout. (**c**) Improved layout

Due to the circular pattern of the sensor layout, guideline 1 is strictly speaking fulfilled, even though many of the distance vectors are very similar. Guideline 2 is not fulfilled: the distance vectors are all close to being solely along the x-direction, and many have equal lengths. The largest peak frequency is close to $1.6\frac{rad}{s}$, implying that the shortest distance should be below 6 m, per guideline 3, something the layout does not fulfil.

Both a reference layout suggested by Donelan et al. (1985), the existing sensor layout, and a modified sensor layout will be considered. All layouts are depicted in Fig. 26.6. The cited reference layout was originally designed for common deep-water waves with peak periods less than 4s, and consists of 14 sensors in total. A comparison with a 6-sensor layout is therefore not fair, but is merely included to illustrate the possibilities of the methods.

26.4.3 Fourier Expansion Method

The Fourier Expansion Method (FEM) was first introduced as a method for the characterization of the directional wave spectral density by Longuet-Higgins et al. (1963), and has been discussed, reformulated and applied in numerous cases since then (Barber 1961; Ochi 2005; Panicker and Borgman 1970; Young 1994). Normally, it is written using the Fourier series defined as sums of sines and cosines. Below, an approach based on the Fourier series as a sum of complex exponentials is presented. The success of this modified procedure relies on proper handling of complex numbers in the numerical pseudo-inverse procedure, but has the benefit that it yields more elegant mathematical expressions.

26.4.3.1 Theoretical Outline

The 2D spectral density is first expressed as a truncated Fourier series, as follows:

$$S(\omega,\theta) = \sum_{n=-N}^{N} c_n(\omega) e^{in\theta} \tag{26.13}$$

By combination with the cross-spectral density between elevations at points p and q found in Eq. 26.3, this gives the following:

$$S_{pq}(\omega) = \int_{-\pi}^{\pi} \sum_{n=-N}^{N} c_n(\omega) e^{in\theta} e^{i\{\kappa\}\cdot\{\Delta r\}} d\theta \tag{26.14}$$

This is written out explicitly, for factorization, as follows:

$$\begin{aligned} S_{pq}(\omega) &= \int_{-\pi}^{\pi} c_{-N} \cdot e^{i(-N)\theta} \gamma_{pq} + c_{-N+1} \cdot e^{i(-N+1)\theta} \gamma_{pq} + \cdots \\ &\quad + c_{N-1} \cdot e^{i(N-1)\theta} \gamma_{pq} + c_N \cdot e^{iN\theta} \gamma_{pq} d\theta \\ &= \int_{-\pi}^{\pi} \gamma_{pq} e^{i(-N)\theta} d\theta c_{-N} + \int_{-\pi}^{\pi} \gamma_{pq} e^{i(-N+1)\theta} d\theta c_{-N+1} + \cdots \\ &\quad + \int_{-\pi}^{\pi} \gamma_{pq} e^{iN\theta} d\theta c_N \\ &= \sum_{n=-N}^{N} \left(\int_{-\pi}^{\pi} \gamma_{pq}(\omega,\theta) e^{in\theta} d\theta c_n(\omega) \right) \end{aligned} \tag{26.15}$$

For illustrational purposes, this is written using matrix notation, with three sensors, and evaluated at the chosen discrete frequency component ω_k:

$$\left\{ \begin{array}{c} S_{12} \\ S_{13} \\ S_{23} \end{array} \right\}_k = \begin{bmatrix} I_{12,-N}(\omega_k) & I_{12,-N+1}(\omega_k) & \cdots & I_{12,0}(\omega_k) & \cdots & I_{12,N}(\omega_k) \\ I_{13,-N}(\omega_k) & I_{13,-N+1}(\omega_k) & \cdots & I_{13,0}(\omega_k) & \cdots & I_{13,N}(\omega_k) \\ I_{23,-N}(\omega_k) & I_{23,-N+1}(\omega_k) & \cdots & I_{23,0}(\omega_k) & \cdots & I_{23,N}(\omega_k) \end{bmatrix} \left\{ \begin{array}{c} c_{-N} \\ c_{-N+1} \\ \vdots \\ c_0 \\ \vdots \\ c_{N-1} \\ c_N \end{array} \right\}_k \tag{26.16}$$

Here, $I_{pq,n}(\omega) = \int_{-\pi}^{\pi} \gamma_{pq}(\omega,\theta) e^{in\theta} d\theta$ is introduced for convenience. This equation system can be solved as a linear least squares problem, using pseudo-inverse when underdetermined. It can be shown that this integral can be solved using Bessel integrals of the first kind $J_n(z)$, as follows:

$$I_{pq,n}(\omega) = e^{in\beta} i^n 2\pi J_n\left(\frac{\omega^2}{g}l\right) \qquad (26.17)$$

where β is the angle and l is the length of the vector $\{\Delta r\}$ between locations p and q, as indicated in Fig. 26.3. This ensures a more robust numerical computation of the integral, avoiding accuracy problems associated with the rapid oscillation of the exponential integrand with respect to the angle, as reported by Giske et al. (2017). The choice of the number of Fourier coefficients included, defined by the constant N, affects the resulting 2D wave spectral density. N was set to 4 for all following applications.

26.4.3.2 Application of the FEM on Simulated Data

Figure 26.7 shows directional wave spectral densities estimated using the FEM, based on cross-spectral densities established from time simulations with the reference layout. When the cross-spectral densities are computed directly, the resulting 2D spectral density match nearly perfectly with the source. Thus, the discrepancies observed in the figure are assumed to be artefacts originating from the spectral estimation. The overall agreement is good, but as reported in the literature, the FEM method yields results with a narrower directional distribution, and thus indicating more spreading, than what is correct.

The current layout proves unable to yield any results when relying on cross-spectral densities estimated from time simulations. In Fig. 26.8, the FEM is therefore applied on the directly computed cross-spectral densities for points corresponding to the current layout. For case 1, the results are very poor, and the only conclusion that can be drawn from it is that the sea state represents a head sea, possibly approaching either from land or sea. For case 2, however, a decent result is obtained.

26.4.4 Maximum Entropy Principle

The Maximum Entropy Principle (MEP), also commonly referred to as the Maximum Entropy Method (MEM), was first applied for the determination of directional wave spectral density by Kobune and Hashimoto (1986). The MEP, which originally was developed for three-quantity point measurements, was thereafter reformulated for wave sensor arrays by Nwogu (1989). Hashimoto et al. (1994) introduced the Extended MEP (EMEP), which deals with the errors in the cross-spectral densities and thus makes it more robust. The implementation of the EMEP found in the DIWASP toolbox for MATLAB (Johnson 2002) was used in the current paper,

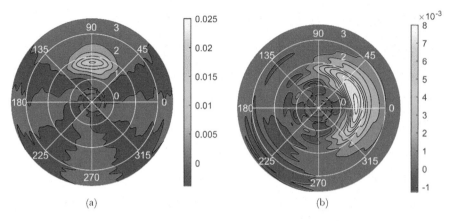

Fig. 26.7 FEM with sensor layout suggested by Donelan et al. (1985). (**a**) Case 1: $H_s = 0.6m$, $s = 30$, $\theta_0 = 90°$. (**b**) Case 2: $H_s = 0.6m$, $s = 3$, $\theta_0 = 15°$

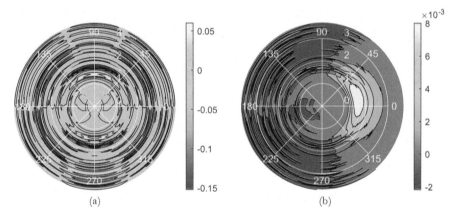

Fig. 26.8 FEM with current sensor layout, based on direct computation of cross-spectral densities. (**a**) Case 1: $H_s = 0.6m$, $s = 30$, $\theta_0 = 90°$. (**b**) Case 2: $H_s = 0.6m$, $s = 3$, $\theta_0 = 15°$

which is based on the mentioned paper by Hashimoto et al. (1994). In the current paper, no further description of the methodology is given.

26.4.4.1 Application of the EMEP on Simulated Data

The directional wave spectral densities estimated from simulated data on the current sensor layout with the EMEP algorithm are illustrated in Fig. 26.9. Due to difficulties in proper estimates of the cross-spectral densities of the wave elevations at the sensor locations, which are less smooth for large distances, gaps are observed for certain

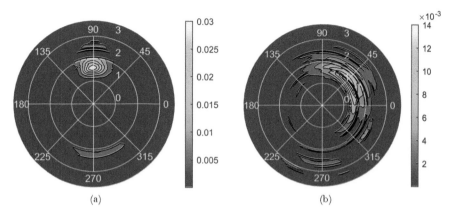

Fig. 26.9 EMEP with current sensor layout. (a) Case 1: $H_s = 0.6m$, $s = 30$, $\theta_0 = 90°$. (b) Case 2: $H_s = 0.6m$, $s = 3$, $\theta_0 = 15°$

frequency ranges in the resulting directional wave spectral density. Furthermore, contrary to what is found when applying the FEM, the sea state indicating head sea (case 1) is much more accurately identified than the one characterized by obliquely approaching waves (case 2). It is noted that the EMEP algorithm is superior to the FEM for the identification of both the simulated cases.

As indicated in Sect. 26.4.2, the sensor layout may be improved. A suggested new sensor layout is depicted in Fig. 26.6c. In the new layout, three aspects are improved: (1) the distance vectors are more uniformly distributed; (2) the anisotropy of the sensor positioning is reduced as the distance vectors are less purely longitudinal; and (3) the sensor distances are reduced. The result from the EMEP algorithm with this layout is depicted in Fig. 26.10. The new layout is in general found to be much more robust for the identification of both simulated test cases. The increased robustness of the spectral estimation, due to shorter distances between the sensors, is believed to be the main cause of this.

26.5 Concluding Remarks

Two well-established methods for determination of the directional wave spectral density, namely the Fourier Expansion Method (FEM) and the Extended Maximum Entropy Principle (EMEP), have been applied on simulated data for the existing wave radar layout on the Bergsøysund Bridge.

The FEM fails to identify the simulated directional spectral density from cross-spectral density estimates based on time simulations, when assuming a sensor layout as is presently installed on the bridge. However, for direct calculation of the cross-spectral densities, the FEM correctly characterizes the sea state when the waves are

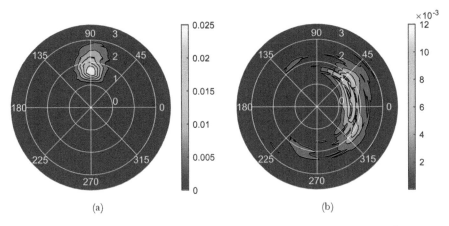

Fig. 26.10 EMEP with improved sensor layout. (**a**) Case 1: $H_s = 0.6m$, $s = 30$, $\theta_0 = 90°$. (**b**) Case 2: $H_s = 0.6m$, $s = 3$, $\theta_0 = 15°$

approaching with an oblique angle to the bridge. This is not the case for a head sea situation, where the method fails also when applying the cross-spectral densities directly. With the selected reference sensor layout, the head sea state (case 1) is estimated as slightly more directionally spread out than correct, an artefact commonly observed with the FEM. The EMEP successfully identifies the head sea state (case 1), but does not reliably identify the sea state characterized by obliquely approaching waves (case 2). However, it performs significantly better than the FEM, also for case 2.

The current layout has a large weakness in its anisotropic distribution, such that mean wave direction plays a major role in the success of the identification. A new layout is suggested to solve this. The suggested new layout results in a more robust estimation of the cross-spectral densities, which in turn improves the quality of the resulting directional spectral densities.

Both methods rely on good estimates of cross-spectral densities, which is a task that requires both experience, skill and sometimes luck. The resulting two-dimensional wave spectral density, and thus directional distribution, is highly dependent on this estimation procedure, i.e., the estimator methods and parameters used.

26.5.1 Future Work

Recordings from the sensors acting on site should be analysed in a similar manner to characterize the wave field on site. The current work represents an important step towards this.

Acknowledgements The research is funded by the Norwegian Public Roads Administration. The authors gratefully acknowledge this. We would also like to thank our dear colleague, the late Prof. Ragnar Sigbjörnsson, for his commitment and inspiration in common scientific endeavours.

References

Barber NF (1961) The directional resolving power of an array of wave detectors. In: Ocean wave spectra. Prentice-Hall, Inc, Englewood Cliffs, pp 137–150

Donelan MA, Hamilton J, Hui WH (1985) Directional spectra of wind-generated waves. Philos Transac Roy Soc Lond Ser A Math Phys Sci 315(1534):509–562. https://doi.org/10.1098/rsta.1985.0054

Giske F-IG, Leira BJ, Øiseth O (2017) Efficient computation of cross-spectral densities in the stochastic modelling of waves and wave loads. Appl Ocean Res 62:70–88. https://doi.org/10.1016/j.apor.2016.11.007

Goda Y (1981) Simulation in examination of directional resolution. In: Proceedings of the conference on directional wave spectra applications, pp 387–407

Hashimoto N, Nagai T, Asai T (1994) Extension of the maximum entropy principle method for directional wave spectrum estimation. Coast Eng Proc 1(24)

Hauser D, Kahma K, Krogstad HE (2005) Measuring and analysing the directional spectrum of ocean waves. Publications Office of the European Union, Luxembourg

Jefferys ER (1987) Directional seas should be ergodic. Appl Ocean Res 9(4):186–191. https://doi.org/10.1016/0141-1187(87)90001-0

Johnson D (2002) DIWASP, a directional wave spectra toolbox for MATLAB®: User Manual. Center for Water Research, University of Western Australia

Kobune K, Hashimoto N (1986) Estimation of directional spectra from the maximum entropy principle. In: Proceedings of the fifth international symposium on offshore mechanics and arctic engineering, vol. 1, pp 80–85

Kvåle KA, Øiseth O (2017) Structural monitoring of an end-supported pontoon bridge. Mar Struct 52:188–207. https://doi.org/10.1016/j.marstruc.2016.12.004

Kvåle KA, Øiseth O, Rønnquist A, Sigbjörnsson R (2015) Modal analysis of a floating bridge without side-mooring. In: Dynamics of civil structures, vol 2. Springer, New York/London, pp 127–136. https://doi.org/10.1007/978-3-319-15248-6_14

Kvåle KA, Sigbjörnsson R, Øiseth O (2016) Modelling the stochastic dynamic behaviour of a pontoon bridge: a case study. Comput Struct 165:123–135. https://doi.org/10.1016/j.compstruc.2015.12.009

Kvåle KA, Øiseth O, Rønnquist A (2017a) Operational modal analysis of an end-supported pontoon bridge. Eng Struct 148:410–423. https://doi.org/10.1016/j.engstruct.2017.06.069

Kvåle KA, Øiseth O, Rönnquist A (2017b) Covariance-driven stochastic subspace identification of an end-supported pontoon bridge under varying environmental conditions. In: Caicedo J, Pakzad S (eds) Dynamics of civil structures, vol 2. Conference proceedings of the society for experimental mechanics series. Springer International Publishing, Cham, pp 107–115. https://doi.org/10.1007/978-3-319-54777-0_14

Longuet-Higgins MS, Cartwright DE, Smith ND (1963) Observations of the directional spectrum of sea waves using the motions of a floating buoy. In: Ocean wave spectra. Prentice-Hall, Inc, New Jersey, pp 111–136

Massel SR, Brinkman RM (1998) On the determination of directional wave spectra for practical applications. Appl Ocean Res 20(6):357–374. https://doi.org/10.1016/S0141-1187(98)00026-1

Nwogu O (1989) Maximum entropy estimation of directional wave spectra from an array of wave probes. Appl Ocean Res 11(4):176–182. https://doi.org/10.1016/0141-1187(89)90016-3

Ochi MK (2005) Ocean waves: the stochastic approach, vol 6. Cambridge University Press

Panicker NN, Borgman LE (1970) Directional spectra from wave-gage arrays. In: Coastal Engineering Proceedings; No 12 (1970): Proceedings of 12th conference on coastal engineering, Washington, DC, 1970. Retrieved from https://journals.tdl.org/icce/index.php/icce/article/view/2612

Pierson WJ, Moskowitz L (1964) A proposed spectral form for fully developed wind seas based on the similarity theory of S. A. Kitaigorodskii. J Geophys Res 69(24):5181–5190. https://doi.org/10.1029/JZ069i024p05181

Sigbjörnsson R (1979) Stochastic theory of wave loading processes. Eng Struct 1(2):58–64. https://doi.org/10.1016/0141-0296(79)90014-2

Young IR (1994) On the measurement of directional wave spectra. Appl Ocean Res 16(5):283–294. https://doi.org/10.1016/0141-1187(94)90017-5

Chapter 27
Identification of Rational Functions with a Forced Vibration Technique Using Random Motion Histories

Bartosz Siedziako and Ole Øiseth

Abstract Rational Functions are used to describe the self-excited forces acting on the bridge deck in the time domain. They can be identified indirectly based on aerodynamic derivatives or directly with the free (E2RFC method) or forced vibration technique, which can significantly decrease the testing time. The approach presented herein enables the extraction of Rational Function Coefficients by testing the section model at only one wind speed. This aim is achieved by increased complexity of the forced motion compared to the previous tests, which made it possible to test a wider range of reduced velocities by adjusting the motion frequency. In this study, motion histories generated from the assumed flat spectra are used. Wind tunnel tests on a streamlined section model utilizing simultaneous vertical, horizontal and torsional vibrations were performed to extract Rational Function Coefficients associated with 3-degree-of-freedom motion. Restrictions and improvements arising from the proposed methodology are described.

Keywords Rational functions · Forced vibration · Section model · Arbitrary motion · Bridge aeroelasticity

27.1 Introduction

Slender structures such as suspension and cable-stayed bridges are especially vulnerable to wind-induced phenomena, namely flutter, buffeting and galloping. Scanlan and Tomko (1971) introduced aerodynamic derivatives (ADs) that characterize the aerodynamic performance of the bridge deck and enable detailed analysis of the bridge's in-wind behavior in the frequency domain. The aerodynamic derivatives that define self-excited forces are most commonly derived experimentally in a

---

B. Siedziako (✉) · O. Øiseth
Department of Structural Engineering, Norwegian University of Science and Technology, Trondheim, Norway
e-mail: bartosz.siedziako@ntnu.no; ole.oiseth@ntnu.no

series of wind tunnel tests with a section model of the bridge deck using the free or forced vibration technique. They can be identified at discrete reduced velocities often within a limited range, depending on the frequencies and velocities tested during the experiments.

Current technological and engineering advances have made it possible to build increasingly slender bridges with very light road decks, leading to the construction of possibly highly nonlinear structures. Moreover, the lower damping of the structure due to the reduced mass emphasizes the significance of aerodynamic damping. Therefore, time-domain flutter and buffeting analyses, which can incorporate structural and aerodynamic nonlinearities, have become more common in recent years (Salvatori and Borri 2007; Øiseth et al. 2011). Formulated in the Laplace domain by Roger (1977), the Rational Function Approximation (LS-RFA) using least squares enabled the time-domain modeling of the frequency dependent self-excited forces. Later, (Karpel 1981) introduced the Minimum State Rational Function Approximation (MS-RFA), which improved the accuracy and decreased the computational time compared to LS-RFA. The main objective of these RFA formulations is to identify the Rational Function Coefficients (RFCs) that define the motion to self-excited forces continuous transfer functions. However, this approximation involves experimentally obtained aerodynamic derivatives in the process of linear and nonlinear optimizations (Neuhaus et al. 2009). This motivated other researchers to find a more direct method to obtain RFCs from wind-tunnel measurements that would make it possible to skip the process of extracting aerodynamic derivatives. Chowdhury and Sarkar (2005) proposed a method to directly extract the RFCs from free vibration tests, while Cao and Sarkar (2012), to overcome some limitations of the free vibration technique, developed a similar algorithm for the forced vibration testing technique. In both methods, the RFCs can be extracted directly from time series recorded during wind tunnel experiments at only a few wind velocities (a minimum of two wind speeds), which can significantly decrease testing time compared to the standard approach with aerodynamic derivatives. However, in the method proposed by Cao and Sarkar (2012), simultaneous pitching and heaving harmonic oscillations of the section model were considered. In this study, a more general, three-degree-of-freedom random motion generated from flat motion spectra is used to identify RFCs. It is shown that through this approach, a bridge deck section model needs to be tested at only one wind speed to extract the full set of RFCs.

27.2 Experimental Setup

27.2.1 Forced Vibration Rig

The forced vibration setup developed at the Norwegian University of Science and Technology has been used in this study (Siedziako et al. 2017). This setup was especially designed to be capable of forcing arbitrary motion histories of the bridge deck section model in heaving, swaying and torsional directions simultaneously.

Fig. 27.1 Experimental forced vibration setup at NTNU (Siedziako et al. 2017)

Figure 27.1 shows the segment of the wind tunnel with the main construction of the setup. The section model of the bridge is attached between the two actuators placed outside on both sides of the wind tunnel. Inside each of the actuators reside two ball screws for the vertical and horizontal motion and a planetary gear for the torsional motion. Two high-sensitivity load cells measure 3 force and 3 moment components acting on the section model during the experiments. The actuators are supported by the steel frame outside the wind tunnel, while the load cells are mounted between the section model and actuators in the centers of two circular holes made in the wind tunnel walls.

The described setup makes it possible to move the section model arbitrarily according to the uploaded motion histories. Data transfer with the time series of displacement is managed using the LabVIEW program, which is also responsible for triggering motion, monitoring, controlling algorithms and acquiring data. In this study, the uploaded motion time series were generated with a time step of 1 ms, while a sampling rate of 250 Hz was set for the data acquisition.

27.2.2 Wind Tunnel

The wind tunnel tests were conducted in the largest wind tunnel in the Fluid Mechanics Laboratory at NTNU. It is a closed loop wind tunnel with a test section 11 m long, 2 m height and 2.7 m wide with a maximum speed of 30 m/s. Temperature inside the wind tunnel was measured with a thermocouple to account for changes in the air density, while to measure the air velocity static,

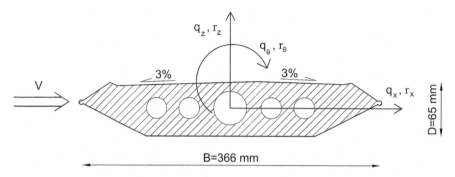

Fig. 27.2 The cross-sectional dimensions of the bridge deck used in this study

a pitot probe was placed 6.10 m in front of the section model. All the tests presented in this paper were performed in a smooth flow.

27.2.3 Bridge Deck Section Model

The bridge deck of the currently longest suspension bridge in Norway, Hardanger Bridge, was used in this study. The geometric shape of the bridge deck allows it to be considered as a perfect example of a streamlined section. The cross-sectional dimensions of the model are shown in Fig. 27.2 together with the coordinate system applied. Thanks to additional holes and very light filling material, the model is very light. With a length of L = 2.68 m, it weighs only 5.45 kg. The high aspect ratio L/B = 7.32 and the fact that the model is only 3 cm shorter than the width of the wind tunnel, eliminated the need to use additional end plates.

27.3 Identification Algorithm

An algorithm used in this study, adapted to the forced vibration technique, has been proposed by Cao and Sarkar (2012) and is based on the previous work by Roger (1977) and Karpel (1981) in the field of aeronautics; therefore, the authors refer to those publications for more details on its derivation. Following Roger (1977), the self-excited forces in the 3-DoF system can be expressed in the Laplace domain as follows:

$$\begin{bmatrix} \widehat{q}_x \\ \widehat{q}_z \\ \widehat{q}_\theta \end{bmatrix} = \frac{1}{2}\rho V^2 B \begin{bmatrix} 1 & 0 & 0 \\ 0 & 1 & 0 \\ 0 & 0 & B \end{bmatrix} \mathbf{Q} \begin{bmatrix} \widehat{r}_x/B \\ \widehat{r}_z/B \\ \widehat{r}_\theta \end{bmatrix} \quad (27.1)$$

Here, ρ is the air density; V denotes the mean wind velocity; B is the bridge deck width, and '^' indicates that the variable is in the Laplace domain. Similarly to the description given by Scanlan and Tomko (1971), Eq. (27.1) presents a linear relation between aeroelastic forces (q_x – drag, q_z – lift, q_θ – pitch) and the horizontal (r_x), vertical (r_z) and torsional vibrations (r_θ) of the bridge deck. The matrix **Q** of Rational Functions is the transfer function in the Laplace domain given by:

$$\mathbf{Q} = \begin{bmatrix} (A_0)_{11} + (A_1)_{11}p + \dfrac{F_{11}p}{p+\lambda} & (A_0)_{12} + (A_1)_{12}p + \dfrac{F_{12}p}{p+\lambda} & (A_0)_{13} + (A_1)_{13}p + \dfrac{F_{13}p}{p+\lambda} \\ (A_0)_{21} + (A_1)_{21}p + \dfrac{F_{21}p}{p+\lambda} & (A_0)_{22} + (A_1)_{22}p + \dfrac{F_{22}p}{p+\lambda} & (A_0)_{23} + (A_1)_{23}p + \dfrac{F_{23}p}{p+\lambda} \\ (A_0)_{31} + (A_1)_{31}p + \dfrac{F_{31}p}{p+\lambda} & (A_0)_{32} + (A_1)_{32}p + \dfrac{F_{32}p}{p+\lambda} & (A_0)_{33} + (A_1)_{33}p + \dfrac{F_{33}p}{p+\lambda} \end{bmatrix}$$

(27.2)

Here, A_0 and A_1 and F are, respectively, the stiffness, damping and lag matrices, all of order 3×3 that contain unknown RFCs. The value λ denotes an unknown lag term, while $p = iK$ represents the dimensionless Laplace variable, where $K = B\omega/V$ is the reduced frequency, and ω is the circular frequency of motion. The expression approximating the Rational Function in Eq. (27.2) can be further extended by including additional lag terms and lag matrices, but previous studies have shown that the Rational Function Approximation with one lag term as presented herein is sufficient (Cao and Sarkar 2010, 2012; Chowdhury 2004; Chowdhury and Sarkar 2005; Neuhaus et al. 2009) in the case of bridge decks. By multiplying Eq. (27.2) by $p + \lambda$ and applying the inverse Laplace transform, the following time-domain equations for the self-excited drag, lift and pitching moment can be obtained:

$$\begin{aligned} \dot{\mathbf{q}}_x + \lambda_x \frac{V}{B}\mathbf{q}_x &= \frac{1}{2}\rho V^2 B \left[\frac{V}{B}\boldsymbol{\psi}_1 \mathbf{r} + \boldsymbol{\psi}_2 \dot{\mathbf{r}} + \frac{B}{V}\boldsymbol{\psi}_3 \ddot{\mathbf{r}} \right] \\ \dot{\mathbf{q}}_z + \lambda_z \frac{V}{B}\mathbf{q}_z &= \frac{1}{2}\rho V^2 B \left[\frac{V}{B}\boldsymbol{\psi}_4 \mathbf{r} + \boldsymbol{\psi}_5 \dot{\mathbf{r}} + \frac{B}{V}\boldsymbol{\psi}_6 \ddot{\mathbf{r}} \right] \\ \dot{\mathbf{q}}_\theta + \lambda_\theta \frac{V}{B}\mathbf{q}_\theta &= \frac{1}{2}\rho V^2 B^2 \left[\frac{V}{B}\boldsymbol{\psi}_7 \mathbf{r} + \boldsymbol{\psi}_8 \dot{\mathbf{r}} + \frac{B}{V}\boldsymbol{\psi}_9 \ddot{\mathbf{r}} \right] \end{aligned}$$

(27.3)

Here, 1×3 size vectors $\boldsymbol{\psi}_i$ i = 1,2 ... 9 contain the unknown RFCs; **r** is the vibration matrix consisting of horizontal vertical and torsional vibrations $\mathbf{r} = [r_x/B\ r_y/B\ r_\theta]^T$; and $\dot{\mathbf{r}}$ and $\ddot{\mathbf{r}}$ are, respectively, the first and second derivatives of the displacements. After a slight modification, Eq. (27.3) can rewritten into the following expression:

$$\mathbf{A}_n \mathbf{C}_n = \dot{\mathbf{q}}_n \quad n \in \{x, z, \theta\}$$

(27.4)

where matrices $\mathbf{A}_n$ and $\mathbf{C}_n$ are given by Eq. (27.5):

$$\mathbf{A_x} = \begin{bmatrix} \psi_1 \\ \psi_2 \\ \psi_3 \\ -\lambda_x \end{bmatrix}^T \quad \mathbf{A_z} = \begin{bmatrix} \psi_4 \\ \psi_5 \\ \psi_6 \\ -\lambda_z \end{bmatrix}^T \quad \mathbf{A_\theta} = \begin{bmatrix} \psi_7 \\ \psi_8 \\ \psi_9 \\ -\lambda_\theta \end{bmatrix}^T \quad \mathbf{C_x} = \begin{bmatrix} 0.5\rho V^3 \mathbf{r} \\ 0.5\rho V^2 B \dot{\mathbf{r}} \\ 0.5\rho V B^2 \ddot{\mathbf{r}} \\ \mathbf{q_x} V/B \end{bmatrix} \quad \mathbf{C_z}$$

$$= \begin{bmatrix} 0.5\rho V^3 \mathbf{r} \\ 0.5\rho V^2 B \dot{\mathbf{r}} \\ 0.5\rho V B^2 \ddot{\mathbf{r}} \\ \mathbf{q_z} V/B \end{bmatrix} \quad \mathbf{C_\theta} = \begin{bmatrix} 0.5\rho V^3 B \mathbf{r} \\ 0.5\rho V^2 B^2 \dot{\mathbf{r}} \\ 0.5\rho V B^3 \ddot{\mathbf{r}} \\ \mathbf{q_\theta} V/B \end{bmatrix} \quad (27.5)$$

To find matrices A_n that contain RFCs, an algorithm that minimizes the sum of squares can be applied:

$$\mathbf{A}_n = \left(\dot{\mathbf{q}}_n \mathbf{C}_n^T\right)\left(\mathbf{C}_n \mathbf{C}_n^T\right)^{-1} \quad n \in \{x, z, \theta\} \quad (27.6)$$

In this study, the derivatives of the drag, lift, pitching moment and displacements were obtained by applying the finite difference algorithm to the recorded time histories. Since the motion considered herein is a combination of horizontal, vertical and torsional vibrations, all the RFCs can be identified using the data from a single forced vibration test at a particular wind speed.

27.4 Random Motion Histories

The random motion histories used in this study were generated by Monte Carlo simulations (Aas-Jakobsen and Strømmen 2001; Øiseth et al. 2011) from an assumed cross-spectral density matrix of the response $S_r(\omega)$. To achieve the maximum possible randomness of the time series and prove that the experimental setup can induce arbitrary motion of the section model, flat spectra in the range of 0.3–2.5 Hz have been used to generate histories of displacements for later upload to the actuators. The amplitudes of the spectra $S_r(\omega)$ have been scaled to obtain standard deviations of the horizontal, vertical and torsional responses, respectively, 6.5 mm, 6.5 mm and 1.4°. The time series for the degree of freedom $m \in \{x, z, \theta\}$ were obtained using Eq. (27.7):

$$x_m(t) = \sqrt{2\Delta\omega} \operatorname{Re}\left(\sum_{l=1}^{m} \sum_{k=1}^{N} L_{ml}(\omega_k) \exp(i(\omega_k t + \phi_{lk}))\right) \quad (27.7)$$

where $L_{ml}(\omega_k)$ denotes the elements of the lower triangular matrix obtained by factorizing the cross-spectral density matrix according to the relation given in Eq. (27.8).

27 Identification of Rational Functions with a Forced Vibration...

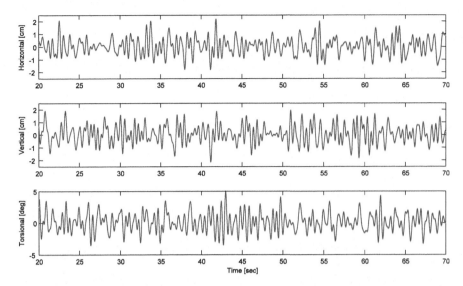

Fig. 27.3 Part of the time series of the section model used for wind tunnel testing generated from assumed flat spectra in the range of 0.3–2.5 Hz

$$\mathbf{S_r}(\omega_k) = \mathbf{L}(\omega_k)\mathbf{L}^*(\omega_k) \tag{27.8}$$

Figure 27.3 presents part of the time series induced on the section model during wind tunnel testing, generated using Eq. (27.7). It can be seen that the created motion histories are very chaotic and simulate a white-noise stochastic process well. The experimental rig used in this study has been designed to address much larger amplitudes and motion frequencies than used herein, and therefore the motion during experiments was very smooth, and the actuators perfectly followed the uploaded motion history.

27.5 Experimental Results

To compare the results obtained in this section with Rational Functions, aerodynamic derivatives of the Hardanger Bridge section model are needed. The aerodynamic derivatives identified in a standard forced vibration procedure with that section have been presented in (Siedziako et al. 2016, 2017). Those two references provide more information about the amplitudes, frequencies and wind speeds tested and also describe the methodology used for extracting self-excited forces, which requires measuring forces for the same motion in still-air and in-wind conditions. The same methodology has been applied herein, considering tests with random motion histories. The duration of each test was taken to be 100s. Tests have been performed at three wind speeds, V = 4, 8 and 10 m/s.

To evaluate the identification algorithm described in Chap. 3 and determine the accuracy of the fit, the extracted RFCs can be used to predict the self-excited forces. Cao and Sarkar (2012) used for this task an expression that contains a convolution integral; however, it has been shown that it can be conveniently replaced with a state-space formulation (Chen et al. 2000; Høgsberg et al. 2000; Mishra et al. 2008). The second approach has been used in this study – see (Øiseth et al. 2012) for more details. Example time series of recorded and predicted self-excited forces are shown in Fig. 27.4. Forces have been calculated based on the Rational Functions identified using the data from the test conducted at V = 4 m/s. Table 27.1 presents collected information about the correlation coefficients between measured and predicted forces together with their standard deviations.

It can be seen that a perfect match between measured and predicted with RFC self-excited forces has been achieved for the pitch and lift. However, in the case of the self-excited drag, the calculated correlation between the measured and predicted forces is significantly lower than for the lift and pitch. Recent studies

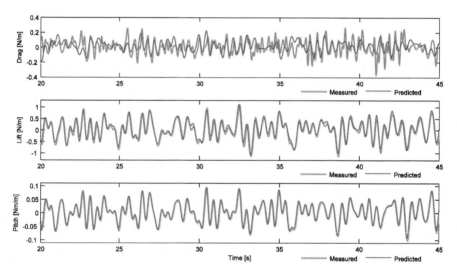

Fig. 27.4 Measured forces vs. forces predicted with RFCs induced during execution of the random motion at V = 4 m/s

Table 27.1 Correlation coefficients and standard deviations of measured (σ_M) and predicted (σ_P) self-excited forces

Wind speed	Drag			Lift			Pitch		
	ρ_{xy}	σ_M [N/m]	σ_P [N/m]	ρ_{xy}	σ_M [N/m]	σ_P [N/m]	ρ_{xy}	σ_M [Nm/m]	σ_P [Nm/m]
V = 4 m/s	0.514	0.087	0.066	0.981	0.395	0.381	0.998	0.033	0.033
V = 8 m/s	0.300	0.082	0.043	0.998	1.670	1.550	0.999	0.131	0.131
V = 10 m/s	0.531	0.115	0.080	0.996	2.77	2.526	0.998	0.209	0.208

by Xu et al. (2016) have shown that the self-excited drag is prone to higher-order contributions that cannot be captured by linear load models and can be especially large when considering streamlined sections, as in this study. This finding agrees with the results presented herein, as the drag force is clearly underestimated in all tests when comparing the standard deviations of the measured and predicted drag.

Knowing that the matrix of Rational Functions $\mathbf{Q}$ can also be described by Eq. (27.9), the relations between particular aerodynamic derivatives and RFCs can be established to allow the direct comparison of the results obtained here with the ones presented in (Siedziako et al. 2017).

$$\mathbf{Q} = \begin{bmatrix} K^2\left(P_1^*i + P_4^*\right) & K^2\left(P_5^*i + P_6^*\right) & K^2\left(P_2^*i + P_3^*\right) \\ K^2\left(H_5^*i + H_6^*\right) & K^2\left(H_1^*i + H_4^*\right) & K^2\left(H_2^*i + H_3^*\right) \\ K^2\left(A_5^*i + A_6^*\right) & K^2\left(A_1^*i + A_4^*\right) & K^2\left(A_2^*i + A_3^*\right) \end{bmatrix} \quad (27.9)$$

Figures 27.5 and 27.6 compare all 18 ADs obtained herein from Rational Functions with the ones identified in the forced vibration tests using the standard procedure. It can be seen that the ADs match very well, especially the most important ADs, namely A_1^*, A_2^*, A_3^*, H_3^*, and also H_2^* as the torsional motion is responsible for most of the induced self-excited forces. However, the ADs found at the lower reduced velocities seems to correspond better to the original

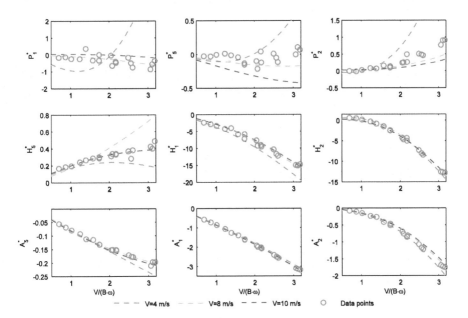

Fig. 27.5 Aerodynamic derivatives related to velocities or angular velocities. Comparison of experimentally obtained ADs (Siedziako et al. 2017) and ADs extracted from rational functions identified at one wind speed

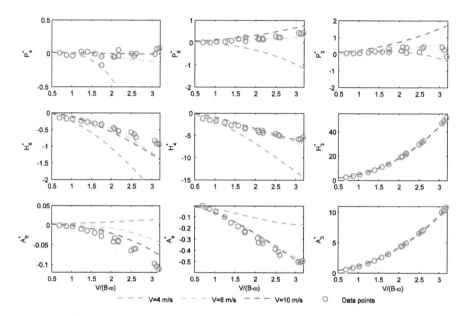

Fig. 27.6 Aerodynamic derivatives related to displacements or rotation. Comparison of experimentally obtained ADs (Siedziako et al. 2017) and ADs extracted from rational functions identified at one wind speed

ones than at the higher reduced velocities, which is especially visible in the case of the ADs extracted from RFCs identified at V = 4 m/s.

27.6 Conclusion

In this paper, a recently developed algorithm for the extraction of Rational Function Coefficients has been used for the first time with a non-harmonic motion pattern. Motion that simultaneously involves horizontal, vertical and torsional vibrations generated from flat motion spectra has been used to measure the self-excited forces induced on the streamlined section model. Preliminary studies showed that the full set of Rational Function Coefficients can be identified from a single test considering only one wind speed. The identified Rational Function Coefficients provided an excellent fit to time series of recorded self-excited lift and pitching moment, demonstrating the high performance of the algorithm used in this study. However, some discrepancies that require separate studies were observed in the drag force.

In the experiments performed, aeroelastic forces related to the torsional motion dominated the measured self-excited drag, lift and pitching moment. Moreover, the motion type used in this study tends to favor the extraction of ADs at the lower reduced velocities. Therefore, suitable design of the motion histories for the wind tunnel testing might be of key importance for the method described in the future.

Additionally, testing section models of the bridge decks considering motions that resemble actual bridge motion would presumably eliminate this problem since the Rational Function Coefficients would be optimized in the range of reduced velocities that correspond to the bridge's natural frequencies. It must be emphasized that the method presented herein assumes that the principle of superposition between the motion and induced self-excited forces is valid. Although the results presented herein strongly suggest that it is, there should certainly be further investigations to assess whether this assumption is well founded.

Acknowledgements This research was conducted with financial support from the Norwegian Public Roads Administration. The authors gratefully acknowledge this support.

References

Aas-Jakobsen K, Strømmen E (2001) Time domain buffeting response calculations of slender structures. J Wind Eng Ind Aerodyn 89(5):341–364
Cao B, Sarkar PP (2010) Identification of rational functions by forced vibration method for time-domain analysis of flexible structures. In: Proceedings of the fifth international symposium on computational wind engineering. Chapel Hill
Cao B, Sarkar PP (2012) Identification of rational functions using two-degree-of-freedom model by forced vibration method. Eng Struct 43:21–30
Chen X, Matsumoto M, Kareem A (2000) Aerodynamic coupling effects on flutter and buffeting of bridges. J Eng Mech 126:17–26
Chowdhury AG (2004) Identification of frequency domain and time domain aeroelastic parameters for flutter analysis of flexible structures. Ph.D thesis, Iowa State University, USA, p 778
Chowdhury AG, Sarkar PP (2005) Experimental identification of rational function coefficients for time-domain flutter analysis. Eng Struct 27:1349–1364
Høgsberg JR, Krabbenhøft J, Krenk S (2000) State space representation of bridge deck aeroelasticity. In: Proceedings of the 13th Nordic seminar on computational mechanics, Oslo, pp 109–112
Karpel M (1981) Design for active and passive flutter suppression and gust alleviation. NASA contractor report No. 3482
Mishra SS, Kumar K, Krishna P (2008) Multimode flutter of long-span cable-stayed bridge based on 18 experimental aeroelastic derivatives. J Wind Eng Ind Aerodyn 96(1):83–102
Neuhaus Ch, Mikkelsen O, Bogunovic Jakobsen J, Höffer R, Zahlten W (2009) Time domain representations of unsteady aeroelastic wind forces by rational function approximations. In: Proceedings of the EACWE 5, Florence
Roger KL (1977) Airplane math modeling and active aeroelastic control design[C]. AGARD-CP-228
Salvatori L, Borri C (2007) Frequency- and time-domain methods for the numerical modeling of full-bridge aeroelasticity. Comput Struct 85(11–14):675–687
Scanlan RH, Tomko JJ (1971) Airfoil and bride deck flutter derivatives. J Eng Mech Div 97(6):1717–1733
Siedziako B, Øiseth O, Rønnquist A (2017) An enhanced forced vibration rig for wind tunnel testing of bridge deck section models in arbitrary motion. J Wind Eng Ind Aerodyn 164:152–163
Siedziako B, Øiseth O, Rønnquist A (2016) A new setup for section model tests of bridge decks. In: Proceedings of 12th UK conference on wind engineering, Nottingham

Xu FY, Wu T, Ying XY, Kareem A (2016) Higher-order self-excited drag forces on bridge decks. J Eng Mech 142(3):1–11

Øiseth O, Rönnquist A, Sigbjörnsson R (2011) Time domain modeling of self-excited aerodynamic forces for cable-supported bridges: a comparative study. Comput Struct 89(13–14):1306–1322

Øiseth O, Rönnquist A, Sigbjörnsson R (2012) Finite element formulation of the self-excited forces for time-domain assessment of wind-induced dynamic response and flutter stability limit of cable-supported bridges. Finite Elem Anal Des 50:173–183

Chapter 28
The Dynamic Intelligent Bridge: A New Concept in Bridge Dynamics

Andreas J. Kappos

Abstract A method is put forward for designing bridges with improved performance under extreme dynamic loadings, such as strong earthquakes. The basic idea is that varying the boundary conditions can lead to an improved structural performance under dynamic actions. The specific goal is to substitute current bridge joints that have a fixed width with variable-width joints, which initially can be either closed or open depending on their length and the serviceability requirements, while under seismic loading their width is optimised either with a one-off adjustment, or continuously varying through semi-active control. In all cases, a novel device is used that permits this improved behaviour of the joints, the moveable shear key (MSK), a device for blocking the movement of the bridge deck, which is not permanently fixed to the seat of the abutment but can slide, hence opening a previously closed gap or closing an existing gap between the deck and the abutment. The performance sought by varying the joint gap depends on the design objectives. A pilot study on the effect of gap size is also presented, which illustrates that it can significantly affect the response quantities of the abutments.

Keywords Bridge design · Seismic loading · Boundary nonlinearity · Semi-active control · Movable shear keys

28.1 Introduction

The paper presents the concept of the 'dynamic intelligent' (DI) bridge that has improved performance under extreme dynamic loadings, such as strong earthquake or strong winds-hurricanes. The basic idea is that varying the boundary conditions can lead to an improved structural performance under different dynamic actions. The

---

A. J. Kappos (✉)
City, University of London, London, UK

Aristotle University of Thessaloniki, Thessaloniki, Greece
e-mail: Andreas.Kappos.1@city.ac.uk

key idea is the control of joint opening through a special type of moveable shear key (MSK) that can be controlled in a number of alternative ways, ranging from simple rough interfaces to semi-active control. The bridge behaves in the most favourable way under dynamic loads due to the presence of joints that can open in either direction (longitudinal or transverse).

Boundary conditions play a key role in the response of bridges to different kinds of loads. Support conditions in bridges are relatively easy to control, as different types of bearings and shear keys that block or release various degrees of freedom are currently available. In the case of environmental loads, the key role of the longitudinal joints is to reduce stresses generated from expansion and contraction due to temperature and time-dependent effects like shrinkage and creep. The usual design approach for joint gaps is to accommodate the full displacement due to the permanent and quasi-permanent actions and a fraction of the displacement due to temperature variations within the end joint (e.g. BSI 2012). The presence of longitudinal and/or transverse gaps at the ends of the bridge may result in more favourable dynamic response, depending on the intensity and frequency content of the dynamic input. Kappos and Sextos (2009) have studied the effect of boundary conditions on the seismic response of a bridge focussing on the effect of closing of gaps at the bridge ends and the changes in the seismic response of the bridge as these gaps closed at earthquake intensities higher than the design one. Neither this, nor any other study has considered alternative solutions for the design of gaps and in all relevant studies the gap size was kept constant during the analysis.

The performance of bridges under low probability loadings like large temperature variations or extreme dynamic loads can be quite poor and points to the need for a proper performance-based design. There are some spectacular cases of bridge collapses due to *extreme winds*, such as that of the Kinzua Viaduct in Pennsylvania that collapsed in 2003 due to a tornado with speeds between 73 and 112 mph, while 45 bridges were damaged during Hurricane Katrina, some of them collapsed (due to span unseating), with a total cost estimated at over $1 billion (Padgett et al. 2008). There are several examples of heavy damage or collapse of bridges from strong *earthquakes*; a well-known one is that of the Hanshin Expressway during the Kobe Earthquake due to failure of some piers (Kawashima and Unjoh 1997), while there were several bridge collapses (Romero, Lo Echevers, La Mochita, Llacolen, Tubul) during the 2010 Chile Earthquake (Yashinsky et al. 2010).

The response of bridges to the aforementioned types of low-probability loadings (normally associated with natural hazards) can be improved by the use of *structural control*. Among the different types of control techniques available, *semi-active* control emerges as a rational combination of efficiency and cost, and was recently implemented mainly as a retrofit measure to control cable or deck vibrations in bridge structures across Asia, Europe and the United States. The specific solutions implemented so far include (Gkatzogias and Kappos 2016): (1) Variable orifice dampers, which are devices that use a controllable, electromechanical, variable-orifice valve to alter the resistance to flow of a conventional hydraulic fluid damper and hence control the damping coefficient; (2) Semi-active stiffness control devices that can vary (either continuously or on an on-off basis) the stiffness of the bracing system of the structure; (3) Friction controlled devices, either in the form of friction dampers (energy dissipators), or as components within sliding isolation systems (coefficient of friction at the sliding bearing interface controlled by adjusting the fluid pressure inside the bearings); (4) Controllable fluid dampers that use controllable fluids instead of electrically controlled valves or mechanisms used in passive fluid dampers; (5) Controllable tuned mass and liquid dampers that consist, in general, of a secondary mass with properly tuned spring and damping elements, providing a frequency-dependent hysteresis that increases damping in the primary structure; (6) Negative stiffness devices which exhibit hysteresis loops with negative stiffness to prevent the transfer of large damping forces developed in long-period base isolated structures with high values of damping ratios to the main structure while maintaining large energy dissipation. All these devices (described in detail in Gkatzogias and Kappos 2016) are typically in the form of dampers or, less often, as braces connected to, or components of, the sliding isolation system.

This paper describes the basic concepts in designing a dynamic intelligent bridge, as well as a pilot study exploring the influence of varying joint gaps on the seismic performance of a typical bridge. If the proposed concept proves to be effective, i.e., if the additional cost required for installing the devices for controlling the opening of the joints is found to be outweighed by the reduced cost of damage due to dynamic loading, then one can claim that a new improved type of bridges can be constructed, with a reduced life-cycle cost.

28.2 Overview of the Proposed System

The proposed novel device that permits varying the joint gap, the MSK, is a stopper, arranged usually internally (Fig. 28.1b), but in the transverse direction, can also be placed externally, to the deck (Fig. 28.1a), which (unlike currently used shear keys) is not permanently fixed to the seat of the abutment but can slide, opening a previously closed gap or closing an existing gap between the deck and the substructure. Vertical support to the deck during the lateral displacement is provided either by the widely used in bridges system of elastomeric bearings or, preferably, by friction pendulum bearings that have the advantage that they restore the bridge to its initial position when the action causing the horizontal displacement ceases.

28.2.1 System Performance Requirements

The performance sought by varying the joint width (gap) depends on the key design objectives of the structural design, i.e.:

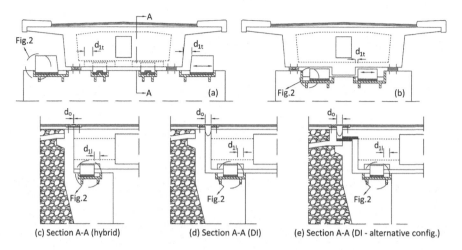

Fig. 28.1 Possible configurations of bridges with adaptive boundary conditions: moveable shear keys in the transverse direction (top); in the longitudinal direction (bottom)

1. When the durability of the bridge and the cost of maintenance are the key considerations, MSKs are provided in the longitudinal direction of the bridge, at one abutment only (Fig. 28.1c), while the other one is monolithically connected to the deck (fully integral). The gap remains closed under normal temperature variations and contractions caused by shrinkage, creep and prestress in concrete decks. The key goal is to further extend the range of spans permitted using the aforementioned techniques by allowing a gap to form at the end of the bridge where the MSKs are located when environmental actions exceed a predefined limit. More details for this solution, which is not the focus of the present paper, can be found in Kappos (2016).
2. In the case that low-probability, high-amplitude, dynamic loads (e.g. strong earthquakes or hurricanes) are a key consideration, an open gap in the transverse direction of the bridge (Fig. 28.1a) reduces the stiffness and results in generally more favourable behaviour under, e.g. medium-intensity earthquakes, whereas a closed gap is preferable for strong earthquakes under which control of displacements and prevention of unseating are the primary performance requirements. The longitudinal gap is also variable, as described in case (1).

Two different solutions are put forward in this case: (i) An *adaptive passive* system, wherein the optimum gap size is applied to the bridge by displacing the shear keys, as soon as an early warning system and/or a measure of ground acceleration in the area of the bridge is transmitted to the actuators of the bridge control system. (ii) A *semi-active* system (primarily meant for major bridges) that entails closed-loop feedback control wherein the position of the shear key can be adjusted in (almost) real-time to obtain the optimum dynamic response of the bridge.

28.2.2 Feasibility and Theoretical Framework

The key issue in the dynamic intelligent bridge (DIB) is to adjust (through the MSKs) the joint gap size in a way that the stiffness of the bridge is optimised with respect to the frequency content of the dynamic loading. The proposed MSKs might be either 'adaptive passive' (no change of device properties during the excitation of the bridge) or semi-active (variation of device properties based on controlling specific response parameters). In the first case the MSK is binary (on-off), in the second it is continuously moveable (resulting in a variable width of the joint to which it is attached). The semi-active solution is primarily meant for the dynamic control of major bridges, due to the higher costs associated with installing and maintaining the control devices; in this case movement of the MSK can only be achieved by a piston or equivalent mechanism, which is feasible but clearly increases the cost of the device. A preliminary study of gap size is presented in the next section.

In the case of the adaptive passive system, the optimum (in a practical context) gap size can be applied to the bridge by appropriate displacing of the shear keys, as soon as the early warning system and/or a measure of ground acceleration in the area of the bridge is transmitted to the actuators of the bridge control system. The "optimum design" of the devices is formulated along the lines: given a bridge structure with certain properties (defined deterministically or as random variables) and a particular intensity/frequency content of the earthquake, what would be the optimal shear key positions such that a particular performance is achieved? This problem needs to be solved "off-line" using classical optimum passive control approaches to provide basic parametric "design charts" from which "optimal" shear key properties are determined for a given bridge (system) and earthquake (input) parameters (or earthquake scenarios) and for a given performance index. In principle, the "adaptive passive" case would be easier and cheaper to implement in practice but requires more off-line "design" work.

The case semi-active system entails closed-loop feedback control in which the position of the shear key can be adjusted in (almost) real-time (i.e. with some minor delay). This involves the regulation of the electromagnetic force/current, continuously or in discrete steps. Various algorithms can be used to achieve effective regulation, e.g. based on classical, optimal, adaptive or model-predictive control. The overall objective of the control design is to achieve an acceptable level of dynamic response, subject to realistic control energy and actuation displacement constraints. Several challenges arise in attempting to analyse the dynamic response of the DIB; for instance, all existing software packages can only deal with link/gap elements with a constant gap size; hence they have to be extended to accommodate gap elements whose size will be continuously updated based on the output of the (semi-active) controller. The controller may be either tuned to the main characteristic frequencies of the bridge or operate over a broader bandwidth. In the case of optimal control, various well-tested design methodologies can be applied to design the controller, e.g. time-domain optimisation based on the maximum principle, LQG or H-infinity optimal control. Whatever method is chosen, the control solution should be robust, i.e. insensitive to unknown spectral characteristics of the seismic signal (which are likely to arise) and to model uncertainty, both parametric and unstructured (e.g. uncertainty arising due to the presence of high-frequency modes that have been ignored, mode interaction, etc.). The choice of appropriate sensors (accelerometers, displacement sensors) and their location is also an important aspect of the control design, along with the effective estimation of dynamic states and parameters, which cannot be directly measured. Finally, the effects of nonlinearities, actuation delays and force saturation constraints need to be fully taken into account during the design and control validation stages, which involve extensive simulation work. Overall, this is a more expensive solution compared to the adaptive passive one, but it requires less work at design stage and is "smart" (i.e. self-adjusted).

Contact interaction (boundary nonlinearity) affects drastically the dynamic/seismic response of both the deck and the MSKs. Given the high mass of the deck, the contact forces/impulses imparted on MSKs, in both directions of contact (normal and tangential) are significant, and should be accounted for. A particular issue to be addressed is to simulate the dynamic interaction between the deck and the shear keys, or between the shear keys and the steel stoppers (Fig. 28.2). A nonlinear dynamic response analysis scheme has to be developed that is capable of capturing the dynamic response of bridges with contact phenomena. The envisaged simulation hinges on the (kinematical) impenetrability constraint between the contacting surfaces and offers a refined treatment of the (frictional) contact interaction. Aim of the simulation is to capture the non-smooth and nonlinear dynamic response of the bridge, and to assess the tendency (if any) of the MSKs to topple and/or twist as a result of contact/impact. The local effects of contact (e.g. the plastic deformation around the interface, shear fracture of the shear key, yielding of the steel stopper or the rail) have to be examined numerically, deploying spatially localised, detailed, nonlinear three-dimensional finite element simulations of the MSK sub-system using explicit time-integration. The numerical simulation of the MSK response is particularly challenging when it is supported on electro-magnetically connected steel plates.

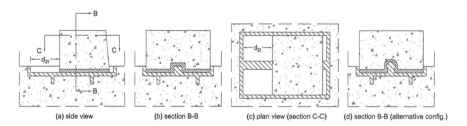

Fig. 28.2 Views of alternative shear key configurations

28.3 Pilot Study of the Effect of Gap Size

As a first pilot study to check the feasibility of the novel concepts described in Sect. 2, the effect of varying the design gap size in an actual bridge on the key response quantities was studied. This sort of analysis will reveal the effect that the gap size can have on response quantities of a bridge such as deck displacements and shear forces in the piers and the abutments.

28.3.1 Description and Modelling of the Selected Bridge

The selected structure (Overpass T7 in Egnatia Motorway, see Fig. 28.3), is of a type common in modern motorway construction in Europe. The 3-span structure with total length equal to 99 m is characterised by a significant longitudinal slope (approximately 7%) of the 10 m wide prestressed concrete box girder deck that results in two single column piers (cylindrical cross section) of unequal height (clear height of 5.9 and 7.9 m). The deck is monolithically connected to the piers, while it rests on its abutments through elastomeric bearings. Horizontal movement in both directions is initially allowed at the abutments, while longitudinal and transverse displacements are restrained whenever a 100 and a 150 mm gap (between the deck and the abutment) is closed, respectively. The bridge rests on firm soil and the piers and abutments are supported on surface foundations (footings) of similar configuration. The above geometrical characteristics (i.e. different pier heights and unrestrained response of the deck at the abutments) result in an increased contribution of the second mode. The bridge generally conforms with EC8 requirements, for a design PGA of 0.16g (return period $T_r = 475$ years.) and subsoil class 'C', for bridges with ductile behaviour.

The analysis of the bridge was carried out using SAP 2000 (CSI 2009); the reference finite element model (Fig. 28.4) involved 32 non-prismatic 3D beam elements. The elastomeric bearings present at the abutments were modelled using equivalent linear springs ('Link elements' in SAP2000 with 6 degrees of freedom). For modelling the closing of the end joints in either direction, gap elements connected to the deck through rigid elements were used (Fig. 28.5). The size of

Fig. 28.3 Studied bridge (Overpass T7, Egnatia Motorway, N. Greece)

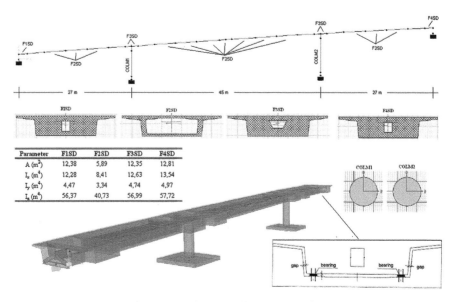

Fig. 28.4 Layout of the bridge configuration and finite element modelling

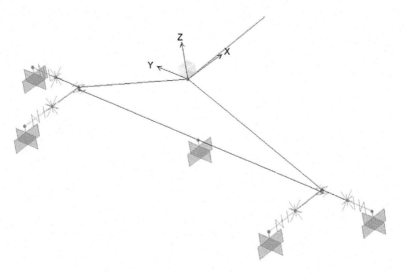

Fig. 28.5 Modelling of abutment area, including gap elements

each gap element corresponds to the actual sizes i.e., 100 mm for the longitudinal gap and 150 mm for the transverse gap. The stiffness of the gap element is assumed 0 as long as the gap is open. Once the displacement in each direction exceeds the gap size, the stiffness will suddenly increase to that of the abutment/backfill system. The latter was estimated using the simplified procedures (based on the limited available test results) adopted by Caltrans (2013), wherein the longitudinal stiffness can be calculated from the initial embankment fill stiffness Ki ≈28.7 kN/mm/(m width of the wall), and this has to be adjusted proportionally to the backwall height. In the transverse direction, a nominal abutment stiffness equal to 50% of the elastic transverse stiffness of the adjacent bent was used (Caltrans 2013). This nominal stiffness has no direct correlation or relevance to the actual residual stiffness but is meant to suppress unrealistic response modes associated with a completely released end condition. In the case studied here, this assumption leads to underestimation of the actual stiffness of the bridge in the transverse direction.

The ground motions considered for the pilot study were five artificial, spectrum compatible accelerograms, which matched the Eurocode 8 design spectrum for the aforementioned PGA and ground conditions. Due to the good match, the variability in the calculated results was quite low.

Three different 'scenarios' were considered regarding the gap sizes, involving values equal to the design ones (actual bridge), one half the design ones, and twice the design ones. It is worth recalling here that the exact size of the joint gaps depends, to a certain extent, on the judgement of the designer, especially as far as the maximum size is concerned.

28.3.2 Preliminary Results from the Analyses

Linear response-history analyses for the 5 accelerograms appropriately scaled to match first the design spectrum and then twice that intensity (corresponding to an earthquake with a return period of at least 2500 years.), showed that the 3rd scenario (gaps equal to twice the value in the actual bridge) was not particularly helpful in assessing the effect of the gap size, as the gaps remained essentially open in both directions even for the 2500 year earthquake. For the other scenario involving half the design gap size (i.e. 75 mm), there was closing of the gap in the transverse direction at Abutment 2 (Fig. 28.2) for the design earthquake and closing of both gaps for twice the design intensity. On the other hand, the longitudinal gap was found to remain open even for the latter case; it is recalled here that longitudinal gaps do not accommodate only the seismic displacement but also that due to temperature variations, creep, shrinkage and prestressing of the deck.

Table 28.1 summarises the key results (pier moments and shears, and abutment shears) for the other scenario involving half the design gap size and the standard case of the as-built bridge. It is noted that the maximum transverse displacement at the abutments is 152 mm in the as-built bridge (i.e. just exceeding the design gap size) and 98 mm in the half-gap size scenario, i.e. well above the assumed gap size, the displacement beyond the 75 mm being due to the deformability of the backfill system. It is clear from these results that while response quantities like pier moments are little affected by the gap size, other quantities, in particular shears in the abutments are very much affected (they are 2.0–2.5 times higher when the transverse gap is equal to one half the design value). Conversely, the displacements of the bridge are smaller when the gap is smaller than the design value.

Table 28.1 Key response quantities for bridge subjected to twice the design earthquake intensity

Response quantities	Member	Original gap size	Half gap size
Shear forces (kN)	Pier 1	11373.6	10635.1
	Pier 2	7910.7	7539.1
	Abutment 1	1406.1	3472.3
	Abutment 2	1579.7	3306.1
Bending moments (kN.m)	Pier 1	54704.1	55469.1
	Pier 2	48571.2	48658.3

It has to be particularly noted that the reported results refer to elastic behaviour of all bridge components (including the piers), hence the only nonlinearity is due to the opening and closing of the gaps. It is clearly important to extend the scope of response-history analysis by introducing nonlinear response of the piers, and even the abutments; this work is currently under way.

28.4 Concluding Remarks

A new method was presented for designing bridges with improved seismic performance. The basic idea is that varying the boundary conditions through the use of the proposed novel devices called moveable shear keys (MSK)s, can lead to an improved structural performance under seismic actions of different levels, including those higher than the design ones. The envisaged variable-width joints are initially either closed or open depending on the bridge length and the serviceability requirements, while under seismic loading their width is optimised either with a one-off adjustment, or continuously varying through semi-active control. The performance sought by varying the joint gap depends on the design objectives; in the case of strong earthquakes, an open gap in the transverse direction of the bridge reduces the overall stiffness, and results in generally more favourable behaviour under, e.g. medium-intensity earthquakes (typically associated with the requirement for uninterrupted operation of the bridge), whereas a closed gap is preferable for strong earthquakes under which control of displacements and prevention of unseating are the primary performance requirements.

A pilot study on the effect of gap size was presented, involving a typical modern overpass with joints in both the longitudinal and the transverse direction. By examining different gap size 'scenarios', it was found that this size can significantly affect some key response quantities like shear in the abutments. Further studies are currently under way to assess the feasibility of the proposed novel ideas.

Acknowledgements The contributions of Kostas Gkatzogias, PhD student, and Le Minh Hoang, MSc student, both at City, University of London, to the numerical part of this study are gratefully acknowledged.

References

BSI (2012) Eurocode 8: design of structures for earthquake resistance – part 2: bridges (BS EN 1998-2:2005 +A2:2011). BSI, London

Caltrans [California Department of Transportation] (2013) Seismic design criteria ver. 1.7, Caltrans Division of Engineering Services

CSI (Computers and Structures Inc.) (2009) SAP2000: three dimensional static and dynamic finite element analysis and design of structures. Berkeley, Computers and Structures Inc

Gkatzogias KI, Kappos AJ (2016) Semi-active control systems in bridge engineering: a review, structural engineering international (IABSE). November 2016 (in press)

Kappos AJ (2016) The dynamically intelligent hybrid bridge. Paper submitted to the WIBE Prize competition

Kappos AJ, Sextos AG (2009) Seismic assessment of bridges accounting for nonlinear material and soil response, and varying boundary conditions. Coupled Site and SSI Effects, Springer, pp 195–208

Kawashima K, Unjoh S (1997) Impact of Hanshin/Awaji earthquake on seismic design and seismic strengthening of highway bridges. J Jpn Soc Civ Eng I(38):1–30

Padgett J, DesRoches R et al (2008) Bridge damage and repair costs from Hurricane Katrina. J Bridge Engng (ASCE) 13(1):6–14

Yashinsky M et al (2010) Performance of highway and railway structures during the February 27, 2010 Maule Chile Earthquake, EE-RI/PEER/FHWA Bridge Team Report

Chapter 29
Systematic Methodology for Planning and Evaluation of a Multi-source Geohazard Monitoring System. Application of a Reusable Template

Fjóla G. Sigtryggsdóttir and Jónas Th. Snæbjörnsson

Abstract In this paper geohazards and their monitoring are considered in the general context of reservoirs and dams. A systematic methodology is used to identify and characterise multiple causes and effects, as well as their interdependencies. This methodology originated at NASA as a part of the safety management of the space program. Here, a recently presented, multi-system expansion of the method is introduced, which has been used to capture complicated data into a reusable template represented by interrelation matrices. This template has a general relevance for monitoring geohazards considering the safety of reservoirs and dams. The methodology and the related system analysis possesses some powerful diagnoses possibilities. The template is for instance, used to reveal the potential safety value of a multi-source system installed for monitoring geohazards in conjunction with a large reservoir in Iceland. The method presents an important step in a holistic risk and safety management of reservoirs.

Keywords Multi-hazards · Geohazard · Systems analysis · Interrelations · Reservoir · Dam

---

F. G. Sigtryggsdóttir (✉)
Department of Civil and Environmental Engineering, Norwegian University of Science and Technology, NTNU, Trondheim, Norway
e-mail: fjola.g.sigtryggsdottir@ntnu.no

J. T. Snæbjörnsson
School of Science and Engineering, Reykjavík University, Reykjavik, Iceland
e-mail: jonasthor@ru.is

29.1 Introduction

Hydroelectric generation requires topographic conditions providing for both adequate storage of impounded water and suitable dam sites, as well as, a significant drop in elevation, between the storage area and a powerhouse over a reasonable distance. Favourable topographic conditions are often found in mountainous regions or highlands with glacially fed rivers and/or significant seasonal and/or annual precipitation. Furthermore, suitable dam sites for impounding of a reservoir are usually found in narrow valleys or canyons.

Mountain slopes, combined with sensitive geology and climatic conditions, are susceptible for landslide occurrence, rock falls and avalanches, while glacially related processes include glacial outburst (jökulhlaup), ice calving and glacier surge which can lead to flooding and debris flow. Moreover, mountainous regions are formed through tectonic forces and volcanism, with rivers and streams in many cases running along valleys or canyons formed through weakness in the geological structure such as faults and lineaments. Hence, it can be concluded that the topographic conditions favouring the construction of a hydropower plant are likely to hold an inherent potential for geohazards.

From a hydropower project perspective, geohazards have been defined (Sigtryggsdóttir et al. 2015) as site specific (in the foundation), local (in the reservoir area) and/or regional depending on the characteristics of the threatening geological condition. Furthermore, a potential geohazard can develop a failure event during the project's lifetime causing one or more of the following effects: loss of life, destruction of downstream property, damage to environment, damage to hydroelectric facilities, loss of sustainability, project benefits and/or socio-economic impacts.

The assessment of relevant geohazards is thus imperative for the planning, design and operation of reservoirs and dams. Furthermore, one has to consider the potential of the reservoir itself to induce or trigger different geohazards during the impounding and operation period. Hence, a comprehensive risk and safety management of reservoirs and dams should consider geohazard/reservoir interrelations in a multi-source monitoring program, linking multi-hazards with the appropriate monitoring parameters. A tool in planning such a program has been developed by Sigtryggsdóttir et al. (2015, 2016), based on an existing method of systems theory, expanded to be applicable for multiple systems. In this paper the methodology will be introduced, and a reusable template of interrelations from Sigtryggsdóttir et al. (2016), used for investigating a multi-source monitoring of multiple geohazards. This template can be further used when planning new reservoirs and dams, or in reviewing existing infrastructure.

29.2 Methodology

The method used is based on an interrelation matrix (an "N-squared" matrix) designed to identify and characterise multiple causes and effects, as well as their interdependencies. This methodology originated at NASA in late 1960s and early

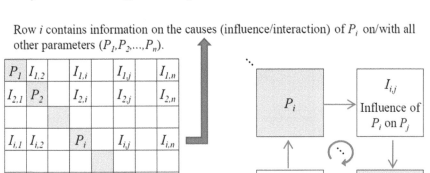

Fig. 29.1 Construction of an interrelation matrix. Parameters are $P_1, P_2, \ldots, P_n$ on the diagonal. The cell $I_{i,j}$ represents the influence of P_i on P_j (cause); vice versa the cell $I_{j,i}$ represents the influence of P_j on P_i (effect). (Sigtryggsdóttir et al. 2016)

1970s as a part of safety management for the space program. Hudson (1992) later developed this method further to solve complex engineering problems in rock engineering systems. Sigtryggsdóttir et al. (2015, 2016) presented a multi-system expansion of the methodology, allowing for a quantitative investigation of the interrelation between individual systems.

The construction of a single system $n \times n$ interrelation matrix is visualized in Fig. 29.1. Here, i and j, respectively, are a row and a column of the matrix. The first step is to identify the components, P_i, of a particular system and align these on the matrix diagonal ($i = j$),. Then the interrelations defined in the off-diagonal cells, I_{ij}, ($i \neq j$) are described and subsequently identified, for example, using binary interactions (1 for existing interrelation, 0 for none). The matrix can also be coded by ranking the interactions in order of importance or strength, using, for example, the numerical values 0, 1, 2, 3, 4 to represent, respectively, "none", "weak", "medium", "strong" and "critical" interrelations (Hudson 1992). Figure 29.1 explains how summation of the interrelations in the same row as parameter P_i, represents its influence or causes, C_i, while the summation of the interrelations in the same column represents the effects, E_i, of the system on P_i. The coordinates, (C_i, E_i), can be plotted in a cause-effect diagram. Furthermore, the interaction intensity ($C_i + E_i$), and dominance ($C_i - E_i$) of parameter P_i within a system, can be plotted.

As previously mentioned, Sigtryggsdóttir et al. (2015 and 2016) expanded the method to include multiple systems. Figure 29.2 describes a multiple system interrelation matrix [MSIM] for N systems. The interrelation matrix of each system J, where $J = 1, 2, \ldots, N$, within the [MSIM] is denoted S_J and is constructed and analysed as described above for one system. The parameters defining each system are aligned diagonally, and the number of these within the respective systems may

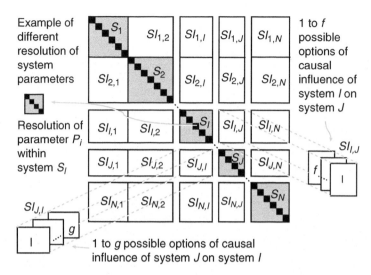

Fig. 29.2 Multiple system interrelation matrix [MSIM] with sub-matrices of system interrelations (S) and sub-matrices of interrelations between systems (SI). (Sigtryggsdóttir et al. 2016)

vary. The interrelation between systems J and I (where $J = 1, 2, \ldots, N; I = 1, 2, \ldots, N$; and $I \neq J$) is described in sub-matrices denoted $SI_{J,I}$ and $SI_{I,J}$. Sub-matrix $SI_{J,I}$ defines the influence of the system defined by S_J on the one defined by S_I and vice versa for sub-matrix $SI_{I,J}$. There may be more than one type of interrelation between systems, that calls for investigation.

29.3 Application of a Reusable Template

29.3.1 Conceptual Model

A conceptual model comprising two systems, GeoRes and SafeMon is outlined in Fig. 29.3. GeoRes is a system of geohazards, a reservoir and large dams, while SafeMon is a system of the monitoring system installed to monitor geohazards within GeoRes. Table 29.1 defines the parameters of each system. Further expansion of the monitoring parameters of specific processes (SpS) and pore pressure (PP) is in Table 29.2. Sigtryggsdóttir et al. (2016) developed the conceptual model, ranking numerically both physical and natural interrelations (Fig. 29.3). The interrelation ranking is based on physical/natural possibility rather than e.g. site specific probability, to allow for general application. Thus, the conceptual model with the interrelation appropriately defined presents a template that can be applied for any reservoir as a first step in a multi-hazard analysis and/or for planning a geohazard monitoring program. However, as later explained, indices must be defined to represent the actual site specific, local and regional settings of a specific reservoir.

The Template with the numerical interrelations is available in Sigtryggsdóttir et al. (2016) and can be copied from there for reuse, or as required. Here, the

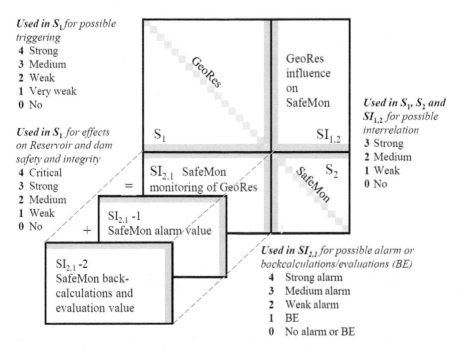

Fig. 29.3 Outline of the conceptual model's multiple systems interrelation matrix and the different SafeMon system interrelations to the GeoRes system. The full model (the Template) with numerical values is in Sigtryggsdóttir et al. (2016). Tables with interrelation categories and ranking used in the model are given

numerical interrelations from the Template are depicted in Fig. 29.4 for the GeoRes system (matrix S_1 of Fig. 29.3) and SafeMon system (matrix S_2). The different shades of grey of the connecting lines in Fig. 29.4 represent strength of interrelation, the darkest shade indicates the strongest interrelation. Similarly, Fig. 29.5(below) depicts the alarm possibilities (matrix $SI_{2,1}-1$) of individual SafeMon components (green) monitoring GeoRes processes. Furthermore, Fig. 29.5(above) presents back calculation possibilities (matrix $SI_{2,1}-2$) provided by the monitoring data collected within the SafeMon system. Finally, Fig. 29.6 shows how actions induced by the GeoRes system components may affect the SafeMon system components (matrix $SI_{1,2}$), i.e. implying that a certain geohazard can trigger changes that influence the monitoring data recorded.

29.3.2 Risk and Susceptibility

The Template from Sigtryggsdóttir et al. (2016) can be applied to an actual reservoir and dams, considering both susceptibility and risk of individual hazards representing this for both site specific, local and regional settings of the particular case. For this

Table 29.1 The conceptual model systems parameters (Sigtryggsdóttir et al. 2016)

System 1: GeoRes		System 2: SafeMon	
Geohazards – Climate – Reservoir		Monitoring parameters	
Climate	Cli	Crustal movement monitoring	Cm
Crustal Movements	CM	Flow measurements	Flo
Surface erosion and scouring	Ero	Fault movement monitoring	Fm
Earthquake	Eq	Groundwater	GW
Flooding out of the reservoir (downstream)	FDs	Leakage/Seepage/Turbidity	L
Flooding into reservoir	FiR	Meteorological data	Met
Fault movement in reservoir/dam foundation	FM	Micro seismicity	MS
Glacial outburst flood	GF	Pore pressure	PP
Glacial – Surface load variations	LV	Reservoir elevation	rEl
Mass wasting into reservoir	MW	Specific Process Monitoring	SpP
Reservoir water body and elevation	REl	Strong EQ Motion	Seq
Reservoir, dams and appurtenant structures	Res		
Rock foundation stability	RfS		
Scouring	Sco		
Sedimentation	Sed		
Soil foundation stability	SfS		
Subsurface hydraulics	SHy		
Subsidence	Sub		
Surge wave on reservoir	SW		
Volcanism	Vol		

Table 29.2 Specific processes (SpP), pore pressure (PP) monitoring and networks (Sigtryggsdottir et al. 2016)

Network #		Specific process	SpP monitoring
PP/SpP	1	Subsidence (Sub)	Geodetic methods to monitor deformation of ground surface. Visual inspection in site visits.
		Soil foundation (SfS)	
PP/SpP	2	Rock foundation (RfS)	Geodetic methods/inverted pendulum (boreholes)/wire alignments (galleries).
PP/SpP	3	Mass wasting (MW)	Displacement/creep monitored by geodetic methods/extensometers.
SpP	4	Sedimentation (Sed)	Sed survey in reservoir. Sediment transport. Remote sensing. Field surveys.
SpP	5	Erosion/scouring (Ero/Sco)	Remote sensing. Topographic/field surveys.
SpP	6	Glacial outburst flood (GF)	Monitoring of glacial lake. Field surveys. Remote sensing. Water level.

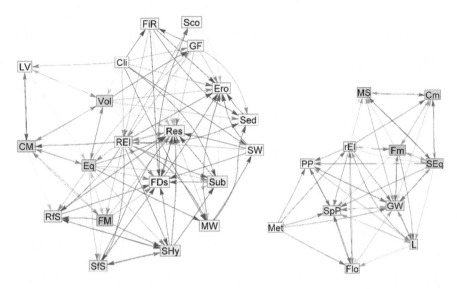

Fig. 29.4 Interrelations within the GeoRes system (Left) (matrix S_1) and SafeMon system (Right) (matrix S_2) depicted. The arrowhead of the connecting lines points to the affected component. The different shades of grey of the connecting lines represent strength of interrelation, the darkest shade for the strongest interrelation. (Legend: Reservoir (blue); Tectonic geohazards (pink); Surficial geohazards (yellow), Site specific geohazards (in the foundation) (green); Site specific geohazard relating to tectonics (brown)). (Sigtryggsdottir et al. 2016)

purpose susceptibility and risk descriptors are defined and suitable indices assigned to each one (See Tables 29.3 and 29.4).

The risk descriptors -very high (VH), high (H), moderate (M), low (L) and very low (VL) – are obtained from Table 29.3 considering both likelihood of a scenario and its consequences (see e.g. AGS 2007). The consequences reflect approximated cost as a percentage of the value of property (order of magnitude) and/or damage to environment. Thus, environmental damages may result in higher consequence class than otherwise appropriate (an example of this is the Sed parameter, presented in the case study in Sect. 29.4). The susceptibility descriptors in Table 29.4 (left) describe the potential threat of individual hazards. In Table 29.4 (right) indices are assigned to the susceptibility and risk descriptors. The interrelations assigned to GeoRes components can be scaled according to either of these descriptors. This is in line with the "expert-semi-quantitative" methodology for quantifying the interrelations.

29.4 Case Study

The case study presented herein, considers the Hálslón Reservoir, the main storage of the 690 MW Kárahnjúkar Hydropower Plant in Iceland. A glacial river fed by the Vatnajökull Glacier and carrying high sediment load, is stemmed with three dams

Fig. 29.5 Back-calculation (BE) (matrix $SI_{2,1}-2$) (above) and Alarm possibilities (matrix $SI_{2,1}-1$) (Below) of individual SafeMon components (green) monitoring GeoRes processes (yellow). The arrowhead of the connecting lines points to the GeoRes component for which an alarm/BE is provided by a SafeMon component

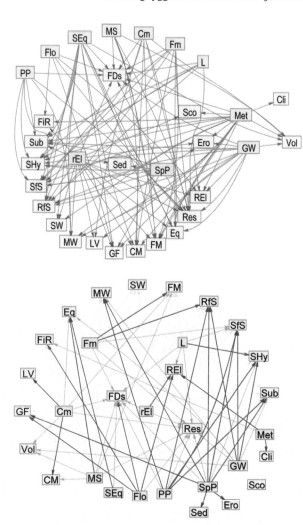

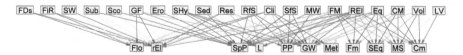

Fig. 29.6 Possible actions applied by the GeoRes system to the SafeMon system (matrix $SI_{1,2}$). The arrowhead points to the SafeMon component reacting to the action imposed by a GeoRes component. (For example FiR influences flow in rivers and the reservoir elevation and thus the relevant monitoring (Flo and rEl)

(Fig. 29.7) to form the reservoir. The highest dam is 198 m high. From the reservoir the water runs through a 53 km long tunnel to an underground powerhouse, from which the tailwater is diverted into another riverbed. Key figures relating to the

Table 29.3 Risk descriptors considering likelihood and consequences

Likelihood				Consequences to property[a] and environment				
	Descriptor	P_A[c]	RP[d] (years)	Cata-strophic 200% (Cat)	Major <60% (Maj)	Medium <20% (Med)	Minor <5% (Min)	Insignificant <0.5% (In)
A	Almost certain	10^{-1}	10	VH	VH	VH	H	M/L[b]
B	Likely	10^{-2}	100	VH	VH	H	M	L
C	Possible	10^{-3}	1000	VH	H	M	M	L
D	Unlikely	10^{-4}	10,000	H	M	L	L	VL
E	Highly unlikely	10^{-5}	100,000	M	L	L	VL	VL
F	Barely credible	10^{-6}	1000,000	L	VL	VL	VL	VL

Adopted from: AGS 2007
[a] As a percentage of the value of the property (order of magnitude) with consideration of environmental damages
[b] L if consequences <0.1% otherwise M
[c] P_A: Indicative annual probability
[d] RP: Indicative return period

Table 29.4 Susceptibility descriptors (left) and indices for the susceptibility and risk descriptors (right)

Potential descriptor	Indicative Annual probability	Susceptibility descriptor		Susceptibility and risk descriptor		Index
Almost certain	10^{-1}	Very High	VH	Very high	VH	1
Likely	10^{-2}	High	H	High	H	0.5
Possible	10^{-3}	Moderate	M	Moderate	M	0.25
Unlikely	10^{-4}	Low	L	Low	L	0.1
Highly unlikely	$<10^{-5}$	Very Low	VL	Very low	VL	0.05

Fig. 29.7 The case study (left (original photo by Emil Thor)) and its location (right) close the the eastern boundary of the volcanic zones (sketches of elliptic circles indicate rough location of volcanic systems). (Sigtryggsdóttir et al. 2016)

reservoir, its dams and waterways are given in Sigtryggsdottir et al. (2016). In addition to the dams, important structures include: a fuse plug and a spillway necessitating a plunge pool.

The Hálslón Reservoir is located close to the South East margin of the Northern Volcanic Zone (NVZ) in Iceland (Fig. 29.7 right). Horizontal crustal movements are continuous in this area, as are post-glacial rebound uplift due to the melting of the Vatnajökull Glacier. The reservoir is in an area generally assessed as a low seismic hazard zone. However, faults with movement in Holocene time have been identified in the reservoir area. Information on the settings of the Hálslón Reservoir can be found in papers and reports referred to in Sigtryggsdóttir et al. (2016).

29.4.1 Results and Discussion

The case study brings forth the difference in considering geohazards from susceptibility versus from risk considerations.

The susceptibility and risk indices for the case study, defined in Sigtryggsdóttir et al. (2016), are presented in Table 29.5. The causal value (Fig. 29.8) and dominance (Fig. 29.9) of tectonic geohazards, other than crustal movements (CM), is high

Table 29.5 Susceptibility and risk descriptors for the case study. (Sigtryggsdottir et al. 2016)

System param.	Susceptibility	Likelihood	Consequences	Risk	System param.	Susceptibility	Likelihood	Consequences	Risk
Cli	H	*B*	*Med*	H	SW	VL	*E*	*Cat*	M
Sed	H	*B*	*Major*	VH	MW	VL	*E*	*Cat*	M
Ero	VH	*A*	*Med*	VH	LV	VH	*A*	*In*	L
Sco	H	*B*	*Med*	H	GF	VL	*E*	*Cat*	M
FDs	L	*D*	*Cat*	H	Vol	L	*E*	*Cat*	M
FiR	M	*C*	*Med*	M	CM	VH	*A*	*In*	L
Sub	VL	*E*	*Cat*	M	FM	L	*D*	*Major*	M
SHy	M	*C*	*Major*	H	EQ	L	*D*	*Major*	M
SfS	L	*D*	*Cat*	H	Res	L	*D*	*Cat*	H
RfS	L	*D*	*Cat*	H	REI	M	*C*	*Med*	M

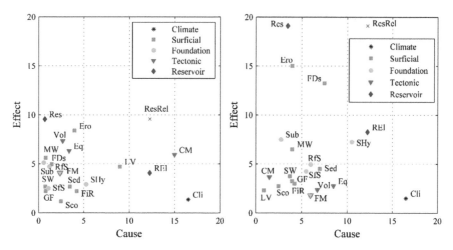

Fig. 29.8 GeoRes system. Causes and Effects of system components from susceptibility (left) and risk (right). (Sigtryggsdottir et al. 2016)

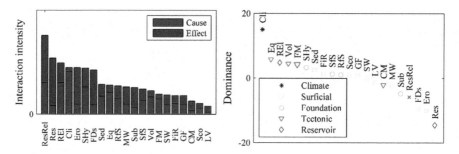

Fig. 29.9 Interaction intensity (left) and dominance (right) within the GeoRes system from risk indices

from a risk perspective but low considering susceptibility/likelihood (Fig. 29.8). A similar situation is observed for subsurface hydraulics (SHy). This results from the fact that potential consequences raise the risk level while the susceptibility index indicates a low likelihood of major tectonic events for the case. Inversely, crustal movements (CM) have a very high susceptibility but a low risk level considering that the consequences are likely to be minor or insignificant. Thus, CM stands out in the susceptibility cause-effect diagram on Fig. 29.8(left) with high causal value, while in Fig. 29.8(right) the causal value is low. The dominance and interaction intensity of processes within the GeoRes system are shown in Fig. 29.9, indicating the dominance of tectonic geohazards within the system from a risk perspective, as well as a high interaction intensity of the reservoir.

In Fig. 29.10 the diagrams labelled (a) represent GeoRes components action imposed on the whole SafeMon system, while the diagrams labelled (b) represent the response of the SafeMon components to the whole GeoRes system. Figure 29.10

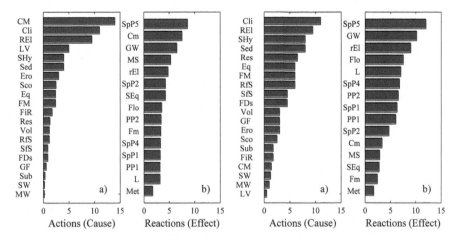

Fig. 29.10 Actions of GeoRes components, and Reactions of the SafeMon components, from susceptibility (left) and risk (right)

presents this from both susceptibility and risk perspective. Considering risk potential, Fig. 29.10 (right) indicates which geohazards induce the most widespread changes in the SafeMon system recordings, and what monitoring parameters are responsive in different hazardous risk situations occurring in the GeoRes system. Conversely, considering the most likely scenarios, Fig. 29.10 (left) indicates which geohazards induce the most widespread changes in the SafeMon system and what monitoring parameters are overall likely to be the responsive during the normal lifecycle phases of the case study.

The calculated causes from $SI_{2,1}$ (see Fig. 29.3) relate to monitoring values of individual SafeMon components for monitoring the whole GeoRes system. On the other hand, the effects represent the value of the whole SafeMon system for monitoring individual GeoRes components. $SI_{2,1}$ is the sum of $SI_{2,1}^{-1}$ for alarm possibilities and $SI_{2,1}^{-2}$ for back-calculation and evaluations (Fig. 29.3). These consider several chains of events. For example, monitoring the development of crustal movement (Cm) and/or microseismicity (MS) may provide early warning of volcanic eruption (Vol). An alarm for volcanic eruption beneath a glacier is also an alarm for impending glacial outburst flood (GF) and consequent flooding into the reservoir (FiR).

The monitoring safety values of SafeMon components for the GeoRes system (causes) are plotted in Fig. 29.11, while those of the SafeMon system for the GeoRes components (effect) are presented in Fig. 29.12. The monitoring values obtained from the three sub-matrices are termed: alarm value (from $SI_{2,1}^{-1}$ and Fig. 29.5 (below)), back-calculation/evaluation (BE) value (from $SI_{2,1}^{-2}$ and Fig. 29.5 (above)) and safety value (from $SI_{2,1}$). A system or component safety value is the sum of the corresponding alarm and BE value. The safety values thus represent the collective monitoring possibilities within the SafeMon system.

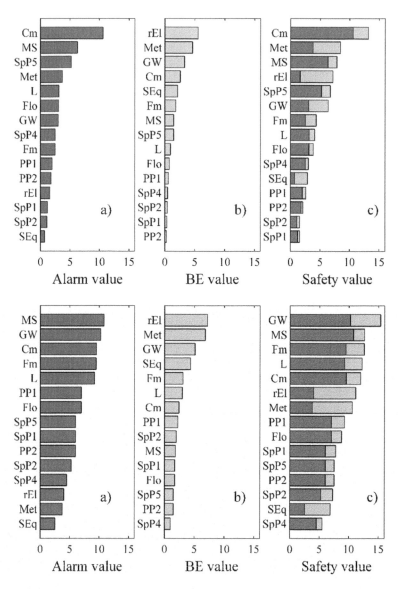

Fig. 29.11 Monitoring values of the case study's SafeMon components based on a susceptibility (above) and a risk perspective (below) (Sigtryggsdottir et al. 2016)

A different order of the SafeMon components, in terms of alarm, BE and safety value, results from considering susceptibility versus risk within the GeoRes system. This can be observed in Fig. 29.11. The susceptibility implies likelihood, and in that sense, the monitoring values in Fig. 29.11 (above) relate to an everyday surveillance scheme. Conversely, Fig. 29.11 (below) brings forth monitoring values of SafeMon components, from a risk perspective, including a warning of potential

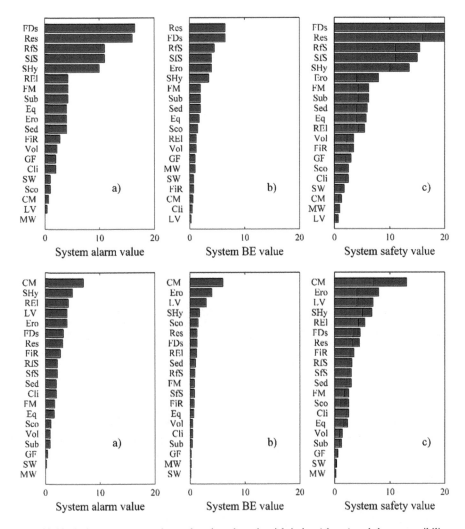

Fig. 29.12 SafeMon system safety values based on the risk index (above) and the susceptibility index (below)

low-probability but high-consequence events. The components monitoring microseismicity (MS) and crustal movements (Cm) are important for providing alarm for hazardous events, from both risk and susceptibility perspective. Groundwater monitoring (GW) is also an important parameter. When considering the safety value, including back-calculation possibilities, the monitoring parameters Met, rEL and L, are among those that score high for the case study. However, all monitoring parameters are important for the overall safety.

Figure 29.12 presents the monitoring systems collective safety values from both a risk and susceptibility perspective. The risk consideration demonstrates how the system alarm, BE and safety values are highest for the most hazardous event in the GeoRes system, namely flooding downstream (FDs). These values are only slightly lower for the Reservoir (Res). This is based on the notion that loss in reliability of the Reservoir (Res) further provides an alarm for flooding downstream (FDs). The SafeMon system also provides relatively high alarm value for the GeoRes processes within the foundation, which may lead to loss of structural safety and consequent flooding.

The GeoRes components with the lowest system monitoring value (Fig. 29.12) from risk perspective are mainly monitored by a single SafeMon component, and/or with a low risk index. This applies to climate and some surficial GeoRes processes, such as scouring (Sco). The GeoRes processes provided with the lowest system safety values generally induce insignificant changes within the SafeMon system (Fig. 29.10 (right)), excluding e.g. climate (Cli). Results presented in Fig. 29.12, deduced from susceptibility, can be considered to represent normal operation. Thus, under normal operation the monitoring system is generally most likely to be recording information on crustal movements (CM), erosion processes (Ero), surface load variations (LV), subsurface hydraulics (SHy) and changes in the reservoir elevation (rEl).

In summary, Fig. 29.12 (Risk) generally indicates the value of the SafeMon system for components of the GeoRes considering the risk associated with individual geohazards. Conversely, Fig. 29.12 (Susceptibility) indicates which geohazard the SafeMon system is likely to provide valuable information for, under normal operational conditions.

29.5 Concluding Remarks

A general template of system interrelations, prepared by Sigtryggsdóttir et al. (2016), has been successfully applied to a study case involving a large reservoir and adjoined dams in Iceland. The general template can be further applied to identify pathways from the interrelations defined for each systems. The pathways within the system of geohazards and reservoir (GeoRes), represent hazard chains for use in risk analysis. Furthermore, the pathways in the safety monitoring system (SafeMon) can be applied to detect precursors for an impending geohazard.

The geohazard monitoring program installed for the case study has been evaluated by using the template. The monitoring of crustal movements is not traditionally a part of a monitoring scheme for reservoirs and dams. However, for the case studied, this, along with monitoring of micro seismicity, is indispensable for a comprehensive safety monitoring.

The analysis confirms the importance of the installed monitoring program for the overall safety of the infrastructure studied herein. The safety value of the monitoring system investigated here depends on expert review of the instrumentation networks and monitoring data. Considering the complex interrelations of geohazards and their monitoring, a multidisciplinary approach in reviewing the monitoring data is essential. Such line of attack was ambitiously taken during the impounding of the reservoir as well as the first years of operation. Normal operations may give the sensation of stable conditions and safe structures. Consequently, less attention may be given to the monitoring of low-probability high-consequence geohazards as time passes. In such cases, it is important to note that the safety value of a monitoring system is non-existent without continuous review.

Acknowledgements At the initiation of the study reported on in this paper, the topic and a draft of the methodology was introduced to Prof. Ragnar Sigbjörnsson. He immediately encouraged further work on this, which later resulted in a PhD thesis and two papers published on the subject. His contribution and encouragement is gratefully acknowledged. The permission from Taylor & Francis, for the reuse of figures and tables in the text that refer to Sigtryggsdóttir et al. (2016) published in http://www.tandfonline.com/toc/nsie20/current, is acknowledged.

References

Australian Geomechanics Society (AGS) (2007) Landslide risk Management: Guidelines. ASG 42:13–35
Hudson J (1992) Rock engineering systems: theory and practice. Ellis Horwood, Chichester
Sigtryggsdottir FG, Snæbjörnsson J, Grande L, Sigbjörnsson R (2015) Methodology for geohazard assessment for hydropower projects. Nat Haz 79(2):1299–1331. https://doi.org/10.1007/s11069-015-1906-4
Sigtryggsdottir FG, Snæbjörnsson J, Grande L, Sigbjörnsson R (2016) Interrelations in multi-source geohazard monitoring for safety management of infrastructure systems. Struct Inf Eng 12(3):327–355. https://doi.org/10.1080/15732479.2015.1015147

Chapter 30
How to Survive Earthquakes: The Example of Norcia

Mário Lopes, Francisco Mota de Sá, Mónica Amaral Ferreira, Carlos Sousa Oliveira, Cristina F. Oliveira, Fabrizio Meroni, Thea Squarcina, and Gemma Musacchio

Abstract In this paper lessons are extracted from the comparison between the very different consequences that similar earthquakes had on the neighbouring towns of Norcia and Amatrice during the 2016 seismic crisis of central Italy. It was found that the differences in damage were essentially due to the strengthening of most houses in Norcia done during the previous decades. This is also likely to lead to a much faster recover of the economy and livelihood in Norcia, as Amatrice needs to be entirely rebuilt.

Keywords Earthquake · Damage · Strengthening

30.1 Introduction

On the 30th October 2016 an earthquake of magnitude $M_W 6.5$ hit the town of Norcia, in central Italy. The earthquake was the strongest to hit Italy since 1980; $M_W 6.9$ Irpinia earthquake).

The October 30th earthquake severely impacted the largely-abandoned towns of Norcia, Castelsantagelo, Preci and Visso, where residents (nearly 8000 people) had left their homes to sleep in cars, campers or moved to shelters or hotels following the earthquake sequence that occurred on October 26 ($M_W 5.9$). Due to the reduced epicentral distance the accelerations in Norcia were extremely high, with a value of

---

M. Lopes · F. Mota de Sá · M. A. Ferreira (✉) · C. S. Oliveira
Instituto Superior Técnico, Department of Civil Engineering, Architecture and Georesources, CEris, Lisboa, Portugal
e-mail: mariolopes@tecnico.ulisboa.pt; monicaf@civil.ist.utl.pt; csoliv@civil.ist.utl.pt

C. F. Oliveira
Escola Superior de Tecnologia do Barreiro, Instituto Politécnico de Setúbal, Setúbal, Portugal

F. Meroni · T. Squarcina · G. Musacchio
Istituto Nazionale di Geofisica e Vulcanologia, Roma, Italy
e-mail: fabrizio.meroni@mi.ingv.it; thea.squarcina@ingv.it; gemma.musacchio@ingv.it

© Springer International Publishing AG, part of Springer Nature 2019
R. Rupakhety et al. (eds.), *Proceedings of the International Conference on Earthquake Engineering and Structural Dynamics*, Geotechnical, Geological and Earthquake Engineering 47, https://doi.org/10.1007/978-3-319-78187-7_30

the horizontal PGA of 0.48g registered at the nearest seismic station (Luzi et al. 2016). Two months before, on 24th August, the neighboring village of Amatrice had been shaken by an earthquake slightly stronger, $M_W 6.0$. In this earthquake most of the constructions of Amatrice collapsed, the ones that did not collapse were so damaged that were useless, and 300 people died. Today, it is a dead village, where nobody is allowed to enter freely. How is it possible to explain the differences between Norcia and Amatrice? How is it possible nobody died in Norcia? In order to find and document answers to these questions a KnowRISK team (2016), with members of Instituto Superior Técnico (IST) and Istituto Nazionale di Geofisica e Vulcanologia (INGV) visited the affected zones during the last week of October 2016, during which several earthquakes hit Norcia and Amatrice. Truthfully, the motivation for this field trip arose from the fact that during August earthquakes Norcia had already been strongly hit but had only minor damages. In October, the ground motion was even stronger in Norcia and, again, there were no mortal victims.

30.2 Characteristics of the Seismic Actions

Figure 30.1 shows the spectra for horizontal accelerations recorded in stations located in Amatrice and Norcia. Both spectra refer to the earthquakes of 24th August and 30th October, as these were the strongest ones in each town and the ones that caused more damaged in each.

If we compare the spectra with the highest accelerations in each location, in the east-west direction, we realise that in Amatrice the maximum spectral acceleration was 2.27g for a period of $T = 0.24s$ and in Norcia it was 1.9g for a period $T = 0.3s$. The Amatrice earthquake produced higher spectral accelerations than the Norcia earthquake for periods below $T = 0.46s$, therefore producing stronger effects in stiffer constructions, i.e. low-rise. The differences are stronger below $T = 0.2s$. The Norcia earthquakes produced stronger effects in more flexible constructions than the ones in both Amatrice and Norcia stock of buildings. But the main issue here is that, qualitatively, the two earthquakes are comparable and therefore the large difference in performance cannot be attributed to differences in the seismic actions alone.

30.3 Vulnerability of the Constructions

The comparison between the damages caused in Norcia and Amatrice due to the largest earthquakes of last year in Central Italy shows profound differences. Since the earthquakes themselves cannot explain these differences, they must be attributed mainly to differences in vulnerability between both towns.

Both in Norcia and Amatrice most constructions are/were old, built in periods in which earthquake resistant design was not enforced in codes of practice, and therefore it is thought that original constructions were vulnerable. However, in

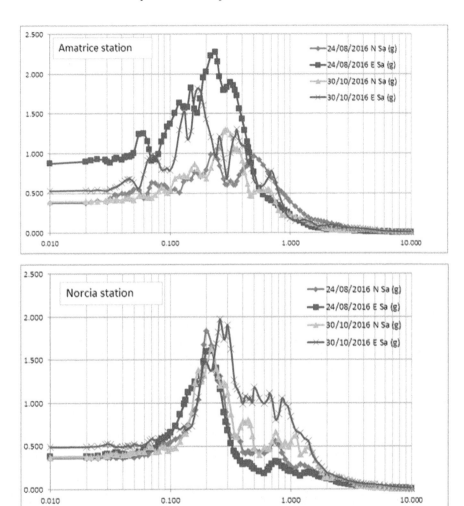

Fig. 30.1 Response spectra for Amatrice and Norcia, east-west direction. (Data provided by Luzi et al. 2016)

Norcia there is a culture of safety in what regards earthquake risk, and due to that the constructions in Norcia were strengthened to provide earthquake resistance during the last 40 years – following the 1979 and 1997 Norcia earthquakes. Higher priority was given to houses where people live permanently, with second houses receiving a lower priority. The main strengthening techniques used in Norcia were i) confinement of masonry walls by a layer of mortar with a pre-fabricated steel welded mesh inside, in both faces, connected by steel bars at a given spacing, and ii) introduction of steel cables connecting parallel walls in order to prevent the out-of-plane movement of exterior walls to the outside of the construction.. This information was

Amatrice, after the 24 August 2016 earthquake Norcia, after the 30 October 2016 earthquake

Fig. 30.2 Amatrice and Norcia after the earthquakes

transmitted to the KnowRISK team by the *Sindaco* (Mayor) of Norcia in a meeting in the morning of 26th October, a few hours before the earthquakes of that day.

The result of the seismic crisis of August and October 2016 was the destruction of Amatrice. However, in Norcia, despite damages in many houses and some collapses of historical constructions, as the exterior walls and churches, most of the houses are standing and, above all, nobody died. Figure 30.2 provides a good comparison of the state of both villages after the earthquakes.

In fact, in Amatrice around 300 people died while in Norcia there were no mortal victims after August and October earthquakes. The policy of strengthening buildings applied in Norcia for four decades is strongly responsible for this significant difference. Yet, another important factor also contributed to this outcome. In October, Norcia had already felt the shaking due to the earthquake of Amatrice on 24th August, immediately after which some of Norcia population was reallocated in order to evaluate structural stability of their homes. Before the last and stronger earthquake in October, Norcia was also shaken two times in the previous days, by two other earthquakes on 26th October not so strong, although causing some minor damages, mostly non-structural. The population was, therefore, on alert, and because of that, when the main earthquake took place on 30th October, at 7 h 41 m am, many people were sleeping on cars and not in their homes.

30.4 Lessons Learned

30.4.1 Rehabilitation of Constructions

The first major lesson from the comparison of the different performances of Norcia and Amatrice constructions is that, in seismic areas, prevention pays off. The seismic strengthening of constructions is able to avoid major collapses and, most important, to save lives.

Other comparisons between the seismic performance of old strengthened and unstrengthened constructions, lead to the same conclusion. Figure 30.3 shows one of

Fig. 30.3 Difference of seismic performance between adjacent constructions

those examples: two adjacent houses in the island of Faial (Azores, Portugal) stroke by a violent earthquake on the 9 July 1998 (M_W6.1). Both houses have similar systems and materials (old masonry) and none of the original constructions had been designed to resist earthquakes. However the left hand side house had been strengthened against earthquakes and the other had not.

The houses were so close that there were no geotechnical differences between their locations and both were subjected to the same seismic action. The comparison is clear and straightforward: the figure shows that the strengthened house resisted the earthquake with minor damage and the unstrengthened house collapsed.

Therefore, it is important to draw attention of decision makers and managers of programs of urban rehabilitation in seismic zones for the importance of seismic strengthening in the rehabilitation of constructions. In seismic zones, improvement of aesthetics and living conditions of old and unsafe houses should in general be accompanied by seismic strengthening.

30.4.2 Non-structural Elements

Even though properly strengthened houses survive strong earthquakes, they vibrate and deform during the vibration. These may introduce relevant non-structural damage, part of which can be avoided by appropriate measures taken by common citizens, which is the subject of the KnowRISK project. Moreover, reducing non-structural damage reduces the probability of people getting injured by falling objects and reduces economic damage. Note that the reduction of economic damage is also important for the affected populations to resume their normal life. Figure 30.4

Fig. 30.4 Televisions connected to walls by chains to avoid toppling

shows a recent example in Norcia: during the August earthquake the television fell down and broke. After that a new television was bought to replace the broken one, but was fixed with chains, as shown in Fig. 30.4 (photo shot by the KnowRISK team on 28 October). The result was that the television suffered no damage during the 26 October earthquakes. The above example has already been used by the Portuguese team in KnowRISK actions in schools, during which young students are taught on how to reduce seismic risk from non-structural elements at school and at home.

30.5 Resilience and Recovery

Most of Amatrice constructions collapsed or were severely damaged during the 24th August earthquake. The town has been closed to the general public, and could only be visited under the supervision of the Italian fire brigades. This included the KnowRISK structural engineers that only got permission to visit the town when the fire brigades could receive them. After the 24th August earthquake, the survivors of the earthquake had no alternative than to leave town and stay in hotels, some in nearby villages, others more far away, mainly in the Adriatic coast, while others went to live with relatives or friends. The Italian government declared that the affected towns will be completely rebuilt. But that will take years, changing completely the livelihood of the population. Even if the streets and constructions are rebuilt keeping the same architectonic and urban characteristics as before the earthquake, the urban environment may be different, as the people will be different. It cannot be taken as granted that the culture, traditions and other factors that are part of the identity of Amatrice before the earthquake will be re-established in the future.

Norcia was strongly hit by the sequence of October earthquakes, mainly by the one of 30th October. However, damage was much less extensive than in Amatrice.

Fig. 30.5 Damage in Norcia after the October 2016 earthquakes

Some monumental constructions suffered partial collapses and several constructions inside town were damaged, a few ones strongly. Figure 30.5 shows some of those cases. The photos were taken by the KnowRISK team on 19th December 2016, one month and a half after the major earthquake that hit the town.

In general, in Norcia, most constructions appeared, from the outside, to have no damage or slight damage, but there were also a few with parts in risk of collapse. In these conditions, any aftershock could lead to more damage, making unsafe to walk normally in the streets. Therefore, the centre of Norcia was closed to the public, until conditions to walk safely in the streets are re-established. This comprises essentially two conditions: (i) all constructions in which there is partial risk of collapse to the streets must be braced to avoid that risk, and (ii) the seismic crisis must be over. When the KnowRISK team visited Norcia in December, together with members of the Italian fire brigades, those works of bracing unstable structures were going on in several parts of the town, which may take a few months to complete. Figure 30.6 illustrates some of those cases.

The second criterion implies a very difficult decision. When the KnowRISK team visited Norcia and Amatrice by the end of October, 2 months after the major earthquakes of August that were followed only by low intensity aftershocks until October, it was thought that the seismic crisis could be fading away. However the October earthquakes, as well as the ones that took place in January 2017, cast high uncertainties on the assessment of the situation from the seismological point of view. In this situation it is likely that only after several months of lack of relevant

Fig. 30.6 Bracing of unstable constructions in Norcia by the fire brigades

earthquakes the crisis will be declared as finished. Therefore, it is unlikely that the centre of Norcia will be accessible to the general public before the Summer of 2017 (and this may be very optimistic), which seriously compromises the normalization of economic and social life.

Many buildings in the centre of Norcia are in safe conditions and could be used if people could access them, allowing the revival of the local economy. Therefore, the lack of conditions to move safely in the streets is a major factor hindering economic recovery. Also, the longer the period in which the centre of the town is closed, the higher becomes the probability that some people will not return.

The KnowRISK team was in Norcia three times, on the 26th of October, from which we left 4 h before the first earthquake of that day, on the 28th October between the major earthquakes of that week and on the 19th December. In the first stay, the team members got acquainted with several inhabitants of Norcia, which we also met in the two following visits, allowing a closer involvement with the reality of the recovery process, including economic, social and psychological aspects of the situation. During the third visit, after the major and more damaging earthquake of 30th October, we were asked by a Norcia resident: "can we live here in peace and safety or should we move to a less earthquake prone part of the country?" The answer given was "you don't need to abandon your town, the place you love and where you always lived, but to live safely here you should take care about the quality and seismic resistance of your constructions, as well as the prevention of non-structural damage, to reduce economic damage and the likelihood of being injured or killed by falling objects and damage to building contents."

For the recovery of a zone affected by an earthquake the time that lifelines and roads and railways are not operational affects all livelihood, including economic activities, in the affected areas, delaying the recovery. The longer this period is, the higher the likelihood that people won't come back. This may change social characteristics typical of some areas, losing part of the "character" of those zones. After many years into the future, when Norcia is repaired and Amatrice is rebuilt, and life is normalized in both towns, there are higher probabilities that social changes are higher, as in Amatrice the process will be much longer. In Norcia, these effects, if any, will tend to be much smaller.

30.6 Conclusions

The comparison between the damages inflicted to Amatrice and Norcia during the August and October 2016 earthquakes in central Italy, led to the following conclusions:

- The lower levels of damage in Norcia were due to the fact that during the last four decades the old constructions of Norcia were strengthened to resist earthquakes, while in Amatrice most of the rehabilitation of the buildings was cosmetic, or, in other words, aimed at improving aesthetics and living conditions only.
- Strengthened and retrofitted constructions prevent major damages and save lives.
- It is strongly likely that the bracing of damaged constructions in Norcia will be much faster than the reconstruction of Amatrice, leading to a much faster recover of the economy and livelihood of the town.

    The longer the reconstruction and recovery takes, the higher is the probability of profound social changes, risking the complete loss of traditional traits, uses and customs of the region.

It may also be worth noting that monetary constraints cannot be used to advocate rehabilitation without seismic strengthening. Both villages, Norcia and Amatrice, were subjected to rehabilitation programs. The decision of including seismic strengthening in the rehabilitation is, most of the time, mainly related, not with monetary constraints, but with incorrectly defined priorities, possibly due to the lack of knowledge that retrofitting represents a negligible increase in rehabilitation costs when compared with simple, cosmetic rehabilitation.

The above conclusions can and should be used for pedagogic purposes in other earthquake-prone areas to demonstrate that seismic strengthening pay is effective and it is worth the investment. _In reality, strengthening not only saves lives and reduces economic damage, but also contributes to the preservation of the cultural characteristics of the building stock and of the population that lives in the affected areas as it allows a much faster recovery of the economy and faster resume of everyday life.

Acknowledgements The support of the Civil Protection headquarters (Dicomac – Direzione di comando e controllo – Protezione Civile) and INGV are gratefully acknowledged. The authors also wish to thank to the *Sindaco* of Norcia, Nicola Alemanno, and the Assessore alla Cultura Giuseppina Perla, for the hospitality and information on the strengthening process of Norcia during the last decades. Special thanks to Agostino Goretti for his suggestions.

This study was financed by the European Commission's Humanitarian Aid and Civil Protection (Grant agreement ECHO.A.5 (2015) 3916812) – KnowRISK (Know your city, Reduce seISmic risK through non-structural elements).

References

KnowRISK (Know your city, Reduce seISmic risK through non-structural elements) (2016–2017) project. Co-financed by European Commission's Humanitarian Aid and Civil Protection Grant agreement ECHO/SUB/2015/718655/PREV28

Luzi L, Puglia R, Russo E, ORFEUS WG5 (2016) Engineering strong motion database, version 1.0. Istituto Nazionale di Geofisica e Vulcanologia, Observatories & Research Facilities for European Seismology. https://doi.org/10.13127/ESM

Chapter 31
KnowRISK on Seismic Risk Communication: The Set-Up of a Participatory Strategy- Italy Case Study

Gemma Musacchio, Susanna Falsaperla, Stefano Solarino,
Giovanna Lucia Piangiamore, Massimo Crescimbene,
Nicola Alessandro Pino, Elena Eva, Danilo Reitano, Federica Manzoli,
Michele Fabbri, Mariangela Butturi, and Mariasilvia Accardo

Abstract KnowRISK (Know your city, Reduce seISmic risK through non-structural elements) is a European project that addresses prevention measures to reduce non-structural damage caused by earthquakes. It is built on risk communication and takes action on pilot areas of the three participating countries: Portugal, Iceland, and Italy. The setting up of risk communication strategies in the project stands on the understanding local communities fragility, on their direct engagement, and on a holistic approach to vulnerability. The level of relevance of seismic compared to other hazards, the understanding, the memory of past disasters are indicators that affect the way a risk is perceived and preventive measures are taken. Similarly, the level of education, wealth, exposure to other,

---

G. Musacchio (✉)
Istituto Nazionale di Geofisica e Vulcanologia, Roma, Italy
e-mail: gemma.musacchio@ingv.it

S. Falsaperla · D. Reitano
Istituto Nazionale di Geofisica e Vulcanologia, Osservatorio Etneo, Catania, Italy
e-mail: susanna.falsaperla@ingv.it; danilo.reitano@ingv.it

S. Solarino · M. Crescimbene · E. Eva
Istituto Nazionale di Geofisica e Vulcanologia, Centro Nazionale Terremoti, Roma, Italy
e-mail: stefano.solarino@ingv.it; massimo.crescimbene@ingv.it; elena.eva@ingv.it

G. L. Piangiamore
Istituto Nazionale di Geofisica e Vulcanologia, Roma 2, Roma, Italy
e-mail: giovanna.piangiamore@ingv.it

N. A. Pino
Istituto Nazionale di Geofisica e Vulcanologia, Osservatorio Vesuviano, Neaples, Italy
e-mail: alessandro.pino@ingv.it

F. Manzoli · M. Fabbri · M. Butturi · M. Accardo
University of Ferrara, Ferrara, Italy
e-mail: fbh@unife.it; mariasilvia.accardo@unife.it

© Springer International Publishing AG, part of Springer Nature 2019
R. Rupakhety et al. (eds.), *Proceedings of the International Conference on Earthquake Engineering and Structural Dynamics*, Geotechnical, Geological and Earthquake Engineering 47, https://doi.org/10.1007/978-3-319-78187-7_31

social, risks are aggravation parameters in risk computation to be accounted for when we communicate risk. Strategies for risk communication in KnowRISK rely on schools and citizen's engagement, citizen's science activities, tools for raising awareness.

Keywords Risk communication · Non-structural components · Earthquake hazard · Seismic risk reduction

31.1 Introduction

Risk communication is a process of exchanging and sharing information about risks among experts, managers, news media, and the general public. These stakeholders do not necessarily share common motivations upon the knowledge of risks; the general public is the most multispectral entity among them all.

Risk includes attributes, such as being unknown, uncertain, unfair, dreaded, towards which human beings react in different ways. There are a number of concerns related to risk (Covello 2003; Fischoff 1995) that need to be taken into account when setting a communication strategy; some of them include – beside safety, the most obvious – the economics, trust, benefits, control, fairness respect, accountability. Since risks never come alone, a fundamental aspect of communication involves the risk relevance: how much a certain risk might prevail in terms of occurrence, magnitude, societal and cultural value and to what extent it might affect highly vulnerable groups (i.e. children, elder, low-income people, foreigners).

In the old approach to risk communication, the only valuable point of view was that of the sender: experts were considered to be enlightening or persuading the uninformed public. Communication was only to correct the distorted perceptions of lay people, providing information about the cost and benefits of risk reduction (Hobson-West 2003).

There are several problems concerning this top-down approach. Providing only information about risk proved to fail to capture the complex and evolving social perceptions, and the interpretations of risk influencing both individual and public decisions (Infanti et al. 2013 and references therein). In some cases increasing the trust towards experts caused communities to overlook personal protection measures and delegate responsibility for safety to appointed agencies (Kuhlicke et al. 2011; De Marchi and Scolobig 2012; De Marchi 2015). In designing communication without input from the audience, experts might use difficult or ineffective wording and omit information their audience might need in implementing mindful decisions (Bruine de Bruin and Bostrom 2012). The needs of recipients can help in understanding why people do not agree upon the harm should be avoided at most and how costs and benefits of risk management are distributed across society (Fischoff et al. 2011).

A participatory process of risk communication and management accounts for all the actors involved: experts, policy makers, and the public. Experts and members of the public are involved in a process of mutual learning that foresees not only the improvement of knowledge, but also the change of perspective and view. This stands on the recognition of the importance of multiple perspectives and domain of knowledge that profit not only from expertise, but also from experience. A dialogic approach may result into an increase of mutual awareness: experts collect inputs to tune their actions and improve their communication skills, while the public participation ensures the establishing of shared rules and reinforces active citizenship. It does also act on the building of thrust among all the actors involved.

Such participatory strategy to risk communication is the approach undertaken within KnowRISK (Know your city, Reduce seISmic risK through non-structural elements). This project, financed by the European Commission under the General Directorate of Civil Protection and Humanitarian Operation, implements actions in pilot areas within the three participating countries, namely Portugal, Italy, and Iceland. The risk addressed within this project is that posed by seismic vulnerability to non-structural elements of building and it is worldwide underestimated. The aim is to facilitate local communities' access to expert knowledge on non-structural seismic protection solutions through risk communication initiatives, and the building of a strategy that can be replicated in other European countries.

In this paper we discuss the rationale behind the set-up a Communication Strategy in the KnowRISK project, assessing more extensively what concerns the profiling of the target audience and presenting some media and tools used and being tuned up during the communication task. We will present the Italian Case study, only mentioning the other pilot areas. We address the reader interested in detailed description of specific actions of the Communication strategy to papers included in this volume (Crescimbene et al. 2017; O'Neill et al. 2017; Piangiamore et al. 2017; Platt et al. 2017; Reitano et al. 2017).

31.2 Establishment of a Communication Strategy

The set-up of a KnowRISK communication strategy passes through seven steps that begin with approaching target communities, ends with the assessment of effectiveness to trigger, and establish prevention of non-structural damage (Fig. 31.1). Assessment of efficacy, rarely taken into account within risk communication actions (Infanti et al. 2013), is a fundamental novelty of the KnowRISK communication strategy and is aimed at providing a replicable tool for other European countries.

Local communities are both the primary victims and the first to respond to emergencies when a disaster strikes mostly because of lack of knowledge or underestimation on protective measures. *Getting to target communities* implies assessing the level of vulnerability in a holistic way to find out the hazards they may be exposed to. In KnowRISK each pilot area has its own peculiarity concerning ranking of hazards, earthquakes recurrence, time elapsed from the last damaging event, building construction, building code enforcement, type and relevance of damage, level of implementation of protective measures and cultural attitudes towards prevention. The KnowRISK communication stands on a Knowledge-

Fig. 31.1 A concept map of the KnowRISK Communication strategy

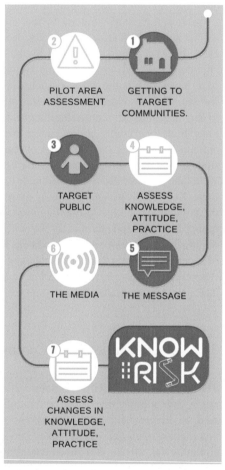

Attitude-Practice (KAP) approach that is the understanding about what the target audience already knows and does: a successful action is the one that reveals changes in the KAP parameters (NSET 2017; Platt et al. 2017). The "Knowledge" is the understanding on earthquake and the associated risk; the "Attitude" refers to feelings and preconceived ideas towards it; the "Practice" refers to the ways in which communities demonstrate their knowledge and attitudes through their actions (NSET 2017).

Table 31.1 The KnowRISK communication plan

The steps of a communication plan	In KnowRISK
WHERE	In pilot areas
WHO	Schools, citizens, building stakeholders
WHEN	Not during emergency
WHY	Reduce non structural vulnerability
WHAT	Little effort and high benefit
HOW	The engagement model

Target public is chosen to be (1) schools in the first place, then (2) citizens, mostly as a by-product of communication in schools, and (3) building stakeholders. Here we mostly discuss the development of communication action for schools and citizens.

KAP assessment is done ex-ante and ex-post with a questionnaire the project has developed to be used in other European countries (Platt et al. 2017).

The setting up of the strategy comes along with the building of the communication plan (Table 31.1). Because communication will occur in pilot areas where research action – evaluation of hazard, risk and social parameters assessment – are developed, the first step of the communication plan has to be the "Where". Particularly, we will focus on the profiling of target audience and discuss "where it lives" and "who is it". The "When" is not during a crisis and, for this reason, we can use an approach that takes into account the KAP criteria. The "How" is an engagement model: the public participate in a co-produced plan for the mitigation of non-structural vulnerability, with focus groups that analyse in detail their thoughts and needs. The target public assesses all the tools for communication, which are produced within the project.

31.3 Profiling the Target Audience: Where & Who?

The public, far from being a monolithic entity, responds to risk and uncertainty differently, depending on culture, personal beliefs and attitudes, and has a personal risk estimation that might not be properly related to its actual value. To understand the needs of recipients, we have investigated in specific communities the level of awareness of non-structural damage.

31.3.1 Profiling the "WHERE"

Based on the Risk Communication Strategy (RCS) in the KnowRISK project, the target audiences in the pilot areas (Fig. 31.2) were chosen to implement communication strategies tailored on the understanding local communities' fragility. The

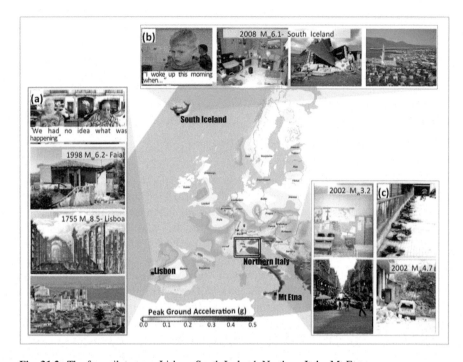

Fig. 31.2 The four pilot-areas: Lisbon, South Iceland, Northern Italy, Mt Etna area
The map is the 2013 European Seismic Hazard Model (ESHM13).
Snapshots from the videos "Before it's too late" (**a**), "In compliance with nature" (**b**) and "Mt Etna" (**c**) – https://www.youtube.com/channel/UCg0VxYGPYa2bUGXIhZl35zQ. Photos of non-structural damage at Mt. Etna pilot area are from Azzaro et al. (2016)

selection stood on two criteria: (i) ensure a broad range of seismic hazard; (ii) areas where it was possible to have a high range of target public.

A holistic approach to community vulnerability takes into account the level of relevance of seismic compared to other hazards, the understanding, the memory all are indicators that can affect the way a risk is perceived and preventive measures are taken. Level of education, wealth and exposure to other social risks need to be taken into account as aggravation parameters in risk computation and in strategies for communication. All these indicators were addressed to draw a general picture of the pilot areas and the target societal groups, which were chosen to be students, citizen living in historical downtowns and building stakeholders.

The pilot area in Portugal is the city of Lisbon. Seismic risk in Portugal derives partly from offshore sources that can cause large events, such as the catastrophic 1755 Great Lisbon Earthquake. Although the Portuguese people have heard about this event at least once in their lifetime, they do not have a clear idea of the likelihood potential future events that can have disastrous effects in their or their children's live. The Portuguese case study differs from the other two, Italy and Iceland cases, by its low disaster experience. Earthquakes in Lisbon are a distant hazard and have a low degree of intrusiveness in people daily lives. In this framework communication should put great emphasis on the raise of awareness, on the fact that simple and low-effort measures can turn in a high benefit.

The pilot-area in Iceland is within the South Iceland Seismic Zone (SISZ). In this area earthquakes tend to occur in sequences, typically every 100 years. It covers the largest agricultural region in Iceland. Several small towns or villages, schools, medical centers, industrial plants, geothermal and hydropower plants, and several major bridges are within this area. In fact, it contains the entire infrastructure that characterizes modern society.

There have been many strong earthquakes in the South Iceland lowland in the recent years that keep high the level of awareness of local communities (Bernharðsdóttir et al. 2015). During the June 2000 and May 2008 earthquakes in South Iceland, residential buildings and other structures suffered some damage. Most of the residential buildings in SISZ proved to sustain significant ground shaking without major structural damage. However, during the earthquakes in June 2000, inhabitants reported having difficulty or finding it impossible to move to a safe place inside their dwellings (Platt et al. 2017 and reference therein). In this framework, communication should put great emphasis on the fact that even though houses resist, non-structural damage still can cause high costs in terms of injuries, economics and resilience.

In general terms Italy is a country with recurrent earthquakes, but a low level of prevention. Seismic hazard spans low to high PGA. Two pilot-areas were chosen – the Mt Etna volcano region and Northern Italy. In Northern Italy, risk communication focal points are the cities La Spezia, Laveno Mombello (a small town in the Varese province), and Ferrara. They were selected based on two criteria: (i) areas affected by the most common non-structural vulnerability; (ii) areas where it was possible to have a high range of target public. Because schools are a major target of the communication action, pilot areas selection was also based on the level of school board engagement in risk reduction activities implemented in the past by the Istituto Nazionale di Geofisica e Vulcanologia.

31.3.1.1 The Italian Case: Mt Etna Pilot-Area

Mt Etna pilot-area is characterized by low to moderate earthquakes, the magnitude of which rarely exceeds Ml 5. However, the seismic foci are often shallow and can consequently create serious damage in restricted areas. For this reason and for the high occurrence frequency of earthquakes, it is important to disseminate information on seismic hazard among students and laypersons, and explain what each of us can do to reduce the damage of earthquakes.

In Catania, the main town in this pilot area (~300.000 inhabitants), Istituto Nazionale di Geofisica e Vulcanologia hosts every year science outreach events that involve public from all the Mt. Etna area. Schools and stakeholders are involved, and discuss with scientists issues concerning earthquakes.

31.3.1.2 The Italian Case: Northern Italy Pilot-Area

The Northern Italy pilot area was chosen to implement communication in regions where PGA was expected to be lower than 0.15g (G.U. n.108 del 11/05/2006); these being seismic zones where strong earthquakes are rare but non-structural damage, for example during the 2012 Emilia earthquake, can be widespread and cause anxiety for people living all over Northern Italy.

Northern Italy pilot-area encompasses two communities: La Spezia, Laveno Mombello and Ferrara city. Risk communication covers two target-groups, respectively: (i) students from a set of schools located in La Spezia and Laveno Mombello; (ii) citizens living in Ferrara downtown historic center.

In La Spezia earthquakes may be strong but they are rare. In Laveno Mombello they are rare but Seismic Building Code for public buildings (i.e schools) is enforced.

Building stock in both areas is diverse and comprising various typologies with building up to 4–6 stories. In La Spezia the historic down town roads can be narrow and composed by the so-called Ligurian "Carruggi", with 4–6 story buildings at both sides, stone balconies, shutters and various pendings that represent the most frequent architectural part of non-structural elements.

31.3.2 Profiling the WHO

To profile the target audience we evaluated the current knowledge about the risk, what information may gain the greatest interest, the major source of information, the level trust on experts and institutions. We also considered how long time it is likely be spent to pay attention to receiving and assimilating information and how he/she usually receives information.

As students are the citizens who will build the world of tomorrow, their present engagement may promote the future capacity to cope with hazards. Involving students will not only act on raising and spreading awareness, but also will enable the choice of an effective language (words, tools, icons) to communicate to their peers.

Results from a test survey addressed to schools' Directors in the three participating countries can help us to identify the most efficacious media to spread information concerning seismic risk and preventive actions to reduce non-structural damage. The survey shows that paper leaflets are a common vehicle of information; in Italy lectures in schools are also valuable, while in Iceland Internet is the second best tool to be informed (Fig. 31.3).

Although state plans do not include efficient actions towards hazard education (Musacchio et al. 2015), many schools were involved in training and lectures organized by research institutions or individual experts on the topic. In the three countries, non-structural measures in schools are highly neglected (Fig. 31.3).

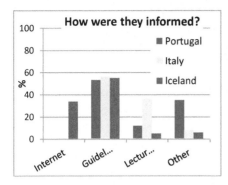 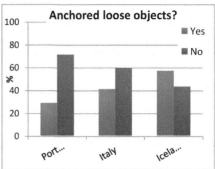

Fig. 31.3 Results from a test survey conducted in schools (Bernharðsdóttir et al. 2015) Left: Return to the question "How did the authorities process information?"; right: Return to the question "Have loose objects, like pictures/frames, statues and shelves, been especially anchored"?

Table 31.2 General safety certification for all school buildings in Italy

	Yes (%)	Not requested (%)	No (%)	No info (%)
Emergency plan	73	0	19	8
Risk assessment certificate	72	0	20	8
Static load assessment	49	7	32	12
Habitability certificate	39	4	45	12
Fire prevention certificate	21	12	54	13

http://www.istruzione.it/edilizia_scolastica/anagrafe.shtml

This applies regardless the experience of recent earthquakes. The earthquake in south Iceland in 2008 highlighted non-structural vulnerability, as the ground motion caused extensive interior damage to buildings. Nonetheless, only 47% of the Icelandic schools had specifically attached loose objects to prevent them from falling in case of an earthquake (Bernharðsdóttir et al. 2015).

31.3.2.1 Profiling the WHO- Schools: The Italian Case

There are about 83.000 school buildings in Italy, public and private, and they host more than 10.6 million students. Most of these buildings are known to be vulnerable to non-structural elements failures even to static load.

According to the official registry database, more than a half of Italian schools (55%) were built before 1976 and around 30% were not originally built to function as schools. A large number of schools (20%) do not have an emergency plan in place and frequently it does not include earthquake emergency. Static load certificate is only available for 49% of the schools (cf Table 31.1) and was only enforced for buildings built after 1971 (Table 31.2).

A recent poll conducted on high-school students (age 13–20) shows Internet, social media and TV as the largest source of information on natural hazards (Musacchio et al. 2016). School is a channel of information only for 15% of students. The poll also assessed students' trust in geoscientists and their science, pointing out a diffuse opinion (56% of respondent) in which scientists are considered not to genuinely be independent from outer urges (Musacchio et al. 2016). Students also believe that media manipulate information with wilful misconduct, to hide inconvenient realities or to get economic advantages. Because the source of information and trust are among of the most important issues in communication, in assessing and going deep in local communities needs, we should take into account the opinion and reliability that students have towards scientists.

31.3.2.2 Profiling the WHO- Citizen's in an Old Downtown: The Ferrara Case Study- Preliminary Results

The town of Ferrara is located in the Northeastern part of Italy. The Emilia earthquake hit the town in 2012, when the memory of past events had been lost over the decades and centuries, and the need to make the citizens active in preventing the damages of possible further earthquakes suddenly became pressing.

Among the other initiatives, the Municipality of Ferrara, with the support of Master Scientific Journalism and institutional communication of science and the participation of the Italian National Agency for New Technologies, Energy and Sustainable Economic Development (ENEA), the Ferrarese Naturalists Society, and the Waseda University of Tokio, organized in 2013 a series of participative events to involve the citizens of the city centre: Laboratories on the prevention of the seismic damage, financed by the Emilia Romagna Region. Major outputs achieved through this initiative were: a series of shared practices to mitigate the non-structural risk, summarized in a booklet called "10 good practices to make our home safer"; a serious game to involve schools goers and citizens, namely the Playdecide "Earthquakes, when and how to communicate an emergency"; a participative proposal, a formal document by the City Council to officially declare the need to develop strong communication and social cohesion actions by the public administration.

For the KnowRISK project we run two focus groups to understand barriers to prevention and tune an efficacious communication strategy. Group 1 had a previous involvement in discussions on seismic risk after the 2012 Emilia earthquake and took part to the 2013 Laboratories (see above), with a focus on the medieval part of the town. Group 2 had no previous involvement in discussions on seismic risk; however, living in the historical downtown, group 2 experienced damage from the 2012 Emilia earthquake in a specific setting. Table 31.3 shows the most representative quotations stated during the two focus groups. Results show that prevention measures and their effective communication are dependent on two main factors: one, wider, based on the societal dimension and one, more specific, based on communication techniques. On the first, prevention can be successful only if citizens participate and, therefore, if social cohesion measures are put in place. On the second, an

Table 31.3 Most representative quotations for each category addressed during the focus groups

Category	Most representative quotations
Memories from the earthquake	My daughter didn't wake up... I went downstairs. All the neighbours were on the street... objects, cornices fell.
	Once back [from the Laboratories], I told myself: what if an earthquake happens now?
	I know I have to wait for the end of the shake and then try to go downstairs and reach the street. However, being so close to the building, a cornice can fall on me. What to do? There are squares a bit bigger and I thought I could try to reach them... I discovered that the urban setting is very important.
Non structural damage	When I think about an earthquake, I do not think about the damage from non-structural elements. I think that if the house doesn't crash, then I will be safe.
	I know that if a shelf falls down I can be in danger. Still, talking with many people after the earthquake, we were all relieved not to see cracks on the walls. The prevention stops there.
The role of the experts	Moderator – Let's start from this claim from our leaflet: "first of all, it is necessary to know our own house; the land where it has been built; the architrave under which to find a shelter; if the electrical, gas, and water systems are safe". Let's also be realistic: is that feasible?
	Participant – experts have to do it, not us.
Prevention, motivating factors: the better motivation is the social cohesion	They said that that neighbours have to find an agreement and help each other [in spreading information and in practice exercises]
	After the Laboratories, I involved also the other people leaving in the building. We live in an ancient building, from the fifteenth century. We all met and checked the roof. I think it is something good to do: to keep a relationship with the neighbours. This is not a joke.
Communicating the best practices – mix and repeat	Undeniable: to repeat what we have to do to protect ourselves is helpful. Newspapers and websites published guidelines for the damage prevention. However, if something as the 29th May [earthquake] happens, I do not honestly know where I have to protect myself... to find the external stairs... under the table... right... we need to have adequate information. And periodical!
	Administrators of buildings, technicians in charge of the buildings' maintenance: they should also be involved in the communication of the seismic prevention. They should be mediators of preventive measures.
	We, as citizens, should involve the people we know, and then somebody has to monitor what happens.
	Regular training at the building and neighbourhood level. Today they are obligatory in schools, public offices, big companies, but not in built up areas and blocks; this kind of events are worth "thousands of folders".

effective communication has to be based on a mix of traditional and innovative techniques, such as leaflets and new technological tools (Reitano et al. 2017). Participants of the focus groups ask for periodical events and trainings, specifically tailoring information delivered by leaflets, apps and websites to local communities situation. More importantly people, even for the simplest checks and for evaluation of possible damages on non-structural elements, ask the consultancy by dedicated professionals. They do not want to be left alone. Moreover there is such a little knowledge on what is structural and non-structural damage that to draw a clear and communicable difference even for the stakeholders closer to technical professions (civil protection, engineer) looks complicated.

31.4 The Setting of Media and Tools

Here we adopt the term media *lato sensu* as channels through which information on measures to reduce risk are disseminated. In this light, the main media are the schools to whom a specific risk communication plan is directed. The approach we have implemented stands on the engagement of students and relies on previous work done with schools in Italy (e.g. Musacchio et al. 2015; Piangiamore et al. 2015). We used a flipped learning strategy (Piangiamore et al. 2016) that is a powerful way to engage students in enhancing the KAP on non-structural risk reduction measures. Students, being part of the whole process, will prepare communication tools to address their peers and act as a vehicle for spreading awareness.

Tools that we developed based on the research we have conducted to profile the target audience. Focus groups with citizens and questionnaire in schools (Bernharðsdóttir et al. 2015; Musacchio et al. 2016) showed that mix of traditional and innovative tools is the best approach: leaflets, "interactive" hands-on products, such as a 3D brochure (Fig. 31.4), and technological, yet highly attractive, portable apps using Augmented Reality applications (Reitano et al. 2017).

Finally, a Practical Guide for citizens suggests solutions for non-structural components vulnerability intended to guide towards the first steps of preparedness, minimizing or avoiding injuries, damage, and long-term financial consequences (O'Neill et al. 2017). Here a clear distinction on what citizens can do by themselves (Do-It-Yourself) and what seek for an expert is drawn.

31.5 Conclusions

This paper presents the rationale behind the implementation of a communication strategy aimed to trigger changes into Knowledge, Attitudes and Practice towards non-structural vulnerability of buildings to earthquakes. The communication relies on the understanding of local communities and target public needs and barriers to the implementation of preventative actions. Students and citizens living in areas with a

Fig. 31.4 The 3D image of the brochure (Orlando et al. 2016). It can be downloaded from http://knowriskproject.com/per-gli-studenti-ecco-la-brochure-di-knowrisk-3d/?lang=it

wide range of seismic hazard and risk in three European countries, namely Portugal, Italy and Iceland, compose the target public. In this paper, we have discussed the importance of profiling the target audience by applying the approach to the Italian pilot areas. These are chosen to be representative of a broad range of seismic hazard levels; in particular, one of these areas (Emilia) has been recently hit by a major earthquake. The profiling of the target audience has been carried out on schools (Bernharðsdóttir et al. 2015) and students all over Italy (Musacchio et al. 2016), and downtown citizens of Ferrara city. The process consisted of several investigations and turned out to be very useful for designing and assembling appropriate tools for communication.

The main results of our study are strictly linked to the approach, in which experts collect inputs before any communication activity to tune their actions and improve their communication skills, while the public participation ensures the establishing of shared rules and reinforces active citizenship. Indeed, the initial effort of designing ad hoc messages and tools reflects on a more efficient communication and significant saving of money and time, as the ex-post questionnaires reveal.

Acknowledgements This study was co-financed by the European Commission's Humanitarian Aid and Civil Protection (Grant agreement ECHO/SUB/2015/718655/PREV28). The 3D brochure was created by Lisa Orlando and Marco Maria Faggioli.

References

Azzaro R, D'Amico S, Tuve' T, Cascone M (2016) Etnean earthquakes, seismic risk of non-structural elements. Istituto Nazionale di Geofisica e Vulcanologia (Catania), KnowRisk Project, 33 pp, ISBN 9788862821797

Bernharðsdóttir AE, Musacchio G, Ferreira MA, Falsaperla S (2015) Informal education for disaster risk reduction. Bull Earthq Eng 14(7):2105–2116. https://doi.org/10.1007/s10518-015-9771-9

Bruine de Bruin W, Bostrom A (2012) Assessing what to address in science communication. Proc Natl Acad Sci U S A 110(3):14062–14068

Covello VT (2003) Best practices in public health risk and crisis communication. J Health Commun 8(1):5–8

Crescimbene M, Pino NA, Musacchio G (2017) Between perception and knowledge: the construction of the Italian questionnaire to assess the KnowRISK project actions. Submitted to this volume

De Marchi B (2015) In: Paleo UF (ed) Risk governance and the integration of different types of knowledge. Springer, Dordrecht, pp 149–165. https://doi.org/10.1007/978-94-017-9328-5

De Marchi B, Scolobig A (2012) Experts' and residents' views on social vulnerability to flash floods in an alpine region. Disasters 36(2):316–337. https://doi.org/10.1111/j.1467-7717.2011.01252.x

Fischhoff B, Brewer NT, Downs JS (2011) Communicating risks and benefits: an evidence-based user's guide. Food and Drug Administration, Washington, DC. http://www.fda.gov/oc/advisory/OCRCACACpg.html

Fischoff B (1995) Risk perception and communication unplugged: twenty years of progress. Risk Anal 15(2):137–145

Hobson-West P (2003) Understanding vaccination resistance: moving beyond risk. Health Risk Soc 5(3):273–283

Infanti J, Sixsmith J, Barry MM, Núñez-Córdoba J, Oroviogoicoechea-Ortega C, Guillén-Grima F (2013) A literature review on effective risk communication for the prevention and control of communicable diseases in Europe. Stockholm, ECDC

Kuhlicke C, Scolobig A, Tapsell S, Steinführer A, De Marchi B (2011) Contextualizing social vulnerability: findings from case studies across Europe. Nat Haz 58:789–810

Musacchio G, Falsaperla S, Bernharðsdóttir AE, Ferreira MA, Sousa ML, Carvalho A, Zonno G (2015) Education: can a bottom-up strategy help for earthquake disaster prevention? Bull Earthq Eng 14(7):2069–2086. https://doi.org/10.1007/s10518-015-9779-1

Musacchio G, Solarino S, Eva E, Piangiamore GL (2016) Students, earthquakes, media: does a seismic crisis make a difference? Ann Geophys 59, fast track 5. https://doi.org/10.4401/ag-7239

NSET (2017) Risk perception survey in Bhimeshwor municipality. National Society for Earthquake Technology-Nepal (NSET) 16WCEE Conference Chile 9 January 2017

O'Neill et al (2017) Geotechnical, geological, vol. 47. In: Rupakhety R et al (eds). Proceedings of the international conference on earthquake engineering and structural dynamics. 978-3-319-78186-0

Orlando L, Faggioli MM, Musacchio G, Piangiamore G (2016) Two concepts of risk communication and learning. Poster presentation at the 2nd general meeting KnowRISK, Catania, Italy, 15–17 December 2016. Vol. 33 MISCELLANEA INGV, ISSN 2039-6651, pp 35–36. http://istituto.ingv.it/l-ingv/produzione-scientifica/miscellanea-ingv/, http://hdl.handle.net/2122/10428

Piangiamore GL, Musacchio G, Pino NA (2015) Natural hazards revealed to children: the other side of prevention. In: Peppoloni S, Di Capua G (eds) Geoethics: the role and responsibility of geoscientists. Geological Society, London, Special Publications. https://doi.org/10.1144/SP419.12

Piangiamore GL, Musacchio G, Devecchi M (2016) Episodes of situated learning: natural hazards active learning in a smart school. In: Hunt LM (ed) Interactive learning strategies, technologies and effectiveness. Nova Science Publishers, New York, pp 21–45. ISBN 97881634841986

Piangiamore et al. (2017) Geotechnical, geological, vol. 47. In: Rupakhety R et al. (eds) Proceedings of the international conference on earthquake engineering and structural dynamics, 978-3-319-78186-0

Platt et al. (2017) Geotechnical, geological, vol. 47. In: Rupakhety R et al. (eds) Proceedings of the international conference on earthquake engineering and structural dynamics, 978-3-319-78186-0

Reitano et al. (2017) Geotechnical, geological, vol. 47. In: Rupakhety R et al. (eds) Proceedings of the international conference on earthquake engineering and structural dynamics, 978-3-319-78186-0

Chapter 32
Seismic Risk Communication: How to Assess It? The Case of Lisbon Pilot-Area

Delta Sousa e Silva, A. Pereira, Marta Vicente, R. Bernardo, Monica Amarel Ferreira, Mario Lopes, and Carlos Sousa Oliveira

Abstract This paper aims to present and discuss the theory and research strategy underlying the assessment of the KnowRISK risk communication in one of the KnowRISK pilot-areas, Alvalade parish in Lisbon city. The theory guiding this evaluation research stands on Precaution Adoption Process Model (PAPM), and is complemented by other proposals. Defining the most appropriate research design required answers to two basic questions: 'what to assess?' and 'how to assess?'. The first-mentioned question implied to take into account KnowRISK risk communication main aims, clarifying which cognitive and behavioral changes were realistically expectable in the context of an earthquake dormant society as is the case of Lisbon. Concerning the second-mentioned question, the assessment of risk communication procedure stands in an evaluation research design that comprehends quantitative and qualitative methods.

Keywords Assessment · Risk communication · Seismic risk

32.1 Introduction

This paper reports a process of assessment of the impacts of a risk communication procedure presently on-going in two schools of Lisbon under the EC project KnowRISK. This is a participatory action research project aimed at transferring expert knowledge on non-structural seismic protection to communities at risk. Schools are relevant targets of risk communication under this project, and were taken as the focus of assessment.

D. Sousa e Silva (✉) · A. Pereira · M. Vicente · R. Bernardo
Laboratório Nacional de Engenharia Civil, Lisboa, Portugal
e-mail: delta@lnec.pt; apereira@lnec.pt; magvicente@lnec.pt

M. A. Ferreira · M. Lopes · C. S. Oliveira
Instituto Superior Técnico, Department of Civil Engineering, Architecture and Georesources, CEris, Lisboa, Portugal
e-mail: monicaf@civil.ist.utl.pt; mariolopes@tecnico.ulisboa.pt; csoliv@civil.ist.utl.pt

Risk communication in Lisbon pilot-area involved two secondary schools–Rainha D. Leonor and Padre António Vieira – and covers approximately 120 students of the 7th and 8th grades with ages between the 12 and the 16 years old. The intervention was structured into a set of activities, which became part of a chair of the 7th and 8th grades curricula, called as *education for citizenship*. KnowRISK intervention programme was the same in the two schools and comprehended a set of sessions/activities. Students and scientists interacted in a regular basis for a period of 2–4 months. In Rainha D. Leonor School, sessions and activities were not sequential resulting in a longer intervention than in Padre António Vieira School, where activities happened once a week and in a sequential way.[1]

Assessing the impacts of a risk communication process, such as the one pursued under KnowRISK project, implies answering two basic questions: (i) *"what is it meant to assess?"*, and (ii) *"how to assess it?"*. The first question sends us to the risk communication aims. The second-mentioned question concerns the research strategy and the most appropriate methodological options taken in order to proceed with the assessment.

Emphasise that the efficacy of education for seismic safety is often inhibited by an incomplete understanding of the process by which individuals decide to protect themselves from harm (Becker et al. 2012a, b). These authors posit that individuals pass through a series of cognitive and social stages until they decide to protect themselves. Those stages are, respectively: *knowledge and awareness* of earthquake hazard and protective measures; *thinking and talking* about the subject with others; *understanding the consequences* of earthquake phenomena in individuals near environment; and *developing skills*. Based on this assumption, KnowRISK risk communication was planned in order to achieve the following aims:

(a) To foster *knowledge and awareness* about risk and protective measures;
(b) To stimulate the formation of favorable beliefs and *behavioral intentions* towards the adoption of protective behaviors.

For several reasons discussed elsewhere (cf. Sousa e Silva 2017), it was found unrealistic to set as an aim the adoption of protective behaviours in the short-term. As mentioned by Weinstein (1988), the adoption of new protective behaviours is usually a relatively long process, made of advances and retreats. Further, it should be taken into account that, in Lisbon, earthquake hazard has a low level of intrusiveness in people's daily lives. Finally, teenagers are our target-group, a social group without full autonomy to make changes in their own homes.

Answering to the question of *"how to assess?"* KnowRISK intervention implied a set of choices concerned with the theory, the research design and underlying methods. This paper aims at presenting the theory underlying the assessment of

---

[1]KnowRISK intervention in Rainha D. Leonor School took place between October 2016 and January 2017. In Padre António Vieira School intervention started in January, 2017 and finished in March 2017.

KnowRISK intervention in Lisbon pilot-area and to discuss some of the methodological challenges that this type of evaluation research (Bryman 2011) poses.

32.2 Theory: The Precaution Adoption Process Model

There is a consolidated volume of accumulated research on protective behaviours and several models co-exist on the process by which individuals decide to act protectively. The *Protection Motivation Theory* (PMT) proposed by Mulilis et al., the *Protective Action Decision Model* (PADM) proposed by Lindell and Perry (2000) are among the most relevant theoretical models. All models share the assumption that decision of acting protectively is influenced by the way people face and assess the following aspects:

- Their own susceptibility to a certain threat (*perceived susceptibility*);
- The severity they assign to a certain event (*perceived severity*);
- The efficacy of a certain precaution measure in reducing the susceptibility to risk (*perceived efficacy*);
- The demands placed by a certain behavior (knowledge, effort, time and money) (*perceived costs*).

In short, individuals' willingness to act protectively is based on the double evaluation they make about their own susceptibility towards a certain risk and the efficacy they attribute to the alternatives of protection. Protection measures related factors, such as time, money and skills required to implement it, usually interfere on individuals' decision (not) to act protectively.[2]

The theoretical framework adopted in KnowRISK assessment stands on Weinstein's model, so-called as Precaution Adoption Process Model (PAPM), with insights coming from other theoretical approaches. Weinstein proposes a dynamic perspective by putting the emphasis on the assumption that the adoption of protective behaviours is process based on the development over time of beliefs and intentions that lead to action. Adopting protection is not a "yes or no" issue. Instead, it is a process concerned with "the initiation of *new* and relatively complex behaviours (...) which are not part of the person's response repertoire" and usually involve high-level cognitive change and advance planning before translated into action.

Weinstein conceptualizes the precaution adoption process into a series of stages defined by beliefs people hold their *susceptibility towards risk* and *perceived efficacy of precaution*. As described in Fig. 32.1, beliefs about *susceptibility towards risk*

---

[2]This is an assumption that stands on the Theory of Reasoned Action, proposed by Fishbein and Azjen (1991). According to this theory, behavioural response towards a certain risk is so dependent on individuals' beliefs about the protective measure (i.e. to fix shelves to the wall) in itself as their interpretation of risk situation (exposure to earthquake threat).

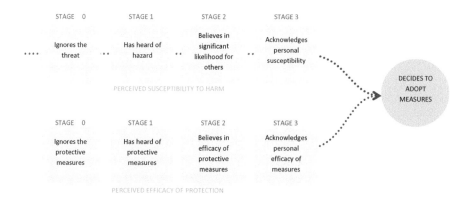

Fig. 32.1 Precaution Adoption Process Model (Weinstein, 1988)

range from "ignorance of the threat" to "belief in personal susceptibility". The criteria that define the stages are, respectively: risk awareness, belief in general susceptibility and belief in personal susceptibility. Changes in individuals' thoughts and behaviour are expected when they surpass a certain threshold and personalize the threat. Generally, risk awareness prevails over the awareness of protection alternatives. It is only after people realize the risks that they feel compelled to find out how can they protect themselves. Beliefs about the efficacy of precaution range from the "ignorance about available precautions" to the belief on individual's own capacity to take precaution.

The transition from one stage to another does not occur in a vacuum. Instead, it is influenced by individuals' experience and particularities of the socioecological context where they live. Disaster experience is usually seen as a major driver of change on human beliefs and practices. Nevertheless, the experience of well-being and stability over long periods may dismiss signals given by disaster experience. Luhmann (1995) emphasises that a certain phenomenon will hardly be perceived as a threat if it does not penetrate in communication system of a society. Given this, information about risk and protection (preferably coming from different sources) needs to circulate on individuals' daily lives for change to take place. Lindell et al. (in ibid) refer that some degree of *hazard intrusiveness* in people's lives, whether through disaster experience or through information, needs to occur in order to stimulate cognitive and behavioral change.

As mentioned elsewhere, the knowRISK intervention was theoretically based on literature about the process by which individuals decide to act protectively expecting to be a catalyst of such process. Level of *intrusiveness* of earthquake hazard in Lisbon inhabitants' daily lives is low. So, it is realistic presuppose that our target-groups are poorly aware of the extent of earthquake threat in the city of Lisbon, where they live. Bridging with Weinstein theoretical framework, we would say that, hypothetically, our target-groups' beliefs on their own *susceptibility towards risk* and *perceived efficacy of precaution* varies between stage 0 through stage

2. Cognitive and behavioral changes are a consequence of KnowRISK intervention, both in terms of *risk personalization* and belief in the efficacy of precaution.

Next, Weinstein model central concepts – *perceived susceptibility towards risk* and the *"perceived efficacy of the precaution–* will be re-worked in order to become operational. Such effort of operationalization will be proceed on the basis of accumulated research on the field, particularly Becker et al. (2012a, b, 2013) research.

32.2.1 Perceived Susceptibility to Harm

As shown by Fig. 32.1, Weinstein (op cit.) depicts *perceived susceptibility to harm* into four major stages, accordingly:

- *Ignorance of the threat* (Stage 0). Individuals that know nothing about the hazard;
- *Heard of the threat* (Stage 1). Individuals who have heard about the hazard but "don't believe in it", "deny its probability of occurrence", "are convinced that risk is insignificant";
- *Believes in significant likelihood for others* (Stage 2). Individuals who believe there is an actual risk but are convinced that they will not be personally affected by it;
- *Acknowledges personal susceptibility* (Stage 3). Individuals who recognize their own susceptibility towards risk.

A person reaches a stage when he or she accepts the idea that defines that particular stage. In the particular case of Stage 4, to achieve this implies to personalize the risk, that is to believe that his or her vulnerability towards a certain threat.

Based on the findings of a research on earthquake beliefs, Becker *et al.* (in ibid) identify a set of risk beliefs favour the adoption of protective measures and beliefs that act as barriers. Both of them are an important contribution to define more accurately each of the previously mentioned stages.

Beliefs that may favour the adoption of protective behaviours are as follows:

- *Perceived probability of disaster occurrence.* Belief that "there is an actual risk" and that "it is significant or high";
- *Degree of eminence.* Belief that the hazard is a phenomena that "may occur anytime without previous warning";
- *Degree of certitude and inevitability.* Belief that the hazard is something "that will inevitably occur, we just don't know when"; "it is not a matter of knowing whether it will happen or not", "because it will inevitably happen, we just do not know when".

These are beliefs that can be shared by people at Stage 2 or 3, being that the only differential aspect is risk personalisation.

Beliefs that may act as barriers to pre-disaster preparation are as follows:

- *Perceived probability of occurrence.* Belief that an earthquake is something "that simply won't happen in one's area of residence"; that "has a low probability of occurrence";
- *Degree of eminence.* Belief that hazard is "something that is not eminent"; that "will not occur in the near future" or "during one's lifetime";
- *Degree of certitude and inevitably.* Belief that "no one knows if an earthquake will occur or not" (uncertainty about disaster occurrence discourages people to perform any action) (Table 32.1).

Table 32.1 Operationalization of perceived susceptibility to harm

	Stage 1	Stage 2	Stage 3
Perceived probability	Risk perceived as improbable or with low level of probability in one's area of residence	Risk perceived as in one's area of residence	Risk perceived as highly probable in one's area of residence
	Possibility of occurrence of an earthquake as highly uncertain	Possibility of occurrence of an earthquake as uncertain, but it may happen	Possibility of occurrence as something certain in the future
Perceived severity	Doesn't know if it will provoke damages in one's area of residence;	It may provoke damages in one's area of residence	It will provoke damages in one's area of residence
Perceived vulnerability/risk personalization	Doesn't know if an earthquake will affect personally or family;	Doesn't know if it will affect personally or family;	An earthquake will affect personally and family
	An earthquake won't affect personally [risk denial]	Own risk is undoubtedly less than risk faced by others;	
Eminence	Earthquake risk pushed to a distant future;	Doesn't know if earthquake is an eminent phenomena;	Na earthquake is a phenomenon that may happen any time, in the near future, and without warning.
		It may happen or not	
Salience	Neve or rarely thinks in earthquake risk and its probability of occurrence in one's area of residence	Sometimes thinks in earthquake risk and its probability of occurrence	Usually thinks in earthquake risk and its probability of occurrence;
	Never or rarely worries with the possibility of occurrence of an earthquake	Sometimes worries with the possibility of occurrence of an earthquake	Usually worries with the possibility of occurrence of an earthquake
Fear	Doesn't feel fear	Sometimes feels fear	Sometimes fears Usually fears

Some of the previously mentioned beliefs are associated with two types of attitude. We refer to the so-called *optimistic bias* and to a general attitude of *fatality* as regards a given phenomenon that is recognised as a threat. Both attitudes are analytically relevant, although distinct, both can act as barriers to the adoption of protective measures.

The *optimistic bias* designates attitudes based on the unrealistic belief that "individual's own risk is undoubtedly less than the risk faced by others". This type of attitude is indeed different from disbelief based on a genuine incertitude about the risk and the likely personal impact. The "unrealistic optimists" are those persons who recognise that there is an actual risk but are convinced that it will not affect them significantly (Becker et al, op cit; Weinstein, op cit). Weinstein mentions that the *unrealistic optimism* may have several causes, which can range from erroneous information about the risk to the personal need in refraining fear. Supposedly, there is a higher tendency for this type of attitudes being observed for risk with a low probability of occurrence and in cases when people are convinced that these can be prevented.

The so-called *fatality* comprises social attitudes based on the belief that the disasters, such as earthquakes, are phenomena that is not within individuals' own possibility of control and, therefore, there is no use in worrying about the subject and to try and find some sort of protection. Turner et al. (1986) outline in some detail the profile of a fatalist individual, based on a study on the social perception about earthquake risk of people living in Southern California. Those authors distinguish from individuals who are fatalist regarding the general impact of an earthquake from those who are fatalist about the efficacy of earthquake protection. Amongst the most fatalist ones, are the individuals who believe that "an earthquake is something that will necessarily cause many damages and fatalities whether we are prepared or not". The less fatalist ones are those who:

- do not show any extreme pessimism as regards the impact of an earthquake but who do not believe in the efficacy of the protective measures;
- show an extreme pessimism as regards the impact of an earthquake, but have some expectations as regards the efficacy of measures.

Interestingly, in Turner et al. (in ibid) study, the people with fatalistic attitudes demonstrated to be less frightened and concerned by seismic issues.

Both the *optimistic bias* and the *fatalism* can act as obstacles to protection. They generate indifference regarding earthquake-related information; they prevent people from thinking about the subject and discussing with others and from trying to obtain further information on the issue; they lead people to distance themselves from the adoption of protective behaviors. This type of people are expected to be also poorly sensitive to the creation of a public risk mitigation policy.

32.2.2 Perceived Efficacy of Protection

In the previous section, emphasis was put on the assumption the process by which individuals decide to take precautions is based on the double evaluation they make about hazard susceptibility and efficiency of precaution measures. As highlighted by Weinstein (1988: 364), beliefs towards risk usually have temporal precedence over beliefs on precautionary behaviors. An individual will hardly search for information concerning protection if not aware of the risks impending on him.

Accordingly to Weinstein model (cf. Fig. 32.1), perceived efficacy of precaution comprehends four major stages, respectively:

- *Ignorance of precaution measures* (Stage 0). Individuals that know nothing about protection and protective behaviors;
- *Heard of* precaution *measures* (Stage 1). Individuals who have heard of, at least, some measures but do not have opinion about it or are not engaged on putting them into practice[3];
- *Belief in precaution measures general efficacy* (Stage 2). Individuals who believe in the efficacy of protection measures, for safety of people and goods, but for some reason postpone protective action[4];
- *Belief in precaution measures' self-efficacy* (Stage 3). Individuals who believe in the efficacy of protection measures and in their own capacity to implement it.

Besides from risk-related beliefs, Becker et al (op cit) identify a set of pre-disaster preparedness beliefs and personal beliefs which may, in our view, be linked with Weinstein's three-stage typology on *perceived efficacy* of precaution. Pre-disaster preparedness beliefs that unfavors protection and may be found in individuals at Stage 0 or 1 are as follows:

- "Investing in pre-disaster preparation will make no difference";
- Disbelief in one's capacity to adopt precautions. Individuals who believe precautions "may not work"; "won't be able to have access to emergency items in case of an earthquake occurrence"; "only at the moment earthquake occurrence one can know what to do";
- Pre-disaster preparedness is excessive. Individuals who regard pre-disaster preparedness efforts as not useful or excessive.

Personal beliefs that may encourage preparedness, and are typical of Stage 3 in Weinstein model, are as follows:

---

[3] In a more recent version of the model, Weinstein et al. (2008) classify these individuals as "aware but unengaged".

[4] Unlike individuals at Stage 2, these individuals are more knowledgeable about the alternatives of protection and have, at some point, considered the possibility of adopting them.

Table 32.2 Operationalization of perceived efficacy of protection

	Stage 1	Stage 2	Stage 3
Perceived efficacy	Disbelief in the efficacy of measures to save people and goods	Belief in the eficacy of measures to save people and goods	Belief in the eficacy of measures to save people and goods
	Pre-disaster preparedness will make no difference in case of earthquake		Pre-disaster preparedness will make a difference in case of earthquake
Self – efficacy	Disbelief in one's capacity to implement protection	Disbelief in one's capacity to implement protection	Belief in one's capacity to implement protection and respond
Values		Safety and survival as important values	Safety and survival as important values
			Pre-disaster preparedness as a way of life
Perceived responsibility	Responsibility for preparedness as something external to the individual		Internalization of responsibility for preparedness

- *Self-efficacy.* Individuals who believe in their own capacity to implement protective measures ("I can prepare myself") and on their own capacity to respond in case of emergency ("I can respond");
- *Perceived responsibility.* Individuals who believe that pre-disaster preparedness is also of his own responsibility (Table 32.2).

32.2.3 Which Factors May Interfere in Decision to Act Protectively?

As emphasized above, adopting protection concerns the initiation of *new* and relatively complex behaviours, which require high-level of cognitive functioning and advanced planning. A variety of factors explain the existing gap between intention and action. Weinstein refers to a variety personal and situational factors that may blockade the initiation of protection, such us: "having to learn what to do", "remembering of doing it", "finding time to do it"; "not having freedom to change one's own house".

There are, also, a set of factors related with the particularities of each protective measure that should be taken into account. Weinstein asserts that it is important to distinguish measures according to its costs and benefits. Low-cost measures, with immediate benefits, tend to be more attractive than the high-cost ones. One specificity of many earthquake-related protective measures concerns the fact that its benefits are not immediate but postponed to a future, often seen as remote and uncertain.

Lindell and Perry (2012) also refer to a set of factors that may interfere on people's decisions to act. They distinguish between *hazard-related attributes* and *resource-related attributes*. The first concern the following aspects: efficacy of protective measure to protect lives and property, protective measure's utility to other aims than not just protection. The second-mentioned factor includes economic cost, knowledge and needed skills; time and effort; degree of dependence of others to implement protection.

32.2.4 Which Actions Are Plausible in Teenagers?

Before discussing which actions are plausible in teenagers, two questions emerge as important to answer, respectively: What does it mean «to be proactive» in terms of protection? Is proactive attitude solely circumscribed to the act of implementing a protective measure?

It has been emphasized, along this paper, that decision to act protectively is the result of a social and cognitive process, often long, complex and marked by drawbacks. Becker et al. (op cit) describe this process as something unchained by social and environmental clues that, in turn, generate the search for more information, dialogue with others and development of skills, based in a realistic vision of what may happen in case of an earthquake. Based on a research on risk earthquake risk perception in Southern California, Turner et al. (1986) advance with a definition of what they mean by "a passive attitude" and a "proactive attitude" towards earthquake risk and protection. Individuals who demonstrate passivity are often those that ignore the threat or got information about it but did not take any action. Proactivity is unchained when the individual develops a series of actions directed 'to confirm the threat' and gain more knowledge about it. Actions such as the 'search for more information about the subject" from credible sources, 'the dialogue and talk with others' (i.e. family, friends, colleagues, neighbours) are signals of a proactive attitude.

Given this, in our assessment of proactivity towards earthquake threat, eventually unchained by KnowRISK intervention, will not be solely circumscribed to the strict act of implementing of protective measures. Actions such as 'to look for more information', 'discuss with others on the subject', 'to influence others to implement protection' and 'to gain skills on protection issues' (cf. Fig. 32.2) will be seen as signals of proactivity.

As mentioned earlier, earthquake risk protection may turn into a demanding task because many measures require skills, time, money and help from others. Such protection-related demands are even more relevant when our target-groups are teenagers. So, it is pertinent to answer to the questioning of which non-structural earthquake protection measures are plausible to expect an adolescent to do by

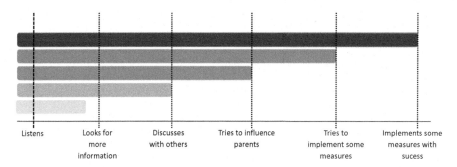

Fig. 32.2 Expectable actions to be taken by teenagers

himself or jointly with his/her family members? We identified emergency preparedness measures and some vulnerability reduction measures, as follows:

(a) Emergency preparedness-related measures:

- To guarantee that house/room exits are unblocked;
- To identify safe places in housing spaces;
- To have water for 2 through 3 days;
- To have food for 2–3 days;
- To have an emergency kit;
- To elaborate an emergency plan with family members and a meeting point;
- To have a battery radio and a flashlight.

(b) Measures for vulnerability reduction

- To reorganize wardrobe and closets with the aim of locating light objects in lower parts;
- To move sofas, beds and tables from areas next to windows.

The act of "fixing shelves and high cabinets' to the walls is among the most commonly recommended non-structural proactive measures. This is an example of a measure that requires time, effort and falls outside teenagers' scope of action. Nevertheless, they can have a crucial role of influencing and stimulating their parents to implement it.

32.3 Research Design: Challenges and Main Options

The assessment of KnowRISK intervention implied the adoption of an evaluation research design. As described by Bryman (2011), evaluation research is concerned with the assessment of social or organizational programmes or interventions and usually follows the principles of experimental design. Methodologically, this corresponds to a type of research that is composed by two groups: one group that is exposed to the intervention (experimental group) and another group that is not (control group).

32.3.1 Is Quantitative Research Enough to Evaluate Impacts?

Evaluation research design was initially structured to stand in a quantitative approach where a questionnaire would be applied to students before the beginning of the Knowrisk intervention (Timing 0–T0) and after it (Timing 1–T1). This type of approach had the advantage of allowing some kind of *measurement* of change. It would allow the identification of the stage at which our target-group is in terms of both *perceived susceptibility towards hazard* and *perceived efficacy of protection* before the intervention and afterwards. Through a comparative data analysis in the two time periods it would be possible to delineate the differences and to gauge eventual changes in terms of knowledge and beliefs.

Along the research process, questionings emerged concerning the suitability of standing the evaluation solely in quantitative methods. Questionnaires hardly capture the particularities of the contexts. Further, questionnaires need to be instruments of easy and fast administration in order to guarantee the maximum quality of data collection, but such demand may turn them into instruments that give nothing a superficial vision of social phenomena (Bryman, in ibid). Under these circumstances, options were made in order to adopt a multi-method approach based on quantitative and qualitative methods. The data obtained through the T0 and T1 questionnaire application, will be complemented by data coming from three other main sources, respectively:

- *KnowRisk notebook* content analysis – this notebook was elaborated for the students to write their impressions and discoveries along the intervention and it will have a function of complementary evaluation;
- *Focus groups* with students – discussions with the students that were involved in KnowRISK intervention are planned to occur 3 months after the intervention and will allows to verify the results of questionnaire data analysis
- Direct observation along the intervention – researchers' evaluation and teachers' assessment will also be taken into account.

32.3.2 The Pertinence of an Exploratory Study

Morgan et al. (2002) posit that too often risk communicators typically ask technical experts what they think people should be told and disqualify the pertinence asking the lay people what they believe and need to know. One of the first findings of knowRISK fieldwork and intervention was that the research team underestimated teenagers' degree of knowledge on earthquake hazard in Lisbon. Many of the teenagers that were part of our sample seemed to know much than it was presumed. Such aspect turned clear the importance of an exploratory study, based on semi-structured interviews, previously the questionnaire elaboration and the KnowRisk intervention. Such exploratory approach would be advantageous for both the questionnaire construction and for the intervention.

32.3.3 How to Administrate the Questionnaire?

This is a fundamental questioning in a quantitative research approach where questionnaires are used as inquiry instruments. In the particular case of KnowRISK evaluation research, this questioning seemed particularly important because our target-groups are teenagers. Should we opt by a self-completion questionnaire or a structured interview? Should we opt by an on-line questionnaire application procedure or by a presential application?

After a ponderation of advantages and disadvantages, an option was made for a self-completion questionnaire, to be applied in the classroom with the supervision of the Knowrisk research team. Structured interviews may allow the researcher to go deeper on the subjects with the interviewees but may favour the so-called social desirability. Through a self-completion questionnaire, our responders would feel more freedom to answer according their own view of the problem. Concerning the issue of presential *versus* non-presential on line questionnaire administration, through this last option we would not have any control over the intrusion of non-respondents (i.e. parents or other member of the household) in the asking of questions.

32.3.4 To Have or Not to Have a Control Group?

The pertinence of having a control group as part of an evaluation research procedure is often questioned. At the basis of such questioning are two main objections. We refer, in the one hand, to the inability of the researcher to control the ecological context and its eventual influence in social and cognitive changes in the period between the T0 and T1 questionnaire application. It may be, on the other hand, difficult to achieve demographic and social similarities between experimental group and control group (Bryman, op cit).

Although the limitations, it was made the option of standing the evaluation research on an experimental design composed by two groups, respectively:

- A experimental group, which embraces all the students from Rainha D. Leonor School and Padre António Vieira School which were part of knowRISK intervention;

- A control group, corresponding to approximately 60 students from the 7th and 8th grades of a neighbour school, Eugénio dos Santos School.

These three schools are part of Alvalade schools' grouping. This means that students often circulate from one school to another along their school cycle.

32.4 Conclusion

The main assumptions the theory underlying the assessment of KnowRISK intervention in Lisbon pilot-area are as follows:

- Both impact assessment procedure and Knowrisk intervention where structured taking into account how social science describe the process by which individuals decide to act protectively.
- KnowRISK intervention intends to be a *stimulus* for unchaining the above-mentioned process and aims at disseminating knowledge on seismic risk and protection, to stimulate the formation of favorable beliefs towards protection and behavioral intentions.
- The theoretical framework guiding impact assessment stands on Weinstein's stage-model, so called as Precaution Adoption Process, where the adoption of protective behaviors is defined as a cognitive and social process where decision to act represents the "arrival point" of a set of cognitive changes and stages.
- The theoretical framework stands on two main concepts – *perceived susceptibility to harm* and *perceived efficacy of protection* based on the assumption that willingness to act protectively stands in the double evaluation individuals make about their own susceptibility and the efficacy of each measure to protect them from harm.
- The sequence of stages concerning *perceived susceptibility to harm* fluctuate between Stage 0, corresponding to ignorance about threat, to Stage 3, where risk is personalized and the individuals envisages himself and his immediate environment as prone to earthquake hazard.
- The sequence of stages concerning *perceived efficacy of protection* fluctuate between Stage 0, corresponding to ignorance about protective measures, to Stage 3, where individuals believe on the efficacy of protective measures and envisage themselves as capable of implementing it.
- Both *perceived susceptibility to harm* and *perceived efficacy of protection* four-stage proposal should be seen as guiding principles, being plausible that research findings to reveal newly and intermediate stages.
- *Perceived susceptibility to harm* will be operationalized through most commonly used variables in research on this domain, more specifically: *perceived probability of occurrence of a disaster*, perceived severity; perceived vulnerability, risk salience and fear; Subjects will be classified in terms of stage accordingly to their position in each item.

- *Perceived efficacy of protection* will be operationalized through their perception of the efficacy concerning a set of pre-established earthquake non-structural measures and their own willingness to implement them; Subjects will be classified in terms of "stage" accordingly to the position they occupy in each item.
- Hypothetically, it is expected KnowRISK intervention to induce on high levels of risk personalization and to allow the formation of beliefs on the efficacy of earthquake protective behaviours.
- Accordingly, to this framework, 'being proactive' is not solely concerned with the implementation of protective behaviours.
- Instead, it includes all actions that an individual usually does along the process by which he looks for "confirming the threat" and considers to adopt protection: 'to look for more information', 'to discuss with others', 'to observe others', 'to influence others to implement protection' and 'to gain skills on protection issues'.

In methodological terms, it was made the option of sustaining evaluation research design in a multi-method approach comprehending both quantitative and qualitative methods.

References

Ajzen I (1991) The theory of planned behavior. Organ Behav Hum Decis Process 50:179–211

Becker JS, Paton D, Johnston DM, Ronan k R (2012a) A model of household preparedness for earthquakes: how individuals make meaning of earthquake information and how this influences preparedness. Nat Hazards 64(1):107–137

Becker JS, Paton D, Johnston DM, Ronan KR (2012b) How people use earthquake information and its influence on household preparedness in New Zealand. J Civil Eng Arch 6(6):673–681

Becker JS, Paton D, Johnston DM, Ronan k R (2013) Salient beliefs about earthquake hazards and household preparedness. Risk Anal 33(9):1710–1727

Bryman A (2011) Social research methods. Oxford Press, London

Lindell MK, Perry RW (2000) Household adjustment to earthquake hazard, a review of research. Environ Behav 32(4):461–501

Lindell M, Perry R (2012, April) The protective action decision model: theoretical modifications and additional evidence. Risk Anal 32(4):616–632

Silva DS (2017) Theoretical framework for risk communication impact assessment. LNEC. KnowRisk Project Task 4 Action D3

Turner RH, Nigg JN, Paz DH (1986) Waiting for disaster, earthquake watch in California. University of California Press, Berkeley

Weinstein N. D. (1988) "The precaution adoption process". Health Psychol, Vol. 7, n°4, 355–386

Weinstein ND, Sandman PM, Blalock SJ (2008). The precaution adoption process model. In: Glanz K, Rimer BK, Viswanath K (eds) Health behavior and health education. Jossey-Bass, San Francisco

Chapter 33
Shaping Favorable Beliefs Towards Seismic Protection Through Risk Communication: A Pilot-Experience in Two Lisbon Schools (Portugal)

Delta Sousa e Silva, Marta Vicente, A. Pereira, R. Bernardo,
Paulo Candeias, Monica Amarel Ferreira, Mario Lopes,
Carlos Sousa Oliveira, and P. Henriques

Abstract Communicating science within disaster risk reduction using methods that encourage two-way dialogue between scientists and laypersons is a challenging task. This paper aims at presenting a methodological strategy of communicating risk and non-structural seismic protection measures through participatory approach. Such methodological strategy is part of a pilot experience of risk communication in two schools in Lisbon (Portugal) under the EU project KnowRISK (Know your city, Reduce seISmic risK through non-structural elements). The efficacy of education for seismic safety is often inhibited by an incomplete understanding of the process by which individuals decide to protect themselves from harm (Becker JS, Paton D, Johnston DM, Ronan KR. Nat Hazards 64(1):107–137, 2012a; Becker JS, Paton D, Johnston DM, Ronan KR. J Civil Eng Archit 6(6):673–681, 2012b). Becker et al. (in ibid) conceive such a process composed of a series of stages: knowledge and awareness, thinking and talking, understanding the consequences, developing skills. The above-mentioned pilot experience of risk communication was designed in order to trigger the cognitive process underlying behavioral change. Lisbon is a dormant society as far as earthquake risk is concerned. Given this, risk communication was firstly designed to generate awareness and knowledge among target-groups.

D. Sousa e Silva (✉) · M. Vicente · A. Pereira · R. Bernardo · P. Candeias
Laboratório Nacional de Engenharia Civil, Lisbon, Portugal
e-mail: delta@lnec.pt; magvicente@lnec.pt; apereira@lnec.pt; pcandeias@lnec.pt

M. A. Ferreira · M. Lopes · C. S. Oliveira
Instituto Superior Técnico, Lisboa, Department of Civil Engineering, Architecture and Georesources, CEris, Lisboa, Portugal
e-mail: monicaf@civil.ist.utl.pt; mariolopes@tecnico.ulisboa.pt

P. Henriques
Proteção Civil Lisboa, Lisboa, Portugal

Keywords Seismic risk · Risk communication · Protective behaviours · Non-structural

33.1 Introduction

Becker et al. (2012a) stress that education on earthquake safety preparedness very often fails due to the lack of knowledge about the way the different individuals apprehend risk-related information and about the way they act preventively.

The intervention foreseen under the KnowRISK project is intended to raise awareness among a specific target public, i.e. the school community, of the seismic risk problem in Lisbon, in particular for a mild to moderate seismic input. The main objective is to eventually change people's attitudes and to stimulate the adoption of protective behaviours. Presently, several models co-exist in social sciences, which are aimed to explain the decision-making processes as regards seismic protection of both individuals and families (Lindell et al. 2009; Lindell and Perry 2012; Becker et al. 2012a, b, 2013). Despite the differences, all of them share a set of general principles, in particular:

- The decision-making process is stimulated by environmental or societal guidelines in the form of disaster experience or risk information, which encourage individuals to ponder the subject;
- Even though risk perception is an important forerunner of protective action, there are other factors that are just as, if not more, important, amongst which we can highlight the individuals' beliefs in each of the protective measures;
- The specific context in which the individuals are integrated plays a part in the individual decision-making process, and it may either be an enabler or a constraint.

The present paper aims to synthesize the intervention that took place within the framework of the KnowRISK project. It begins by presenting the theoretical principles on which the organisation of the intervention was based. Afterwards, it establishes the objectives to be reached with the intervention and, in Sect. 33.4, it presents the actual intervention proposal and the actions that took place in the two schools, between October 2016 and March 2017.

33.2 Principles of Intervention

The KnowRISK intervention is based on the Becker et al. (2012a, b, 2013) model. These authors define the adoption of protective behaviours as a four stages process: (i) *risk knowledge and awareness;* (ii) *reflection and dialogue*; (iii) *vision of the consequences*; e (iv) *development of competences.*

The stage of *risk knowledge and awareness* is triggered either by the exposure of individuals to information about risk or by the direct or indirect disaster experience. One or the other provides clues that act as a warning leading the individuals to ask themselves the following question "*is there an actual threat that must be taken into account?*"

However, *risk knowledge and awareness* will not necessarily lead to the immediate adoption of a protective action. The most likely hypothesis is that they will first lead to *reflexion and dialogue*, as an attempt to confirm the threat and to understand it better. This is the stage in which the individuals feel compelled to search for more information about the subject; to bring the topic up in conversations with friends and relatives; and to contact field-related organisations or experts. The main purpose is the clarification of doubts, such as: "*is the risk truly relevant where I live? What would happen to my town if an earthquake occurred? What should I do in case of disaster?*". Becker et al. (in ibid) refer to the important role played by organisations and experts in this stage, as it regards the clarification of doubts and the encouragement to adopt protective actions.

The interaction with others, especially with credible experts or organisations, can be important to help individuals building a *vision of the consequences* of earthquakes and to weigh in the advantages/disadvantages of the adoption of protective behaviours. Individuals need to have a clear perception of the expectable impact of an earthquake in their homes, neighbourhood and town as to be able to make decisions in terms of protection. "*How damaged will my house be after an earthquake? What might collapse? For how many days should I store water and food?*". In other words, they need to "*personalise the threat*" and also to believe that there are various protection alternatives that are worth exploring and implementing.

Once the individuals are convinced that the risk is real and that some protection alternatives can be adopted, they become more motivated to *develop their skills* in terms of protection. Generally, it is in this stage that the individuals formulate questions such as: "*which are the best protective actions? What should I do to put them into practice?*". Reference must be made to the fact that once this stage is reached, the individuals may only develop *behaviour intents* rather than put actions into practice. The postponing of protective actions, in favour of pressing daily demands, is fairly common, particularly in the cases when no eminent threat exists (Fig. 33.1).

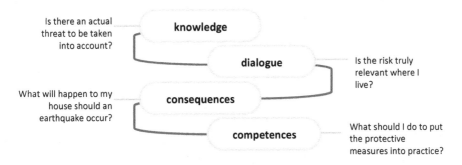

Fig. 33.1 Diagram for the model of protective behaviours adoption (Beker et al., op. cit)

Overall this intervention is taken as an opportunity to disseminate information on risk and seismic protection, and it is expected to eventually lead to risk awareness; to the search for further information; to debates about the subject; to a perception of the consequences and, lastly, to the development of competences from a protection viewpoint.

33.3 Objectives

Lisbon is a city with virtually no *disaster culture*, of which the effects are felt at both the societal and the political level, as well as at a micro level, i.e., families and individuals. Historically unaware of its growing vulnerabilities, Lisbon has failed to adopt and implement a long-overdue policy aimed at seismic risk mitigation. The government, who is unavoidably responsible for promoting a public policy in terms of seismic risk, has not instigated the municipalities to act accordingly. In fact, this is a type of risk that has poor impact on the citizens' daily life, and the protective behaviours that are likely to be implemented by citizens are neither the social norm nor have they been fully assimilated by the individuals.

Therefore, the school community, which is the target of this intervention, might be poorly aware of the seismic risk. Generally, this community is assumed to integrate individuals: with a poor knowledge about Lisbon's seismic risk and non-structural protection alternatives; with no perception about seismic risk or having perceptions based on beliefs that discourage the adoption of protective measures; who are poorly pro-active.

The above referred aspects are important because they establish the starting point for the KnowRISK intervention. According to Eiser et al. (2012:13), pre-disaster preparedness is not an all or nothing process. Some people are expected to remain unaware of the risk, even after being exposed to information. Others may become aware of the risk, but may stay out of preparedness. Others may still feel the need for further guidance so that they can act. Therefore, we are in the presence of a long process, which will inevitably encounter a few ups and downs along the way.

This being said, the KnowRISK intervention is expected to reach the following objectives:

1. To foster *knowledge* about the seismic risk problem, its incidence in the city of Lisbon, and about the protective measures that citizens can adopt to increase their safety;
2. To stimulate the *formation of beliefs* intended to encourage the adoption of seismic protection measures;
3. To foster the development of protective *behaviour intentions*.

Of note is the importance of the dissemination of knowledge, included in objective (1), which is based on information related with seismic risk and protective measures. The studies in this field are unanimous in assuming that the individuals do adopt protective behaviours on the basis of a double evaluation of the risk and of the available protection alternatives (Neuwirth et al. 2000).

The concept of belief, according to objective (2), refers to the information a given individual has – and in which he believes – about a specific subject. This information constitutes a cognitive and rational background for the development of attitudes. The intervention is expected to stimulate the formation of beliefs, intended to encourage seismic protection actions. Becker et al. (2013), based on a study on individuals living in seismic risk areas, in New Zealand, carried out a systematisation of beliefs that either encourage or discourage the adoption of protective attitudes and behaviours.

Regarding objective (3), it is important to clarify the meaning of *behavioural intention*. It refers to the motivation in adopting a given behaviour and it indicates how much effort an individual is willing to put into adopting that given behaviour. Generally, the stronger the *behavioural intention*, the higher the individual's likelihood of transforming it into actions (Ajzen 1991).

In the particular case of KnowRISK, it is methodologically unfeasible to evaluate the attainment of objective (3) based on the direct observation of behaviours. Moreover, it is necessary to take account of the fact that the target-group of the intervention includes 12–15 years-old teenagers. Since this group does not play a major part in decision-making about the use of household space, some protective behaviours (ex. fixing furniture to the wall, fixing TV sets or similar appliances) will not fall within their possibilities of implementation. The expectable scenario is as follows:

- Even though they are not the main decision-makers, teenagers may influence parents' decisions regarding seismic protection (social influence);
- Having no implementation freedom in the present, teenagers may develop strong enough *behavioural intentions* to adopt protective behaviours in adulthood.

33.4 KnowRISK Intervention: Proposal and Actions

By emphasising the vital role of school education in disseminating knowledge and in raising risk awareness, Luna (2012:750) refers to schools' resources and to the fact that these very resources can be mobilised in the framework of a strategy aimed to reduce the risk of disaster. These resources include students, teachers, parents, community associations, the school infrastructure and the endogenous knowledge of the school system.

As previously mentioned, the KnowRISK intervention aims to raise awareness among a specific target, i.e. students and hopefully parents, of the problem of seismic risk and associated safety. The success of such interventions often rests in the participation of teachers in its preparation and implementation. Besides teachers, the school infrastructure is also seen as a helpful resource. The target-schools of this intervention have been recently subjected to seismic strengthening works. Therefore, this must be maximised as an opportunity to raise awareness among the school community, from students to teachers, about the importance of structural safety of buildings.

The total of six actions happened in two schools, School *Rainha Dona Leonor* and School *Padre António Vieira*. Overall 108 students were involved. The intervention comprises a set of actions devised in the light of the previously mentioned Becker et al. (op. cit) four stage model. Accordingly, the actions proposed are aimed to:

1. Promote dissemination of *knowledge* and to raise *awareness of the risk*;
2. Stimulate the *reflexion and dialogue* about the problem;
3. Stimulate students to build a *vision of the consequences* of an earthquake on the place where they live;
4. Promote the *development of competences* regarding non-structural protective measures.

To support the actions foreseen under the intervention, a notebook was designed to guide students through. The KnowRISK notebook main objectives are to consolidate the information given and also to help the evaluation of the intervention. There are several spaces where students are invited to give their opinion on certain topics or actions (Fig. 33.2).

As previously mentioned, we assumed that students would have little knowledge about seismic risk, its level of incidence in Lisbon, and about the protection alternatives available for societies and individuals. Therefore, the initial stage of the intervention included two *knowledge* dissemination actions, with approximate duration of 45 min, intended to stimulate risk awareness:

- **Is there seismic risk in Lisbon?** | First lecture, introducing seismic risk concepts and Lisbon's vulnerability to this type of extreme events;

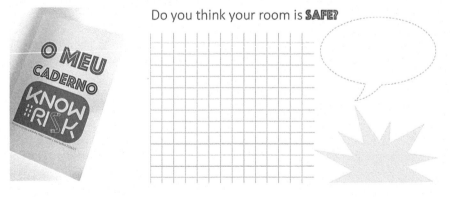

Fig. 33.2 The KnowRISK notebook: cover and page sample

Fig. 33.3 KnowRISK pictograms representing vulnerable situations often found inside buildings

- **What can we do to protect ourselves?** | Second lecture, presenting the various protection alternatives available, with particular emphasis on non-structural measures that can be implemented by citizens. This session finished with the presentation of the "Home Hazard Haunt", a challenge for students and parents to find the non-structural vulnerabilities of their homes (Fig. 33.3).

The knowledge dissemination stage was followed by an action aimed to initiate a dialogue between scientists and students. As previously stated, once exposed to information about risk, individuals are normally led to explore the subject and to dialogue with others to confirm the threat previously presented.

This *reflexion and dialogue* stage was promoted by a field trip to LNEC campus, during approximately 3 h. In this visit students were able to:

- **Visit the shaking platform** at LNEC Seismic Engineering facilities, where they were able to discuss and learn more about the behaviour of structural and non-structural elements during earthquakes;
- Interact with **Civil Protection** specialists, during a presentation on emergency and preparedness and a visit to emergency communications vehicles;
- Participate in the **contest "Who wants to be safe?"**, inspired on the television show *Who wants to be a millionaire?*, specially design to help students consolidate what they already knew and to talk to several specialists about their doubts regarding non-structural protective measures (Fig. 33.4).

According to the Becker model, after getting the chance to confirm the threat, it is important that individuals visualize the possible consequences caused by earthquakes. In this context, under the *visualization of consequences* stage, during a 45-minclass, the following actions took place:

Fig. 33.4 Three key- moments of the field trip to LNEC

Fig. 33.5 Video frame from the moment when Bruno manages to exit the room

- **Bruno's Story** | Short video inspired by an interview from one individual who survived the Amatrice earthquake (Italy, August 2016). The animated video shows a set of events that take place inside a bedroom during an earthquake;
- **Debate** | Conversation with students is triggered by asking them if they think the video represents reality or fiction. Students are invited to try to visualize likely effects of an earthquake on their own room, providing a vision of the problem and an opportunity to personalize the threat (Fig. 33.5).

Fig. 33.6 Assembling of cardboard furniture, scale 1/10

This third stage of the adoption of protective measures process, provides proper visualization of consequences and it is considered to stimulate people to act.

The last stage of the Becker et al. (op. cit) model is the *development of competences*. Once the threat has been interiorized, individuals try to acquire the practical knowledge and the necessary tools for the adoption of protective behaviours. Regarding the intervention, this stage was planned in a way that would allow students to create an object, on which they could be able to develop a set of skills aimed to perceive both the vulnerability and resilience areas of a house; as well as to define and apprehend the attitudes to be adopted in case of earthquake, both inside and outside.

During two 90-min sessions, students were invited to develop a set of skills under the concepts they got to know during the previous stages of the KnowRISK intervention. Each group of three students worked on one specific room (living room, office, kitchen, bedroom), managing a set of tasks:

- Assemble a given set of furniture according to the house division. This activity is intended to allow individuals to have an active role on actually building the house model (Fig. 33.6);
- Display the furniture set freely inside the room. In this stage students were not forced to organize the rooms safely, they could display things as they liked;

Fig. 33.7 House model after shake test

- Assemble the several parts together, in order to have the complete house and shake the model in a way to simulate the action of an earthquake. This shaking test and its consequences allowed students to have a clear vision of the effect of an earthquake inside a regular house. Individuals are invited to take a close look at the house area they were working on and identify the main problems (Fig. 33.7);
- With the effects of the earthquake in mind, students go back to work and are able to rearrange the furniture in a safer way. After a correct display of furniture, students were invited to simulate other protective measures such as fixing tall furniture to the walls; moving heavy objects to lower shelves; moving beds away from windows or, even, place heavy curtains on windows to prevent broken glass to spread through the room (Fig. 33.8).

After building, arranging and handling the décor, students were invited to write signs to place inside their room, displaying three of the protective measures adopted. After a brief presentation of all groups, about the main vulnerabilities and resiliencies found in their house part, Civil Protection took the lead handling a discussion about what to do before, during and after an earthquake. By looking at the model previously built, it was possible to visualise the possible consequences of an earthquake on a house and to address the attitudes that can be adopted in such a situation (Figs. 33.9 and 33.10).

Once the intervention ended the resultant house models are now an object that individuals who took place in the KnowRISK project may use to communicate risk to others. The fact that students took their time to prepare the models and to make

33 Shaping Favorable Beliefs Towards Seismic Protection Through Risk...

Fig. 33.8 Students handle last details of their room. (Photography taken by Nuno Patrício – RTP Notícias)

Fig. 33.9 Signal: tables should be as far away as possible from windows. (Photography taken by Nuno Patrício – RTP Notícias)

Fig. 33.10 A group signals that bookcases were fixed to the walls. (Photography taken by Nuno Patrício – RTP Notícias)

signs with the adopted non-structural protective measures makes possible the communication of vulnerabilities and resiliencies found inside regular household spaces. We believe this a chance not only to empower the individuals responsible by making the model but also to get the information to a larger group of people (Fig. 33.11).

33.5 Discussion

Once the intervention on the two Portuguese schools is now over it is possible to make some considerations on the process. As stated in the beginning of this paper, the KnowRISK was intended to raise awareness among school community on the seismic risk problem in Lisbon. The main objective was to stimulate the adoption of protective behaviours.

Despite the fact that the target group, ages from 12 to 15 years-old, may fail to influence the decision-making process of adopting protective measures inside their household space, after the intervention we tend to believe that the skills developed during the KnowRISK project may be used by the individuals later on in their lives.

Regarding the theoretical model adopted for this intervention, we believe that, if implemented correctly, it provides individuals a set of information and tools that can prepare them to act protectively. Of course, either they act on it or just develop behaviour intentions. During the process in both schools, we realized that by the time students reached stage four, *development of skills*, they already knew and comprehended a given set of non-structural protective measures. In this scenario,

Fig. 33.11 Completed house models during debate on *"Before, during and after an earthquake"*

the house model activity represented a great moment when students were able to act and put into practice the concepts they discovered during previous stages.

Regardless of the successful implementation of all proposed actions, some questions remain to be answered: *is this the way to communicate risk to this target group? Is this the right age to communicate risk? How are these individuals going to adopt such measures?*

Acknowledgements We would like to thank all the students and teachers, from schools Rainha Dona Leonor and Padre António Vieira, involved in the KnowRISK intervention for their commitment. Without it, it would have been impossible to carry out this pilot-intervention.

This project was financed by the European Commission's Humanitarian Aid and Civil Protection (Grant agreement ECHO.A.5 (2015) 3916812) – KnowRISK (Know your city, Reduce seISmic risK through non-structural elements).

References

Ajzen I (1991) The theory of planned behaviour. Org Behav Human Dec Proc 50:179–211

Becker JS, Paton D, Johnston DM, Ronan KR (2012a) A model of household preparedness for earthquakes: how individuals make meaning of earthquake information and how this influences preparedness. Nat Hazards 64(1):107–137

Becker JS, Paton D, Johnston DM, Ronan k R (2012b) How people use earthquake information and its influence on household preparedness in New Zealand. J Civil Eng Archit 6(6):673–681

Becker JS, Paton D, Johnston DM, Ronan k R (2013) Salient beliefs about earthquake hazards and household preparedness. Risk Anal 33(9):1710–1727

Eiser J.R., Bostrom A, Burton I., Johnston D.m., McClure J., Paton D., Van der Pligt J., White M.P. (2012), "Risk interpretation and action: a conceptual framework for responses to natural hazards", Int J Dis Risk Reduc, 1, 5–16

Lindell MK, Perry RW (2012) The protective action decision model: theoretical modifications and additional evidence. Risk Anal 32(4):616–632

Lindell MK, Arlikatti S, Prater (2009) Why people do what they do to protect against earthquake risk: perceptions of hazard adjustment attributes. Risk Analysis, vol. 29, n°8, 1072–1088

Luna EM (2012) Education and disaster. In: Wisner B, Gaillard JC, Kellman I (eds) The Routledge handbook of hazards and disaster risk reduction. Routledge, New York

Neuwirth K, Dunwoody S, Griffin RJ (2000) Protection motivation and risk communication. Risk Anal 20(5):721–733

Chapter 34
The KnowRISK Action for Schools: A Case Study in Italy

Gemma Musacchio, Elena Eva, and Giovanna Lucia Piangiamore

Abstract *"Know your school: be safe!"* is a risk communication campaign, within the KnowRISK (*Know your city, Reduce seISmic risK through non-structural elements*) project, involving schools in three European countries, namely Portugal, Italy and Iceland. Its main aim is to facilitate school communities' access to experts' knowledge on non-structural seismic risk reduction. In Italy we implemented a learning strategy to capture young people attention. We believe students will be influential in turning scientific knowledge to practical know-how. The strategy is based on *Situated Learning Episode* (EAS), where homework is for learning and skills, and classwork is for reworking and understanding. The KnowRISK-EAS starts and ends with two focus groups where students, and experts rework concepts and discuss best practices. Students are asked to implement a communication product addressing their peers. The assignment has a double goal: it helps to activate reflexive learning; it will be a project tool to trigger risk reduction attitude within schools communities.

Keywords Natural hazards · Active learning · Risk communication · Prevention · Schools education

G. Musacchio (✉)
Istituto Nazionale di Geofisica e Vulcanologia, Roma, Italy
e-mail: gemma.musacchio@ingv.it

E. Eva
Istituto Nazionale di Geofisica e Vulcanologia, Osservatorio Nazionale Terremoti, Rome, Italy
e-mail: elena.eva@ingv.it

G. L. Piangiamore
Istituto Nazionale di Geofisica e Vulcanologia, Roma 2, Roma, Italy
e-mail: giovanna.piangiamore@ingv.it

34.1 The Partecipatory KnowRISK Setting in Italy

The KnowRISK project (Know your city, Reduce seISmic risK through non-structural elements) is financed by the European Commission (AGREEMENT NUMBER – ECHO/SUB/2015/718655/PREV28), under the General Directorate of Civil Protection and Humanitarian Operation, with a major goal to help communities mitigate risk posed during earthquakes by non-structural elements. Within the task that implements strategies for risk communication, action E3 – "Know your school: be safe!" – is devoted to schools in pilot areas chosen within the three participating countries. It Italy pilot area for intervention in schools is the northern part of the country. Here schools communities in Varese and La Spezia, and their surroundings, were chosen to implement a participatory action, mostly profiting of previous schoolboard engagement. The action started in Spring of 2016 and it is on-going involving more than 2000 students.

"Know your school: be safe!" stands on a flow process that passes from understanding and ends with building up or reinforcing knowledge, acts on attitude to promotes a shared view of a more resilient society. School education is fundamental in disseminating knowledge and in raising risk awareness. Schools include resources – that can be mobilized to reduce the risk of disaster (Luna 2012). These resources are students, teachers, parents, community associations, the school infrastructure and the endogenous knowledge of the school system.

The KnowRISK campaign addresses last years of the compulsory cycle of education and, more specifically Level 2 and 3 of the 2001 International Standard Classification of Education, ISCED.

In Italy ISCED 2 level refers to Middle Schools and ISCED 3 level to High Schools. A robust engagement of school board prompted the selection of schools in the Northern Italy pilot area. Here Istituto Nazionale di Geofisica e Vulcanologia had implemented outreach and dissemination programs for the past 10 years. This allowed KnowRISK to start the action in Italy at the very beginning of the project.

Our targets audience is 13–15 years old students attending the third class of Middle schools and the first class of High schools in the city of La Spezia and surroundings (Lerici, San Terenzo and Sarzana) and Laveno Mombello (in the province of Varese).

The Middle schools of J. Piaget, in La Spezia, and G. B. Monteggia, in Laveno Mombello (Varese), have been involved in the project interventions, with sixteen classes and about 350 students, for 2 years. The U. Mazzini and the ISA 10 Middle schools in La Spezia joined the KnowRISK action with 7 classes and about 160 students in the second year of project, when the protocol of intervention was ready to be fully applied. In Tables 34.1 and 34.2 are shown details concerning schools involved in Italy up to now.

High schools (Table 34.2) were involved only in the second year of project when some of the communication tools specifically implemented within the project were expected to be ready. The scientific and classic Lyceums of the Parentuccelli-Arzelà high school, in La Spezia surrounding, and the Scientific and Technological

Table 34.1 Italian middle schools (ISCED 2 level) involved in "Know your school: be safe!"

Location	Name	N° Classes	Level	N° Students
La Spezia city	J. Piaget	10	III	201
La Spezia city	U Mazzini	4	III	96
Lerici (La Spezia)	F. Poggi	2	III	45
S. Terenzo (La Spezia)	P. Mantegazza	1	III	21
Laveno Mombello (Varese)	G. B. Monteggia	6	III	130

Table 34.2 Italian high schools (ISCED 2 level) involved in "Know your school: be safe!"

Location	Name	N° Classes	Level	N° Students
La Spezia city	A. Pacinotti	3	I	81
Sarzana (La Spezia)	Parentuccelli-Arzelà	6	I	170
Varese	Sacro Monte	4	III–IV	80

Lyceums of the Sacro Monte high school, in Varese, joined KnowRISK action in the Fall of 2016. We are adapting the protocol of intervention to be ready for specific needs of different levels of schools and setting up a non-structural elements checklist that can be used by students. It will be the results of a participatory research in which experts and student side by side have discussed a swot analysis in the classroom.

Three classes of the A. Pacinotti high school were the control group for the KnowRISK common questionnaire (Platt et al. 2017, this Volume).

34.2 The Approach

The early starting of the action in Italy allowed implement and test a protocol of intervention that could be used in other pilot areas of the project. The approach stands on the engagement of students and relies on previous work done with schools in Italy (Musacchio et al. 2015a, b, c; Piangiamore et al. 2015; La Longa et al. 2012).

It includes the six-steps listed in Table 34.3: (1) Ex-ante survey, (2) T_0-focus group, (3) the first expert Lecture, (4) Home-works, (5) T_1-focus group, (6) Ex-post survey. Experts in the classroom are needed only to lead steps 2–3 and 5. This allows have a more sustainable action, mobilizing the least number of experts that, because of other work related duties, might hold from offering their expertise to the intervention. The protocol, based on Becker et al. (2012), has three main elements of novelty: the assessment of pre-existing situation; the use of a dialogic framework; a flipped up learning strategy.

Seismic risk reduction starts from best practice of safety and educational seismology. The results expected after the KnowRISK intervention are listed as follows:

- Engagement of students in the discovery of vulnerability their school and home;
- Engagement of students in understanding resilience;
- Dissemination of knowledge on non-structural seismic protection measures;
- Spreading Knowledge, raising Awareness and improving Practice

Table 34.3 The protocol of intervention for schools in Italy

Type of action	Action	Description	Support material	Duration	Observations
Assessment of perception	Survey T_0	(At school)	On-line questionnaire	30 min	Filled in by each single student and on his/her own
Dialogue	T_0-Gocus Group	(at school) Assessment of previous knowledge and opinion on hazard and risk. Establish a participate knowledge	Images and project brochure	1 h	The debate should try to understand students opinions and push them to make free observations
Knowledge	Lecture (What kind of damage have earthquakes produced in Italy? or Earthquakes: where and how?)	(At school) Fire Brigade will document disasters caused by. Emotional learning: story-telling based on personal involvement	Miscellanea	1 h	The Fire Brigade's lecture is on damage and earthquake story-telling.
		and/or Scientist runs a hands-on activity			The scientist's lecture is an hands-on and interactive seminar
Vision and development of consequences	Flipped-up learning strategy	(At home) Active learning phase: homework	Video (Shake table video or augmented reality)	Free at home	Students are asked to prepare a product to communicate risk related to non-structural elements to their peers
Dialogue	T_1-Focus Group	(At school) Assessment of effectiveness reinforce of knowledge	Images, videos, music, presentation, serious games, posters, checklist	90 min	Students present their products to the class. Discussions and shared knowledge
Assessment of effectiveness	Survey T_1	(At school)	On-line questionnaire	30 min	Filled in by each single student and on his/her own

34.3 The KnowRISK-EAS

In the protocol of intervention listed in Table 34.3 the action including vision and development of consequences is the core of a method based on Situated Learning Episode, EAS, (Episodio di Apprendimento Situato; Rivoltella 2013, 2014a, b). This is a new learning method that wants to address the low motivation, the fragile attention and the overload of information that nowadays students have. It is related to active learning where the classroom is rethought as a lab with lessons becoming workshops with experts.

Neuroscience suggests that there are three ways our brain learns: repetition, experience and modelling. Repetition is what helps to switch from short to long-term memory that is, in some ways, from memorization to consolidation. Experience acts on emotions, through the so-called somatic marker: it provides the tools to preview what is happening without having to experience it in that precise moment (Frith 2007; Friston 2010, 2012). Modelling triggers learning from the others.

EAS stands on *episodes*, which are micro contents (Patchler et al. 2010) that best suites repetition, and on *situations*, which are real life experience included into a framework that needs to make sense for the students. Situations need to be simple to trigger micro learning. Repetition, experience and modelling are activated within a three-phase procedure (Fig. 34.1): the Preparation phase that stands on a problem solving learning strategy; the Activity phase that is based on learning-by-doing; the Debriefing phase that activates reflective learning.

In the Preparation phase the mentor's action is to build the conceptual framework, presents it to the students in dialogic form, give inputs and assign homework. The inputs need to be short-shoots of videos, presentations, and a text that can catch attention, raise the needed curiosity to do the homework.

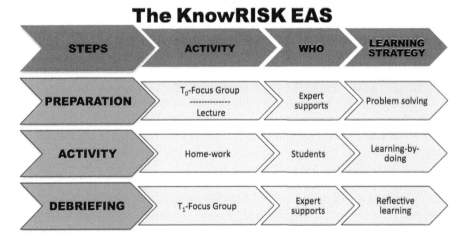

Fig. 34.1 The KnowRISK-EAS scheme. Focus groups embedded in the learning method are its element of novelty

In the Activity phase the mentor sets the activities time and allow students to work on their own. Students are asked to prepare a digital product. They will be given a list of websites where to search for additional information, derive in-depth contents and be inspired for the preparation of the product.

In the Debriefing phase the mentor makes assessments, discusses misconception and defines concepts while students analyse schoolmates products, discusses with them and reflect on products and processes.

EAS conceptual background and operative structure perfectly suites risk education: risk can be segmented into micro-contents, they may affect daily life experience and are suitable to modelling.

The KnowRISK-EAS (Fig. 34.1) is peculiar with respect the activity implemented to trigger problem-solving and reflexive learning. This peculiarity is given by focus groups that set a dialogical approach and build the conceptual framework, in the Preparation phase; they work out misconceptions coming up with shared views, in the Debriefing phase.

Students get acquainted with key-concepts on seismology and safety; they implement safe strategies in case of hazard. The Episode in the EAS methodology is a real life simple situation that students might have experienced at home or at school. In this framework non-structural elements toppling may recall real life situations that students can easily detect. The aim is to help students in better understanding seismic phenomena and their consequences on society. Experience marks learning and it is helped by emotions that lead students to develop proper skills to specific situations. Modelling allows them to be proactive, which is peculiar to resilience: do their best when they are in need (Frith 2007; Friston 2010, 2012; Piangiamore et al. 2016).

34.3.1 Preparation Phase

The KnowRISK EAS Preparation phase starts with a focus group where researchers, fire fighters and students debate the difference between hazard and risk. This is a dialogic type of action where researchers moderate the group and offer stimuli (images, video, ideas) to engage discussion. The focus group is also used to provide a qualitative assessment of the situation prior than the intervention.

Stimuli are video and images recalling situations the students are familiar with (https://www.napofilm.net). They are carefully selected because they must be relevant to students and help them focus on a fundamental concept: this is the difference between hazard and risk, which are normally not clear and often confused.

Once the difference between hazard and risk is understood, the experts ask whether it is possible to remove risk or hazard. Up to this point, no scientific terms but only practical examples from everyday life are used. This makes it easier for the students to understand the concepts and, in turn, raise awareness through an in-depth comprehension (Fig. 34.2).

Later on, we show a seismic map of Italy and introduce the scientific concepts of seismic risk, hazard and vulnerability. Reflection now becomes deeper and the

Fig. 34.2 Photographs of the Focus Groups

students discuss how natural hazards cannot be removed, the need of prevention and what solutions can be adopted to reduce the risk. The end of the meeting makes focus made on what is non-structural vulnerability mostly using products specifically prepared by the project (Fig. 34.3).

It is important that students understand that an earthquake can cause both structural and non-structural damages and that both might cause serious injuries. The discussion on this point is kept simple, only few images are presented because it is necessary that the students make their own assessments.

34.3.2 Activity Phase

The students will be asked to prepare homework in small groups and on their own. The assignment is: prepare a product of science and risk communication having as target public your peers. The mentor provides Internet links that the students can surf to find more information or to get inspired on how to prepare a creative product on non-structural protective measures. We ask them to follow a multidisciplinary approach: they should include, beside science, several other disciplines such as art, music, design, and citizenship. The students are also asked to use open-source software.

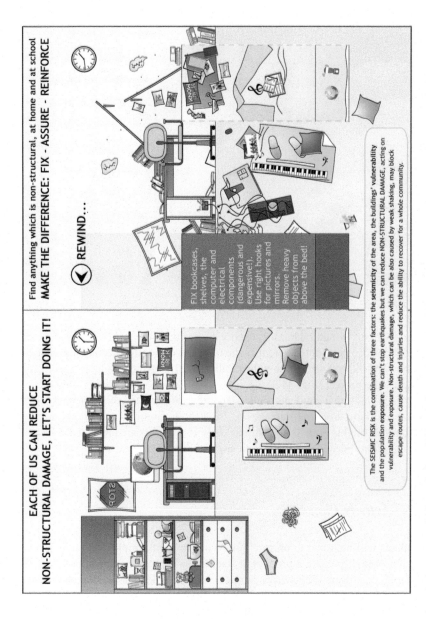

Fig. 34.3 A typical teen-agers bedroom before (left) and after (right) the shaking caused by an earthquake. Images are from the project pop-up brochure. They are used to trigger discussion during T_0-Focus group. The focus is the way furniture are placed in a house and the risk they pose: "Is it a safe room, in case of earthquake?" or "Is it just a messy room?"

34.3.3 Debriefing Phase

In this phase the students present their products to classmates and the discussion is set within a focus group. The same researchers that piloted the T_0-focus group will moderate the discussion, analyse students' products and workout misconceptions (Figs. 34.4a, 34.4b, and 34.4c).

We collected very interesting brief video-reports, interviews, video-pills, serious games, non-structural damage models, comics, music compositions, poems, etc. that will soon be shared on the *KnowRisk* website.

Finally students' creations will be part of a contest entitled "Are you taking too many risks?". The competition will prize the best digital compositions expressing youth point of views, their suggestions to their peers, but with the intention to convince families and/or politicians and administrators of the importance to reduce risk though non-structural elements.

The subjects that can be addressed to participate are the following:

- We know that non-structural building components (furniture, ceilings, partitions, panels, maps) can get damaged by earthquakes;
- We prevent a destructive phenomenon in the place we live in;
- We learn to protect ourselves by adopting proper strategies to secure furniture and proper behaviour of self-protection;
- We are in solidarity: we become young Fire Brigade and Civil Protection operators to help for ourselves and our families by promoting the best practice on non-structural seismic risk reduction.

34.4 Conclusions

"Non-structural elements cause damage. Repairing them takes time and money. If you feel an earthquake it might already be too late to get prepared. You must be proactive". This is in a few words our message to help and reinforce the community on prevention of seismic disaster. To keep the interest in earthquake alive during inter-seismic period we have searched effective ways to communicate seismic safety to school communities. Here young people (the students) but also adults (teachers, school managers, parents and families, too) are the targets of our communication strategy.

A comparison between protocols implemented in Italy and Portugal was performed by having the Middle school J. Piaget (the 3E) meet in Lisbon students from the "Padre António Vieira" School. We tested a peer-education experience that will help validate the effectiveness of the intervention. Students from the two countries exchanged their non-structural elements checklists discussing risk reduction solutions. The Portuguese research team offered Italian students some of the activities included in their protocol of intervention.

Fig. 34.4a Examples of homework: a video-pill of non-structural damage in a butcher's shop in Norcia (left); on a video of likely damage in the case of non-structural risk (right)

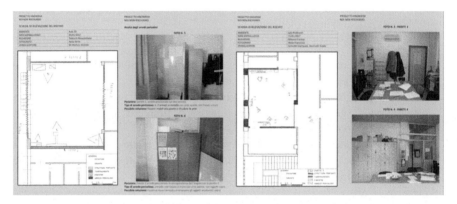

Fig. 34.4b Example of homework: a room-by-room inventory of likely non-structural damage at school

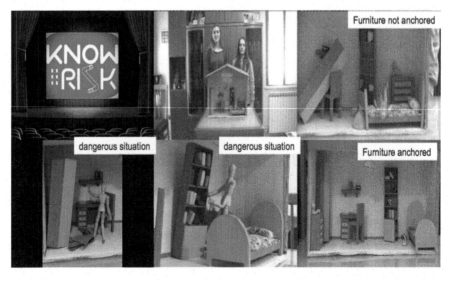

Fig. 34.4c Example of homework: video of the behaviour of a scale model of a bedroom

Because students are citizens that will be protagonist in building the world of tomorrow, to involve them could turn out to be effective in promoting future capacity to cope with hazards.

The student's awareness ensures the establishing of shared rules and reinforces a future active citizenship. In this way we aim to eventually change people's attitude and stimulate the adoption of preventive measure. Young people can also be a valid channel for disseminating information.

The commitment and participation of the students is very high, despite the age switch to attention to other interests, this shows that any topic, with the right approach, can be interesting and easy to understand. With the KnowRisk competition we expect students be more challenged to go more in depth on concepts that we debated with them in classroom (Musacchio et al. 2015c; Piangiamore et al. 2012).

The protocol we have implemented can be considered an example for teachers (also those in subjects other than Geoscience) on how they can introduce earthquake topics and seismic issues in STEM (Science, Technology, Engineering, and Mathematics) curricula. The KnowRisk project would provide the added value of increasing the security protocols in schools.

Acknowledgements This study was co-financed by the European Commission's Humanitarian Aid and Civil Protection (Grant agreement ECHO/SUB/2015/718655/PREV28). We are grateful to Calogero Daidone and Vincenzo Melillo from the Provincial Fire Brigade Department of La Spezia for kindly allowing Alessandro Frione, Paolo Trolese, Simone Pedretti, Fabio Bardelli to be part of KnowRISK team contributing to enrich the focus groups with emotional learning activities (unpublished photos in areas stroke by earthquakes, rescue experience of seismic disasters and precious safety advices). Many thanks to all the school heads, teachers and students of the Middle school "J. Piaget" and "U. Mazzini" of La Spezia, "F. Poggi" of Lerici and "P. Mantegazza" of San Terenzo, "G. B. Monteggia" of Laveno Mombello (Varese), of the Scientific Lyceum and Classic Lyceum "Parentuccelli-Arzelà" of Sarzana and of the Lyceum "Sacro Monte" of Varese. We would thanks also the Scientific Lyceum "A. Pacinotti" of La Spezia for the collaboration in testing the new Ex-ante and Ex-post survey by the Actions D3 and E5. Thanks to the Portuguese KonwRisk team and to the school heads, teachers and students of the "Padre António Vieira" School of Lisbon to support the exchange of the intervention protocol with the Italian students of the Middle school "J. Piaget" of La Spezia.

References

Becker JS, Paton D, Johnston DM, Ronan KR (2012) A model of household preparedness for earthquakes: how individuals make meaning of earthquake information and how this influences preparedness. Nat Hazards 64(1):107–137
Friston K (2010) The free-energy principle: a unified theory of the brain? Nat Rev Neurosci AOP. https://doi.org/10.1038/nrn2787
Friston K (2012) A free energy principle for biological systems. Entropy 14:2100–2121. https://doi.org/10.3390/e14112100
Frith P (2007) Making up the mind: how the brain creates our mental world. Blackwell, New York

La Longa F, Camassi R, Crescimbene M (2012) Educational strategies to reduce risk: a choice of responsibility. Ann Geophys 55:3. https://doi.org/10.4401/ag-5525

Luna EM (2012) Education and disaster. In: Wisner B, Gaillard JC, Kellman I (eds) The Routledge handbook of hazards and disaster risk reduction. Routledge, New York

Musacchio G, Falsaperla S, Bernharðsdóttir AE, Ferreira MA, Sousa ML, Carvalho A, Zonno G (2015a) Education: can a bottom-up strategy help for earthquake disaster prevention? Bull Earthq Eng. https://doi.org/10.1007/s10518-015-9779-1

Musacchio G, Falsaperla S, Sansivero F, Ferreira MA, Oliveira CS, Nave R, Zonno G (2015b) Dissemination strategies to instil a culture of safety on earthquake hazard and risk. Bull Earthq Eng. https://doi.org/10.1007/s10518-015-9782-6

Musacchio G, Piangiamore GL, D'Addezio G, Solarino S, Eva E (2015c) Scientist as a game: learning geoscience via competitive activities. Ann Geophys. https://doi.org/10.4401/ag-6695

Pachler N, Bachmair B, Cook J (2010) Mobile learning: structures, agency, practices. Springer, New York

Piangiamore GL, Pezzani A, Bocchia M (2012) ERiNat project (training on natural risks): from informed children to knowledgeable adults. In: Proceedings of the 7th EUREGEO-EUropean congress on REgional GEOscientific cartography and information systems, Bologna, vol 1, pp 321–322

Piangiamore GL, Musacchio G, Pino NA (2015) Natural hazards revealed to children: the other side of prevention. In: Peppoloni S, Di Capua G (eds) Geoethics: the role and responsibility of geoscientists. Geological Society, London, Special Publications. https://doi.org/10.1144/SP419.12

Piangiamore GL, Musacchio G, Devecchi M (2016) Episodes of situated learning: natural hazards active learning in a smart school. In: Hunt LM (ed) Interactive learning strategies, technologies and effectiveness. Nova Science Publishers, New York, pp 21–45. ISBN 97881634841986

Platt S, Musacchio G, Crescimbene M, Pino NA, Ferreira MA, Oliveira CS, Lopes M, Rupakhety R, Silva DS (2017) Development of a common (European) tool to assess earthquake risk communication. In: Rupakhety R et al (eds) Proceedings of the international conference on earthquake engineering and structural dynamics, Geotechnical, Geological, vol 47. Springer, Spam

Rivoltella PC (2013) Fare didattica con gli EAS (Episodi di Apprendimento situato). La Scuola, Brescia

Rivoltella PC (2014a) Smart Future – didattica, media digitali e inclusione. Franco Angeli, Milano

Rivoltella PC (2014b) Episodes of situated learning. A new way to teaching and learning. Res Educ Media VI(2):79–87. http://ojs.pensamultimedia.it/index.php/rem_en/article/view/1070

Chapter 35
Risk Perception and Knowledge: The Construction of the Italian Questionnaire to Assess the Effectiveness of the KnowRISK Project Actions

Massimo Crescimbene, Nicola Alessandro Pino, and Gemma Musacchio

Abstract In this paper we describe the design of the Italian version of the KnowRISK (EU project *Know your city, Reduce seIsmic risK through non-structural elements*) questionnaire. Purpose of the questionnaire is to evaluate if the actions of the KnowRisk project can promote in students and in the people involved, attitudes and behaviours to reduce seismic risk. In the first months of the KnowRisk project, a questionnaire was designed to assess the starting point (T_0) on seismic risk Perception, Knowledge of risk and Practice (PKP) of students and public in general before the project actions. Practice was meant in terms of intention to act. The first versions of the KnowRISK questionnaire were built around four theme questions: Who are you?; Do you feel safe?; What do you risk? What would you do?. We also present a preliminary data analysis of the answers collected between March 29 2016 and May 12 2016 in three schools (one each) of the Lombardia (N = 127), Lazio (N = 24) and Liguria regions (N = 14) for a total of 165 students. From this experience, we derived the guidelines for the construction of a common questionnaire of the KnowRISK project: (a) to assess in different countries the effectiveness of a project it is preferred to have a common questionnaire; the questionnaire should consider both qualitative and quantitative aspects. In this regard it is helpful to use Likert scales and methods such as the semantic differential that consider both aspects and they allow quantitative scores which are amenable to sophisticated statistical data analysis; (c) there are three important dimensions to assess project effectiveness: perception; knowledge; and intention to act to reduce the risk.

M. Crescimbene (✉)
Istituto Nazionale di Geofisica e Vulcanologia, Centro Nazionale Terremoti, Roma, Italy
e-mail: massimo.crescimbene@ingv.it

N. A. Pino
Istituto Nazionale di Geofisica e Vulcanologia, Osservatorio Vesuviano, Neaples, Italy
e-mail: alessandro.pino@ingv.it

G. Musacchio
Istituto Nazionale di Geofisica e Vulcanologia, Roma, Italy
e-mail: gemma.musacchio@ingv.it

Keywords KnowRisk questionnaire · Seismic risk assessment · Risk perception · Knowledge · Actions

35.1 Introduction

KnowRisk is an EU project that aims to improve local communities' access to expert knowledge on non-structural seismic protection solutions. It involves case study areas in Italy, Portugal, and Iceland. The project, still ongoing, includes risk communication strategies for the general public and schools children through the engagement of students. Augmented reality, shake table tests, and audio-video media have been used as tools to communicate the seismic performance of non-structural elements in common living spaces. Risk maps based on seismic scenarios have been produced and tested in the pilot areas of Portugal, Italy and Iceland. Also, a practical guide is being produced for the dissemination of good practices to reduce the most common non-structural vulnerabilities, in order to promote reduction in casualty, injury and damage (http://knowriskproject.com).

The project incorporates various risk communication and awareness raising activities (collectively referred to as intervention in the following) addressed to school children and the general public. The assessment of the effectiveness of these activities in achieving the envisioned goals of the project is among the main objectives of the project. There are a large number of risk communication strategies that embraced different disciplines and address a broad range of risks, natural, technological and those involving health (Infanti et al. 2013 and references therein). All of them lack of an in-depth assessment of efficacy (Infanti et al. 2013), which instead is a fundamental novelty of the KnowRISK communication strategy.

The assessment requires a tool that can be used to evaluate the outcomes of these interventions and, in part, the methods and tools employed in the intervention process. Unfortunately, as is evident from social sciences research in recent years, a generally accepted method to evaluate the effectiveness of such interventions still does not exist (Chaiken and Trope 1999; Covello and Merkhoher 2013). In the recent years, many methods and different theories have been used to try to observe and to explain the change of thought, opinion and behaviour linked to this type of initiatives (Chaiken and Trope 1999; Covello and Merkhoher 2013). We think that the efficacy evaluation of a risk reduction project should necessarily include three dimensions: risk Perception, Knowledge of risk and Practice (PKP) that is actions to mitigate it, and the will to achieve these actions. The evaluation tool in this project is envisioned in the form of a questionnaire survey, which we planned to submit before and after the interventions in schools, referred to as T_0, and T_1, respectively. It is assumed that the comparison of response to T_0 and T_1 questionnaires provides an indication of not only the change brought about by the intervention activities, but also the effectiveness of the methods and tools used in the intervention. This paper describes background and methods used to construct such questionnaires to assess the efficacy of intervention actions of the KnowRISK project in Italy.

35.2 Previous Experiences: Research on Risk Perception in Italy

In recent years, our research group has conducted several surveys on the perception of seismic risk in Italy. We have designed and developed questionnaire to evaluate seismic risk perception (SRP-Q) that was used to conduct a survey on the web (over 9000 respondents) and a telephone survey CATI (Computer Assisted Telephone Interview) on a statistical Italian sample of more than 4000 people (Crescimbene et al. 2016). The design and construction of these questionnaires was mainly inspired by the theories and methodologies developed in risk perception research (Flischoff et al. 1978; Slovic 1987, 1992, 2000; Slovic and Weber 2002; Wachinger and Renn 2010). The SRP-Q is based on principles and methodologies of Semantic Differential (Osgood et al. 1957) and Likert scale (Likert 1932). The questions seek answers on three risk perception indicators: hazard, exposure and vulnerability. The level of perception can be evaluated by comparing the response of the sample population with actual value of hazard, exposure and vulnerability of a specific geographic area. This approach allows us to estimate the difference between the perceived risk and actual risk: whether the perception scores are close to or far from the actual values of the risk. In addition, from a theoretical point of view, we overcome the impasse between perception and reality by making a direct comparison between what is perceived and what actual risk exists.

Results obtained from the surveys showed that, unfortunately, there is still a big gap between the perception of risk in Italy, as measured by our questionnaire, and the real risk defined by information based on seismology (hazard), the economic values (exposure), and earthquake engineering (vulnerability).

The SRP-Q also contains two other sections: People and Communities and Earthquake Phenomenon. The section People and Community evaluates the level of preparation and attitudes of individuals and communities against seismic risk. For example, it evaluates if the people around us are perceived to be prepared or unprepared if an earthquake were to occur. This indicator, in our view, could be compared to the existing social resilience of a community before earthquake occurs. Such a comparison was, unfortunately, not possible because social resilience to earthquakes is hard to quantify. The section Earthquake Phenomenon is particularly descriptive of the earthquake phenomenon and concerns some cultural and social issues often associated with it. Some questions in this section measure the perception of its predictability vs. unpredictability, if it is considered a natural phenomenon or caused by man, if it is a divine punishment or a natural phenomenon. Further details on the questionnaire and the research results and experience can be found in Crescimbene et al. (2016).

Our experience highlights that research on seismic risk perception provide insight to failure or ineffectiveness actions towards seismic risk reduction conducted in the last decades: perception of seismic risk in Italy continues to be too low compared to actual risk. This assortment is confirmed by the reactions of astonishment and surprise of people who have been victims of the recent earthquakes in central Italy in 2016. In this sense, it is clear that improving risk perception is the first step towards a Copernican revolution to transform the mentality of the "after" an event in a prevention mindset (to do before it happens). We consider this to be an absolutely necessary condition to stimulate a serious and effective seismic safety culture in Italy (Crescimbene et al. 2014).

35.3 The Specificities of the Know Risk Project

The KnowRISK (KR) project lays in a quite different context than the one described above, where the emphasis was mostly on losses due to failure or collapse of load bearing elements in buildings. Risk posed by non-structural elements is the main target of the project and reduction is pursued following a twofold strategy: improve knowledge and perception of risk among the population, and provide them with information and expert knowledge to mitigate risk. It incorporates a range of very different actions that cannot be assessed individually.

At the initial stage of the project, learning from our experience and research in Italy, our emphasis was on risk perception. However, due to the different nature of this project (non-structural versus structural problems) as well as the goals and objectives of the project, it soon became clear that although evaluation of risk perception is a fundamental requirement, it would not be enough to assess all the issues that the project wanted to address. Based on previous research in Italy, it was also evident that a significant change in risk perception involves factors such as cultural and behavioural improvements as well as knowledge and awareness, in addition to knowledge of effort and cost related to mitigation measures, which can either hamper or motivate preventative actions. Such changes require longer time to go deep into people mind, and cannot be expected be completed during the 2 years of the project.

Based on these arguments, we envisioned the tool for assessing the impact of KnowRISK intervention to contain additional dimensions to the sole risk perception. The idea was to build a new questionnaire considering the three PKP dimensions. These dimensions are supposed to be improved by the intervention actions, albeit to varying degree. Our argument is that the knowledge to be delivered to the public is not independent of their level of risk perception and awareness. By the same argument, the communication of preventative actions needs to be designed, at least partially, on the level of risk perception. The structure of the new questionnaire was then envisioned to contain a section to evaluate risk perception, which we believe to be the starting point of risk awareness process, and is also useful to assess the awareness level in different countries participating in this project. To this dimension, dimensions evaluating change in knowledge and attitude to mitigate risk brought about by the intervention actions were to be added, which were designed in collaboration with the partners from Iceland and Portugal.

35.4 Towards the Construction of a Questionnaire to Test the Effectiveness of a Risk Reduction Project

In the first months of the KnowISK project, we designed a questionnaire to assess the starting point (T_0) about PKP of students before the intervention were conducted. Knowledge at T_0 is fundamental to assess impact of the project and may give important information to improve project actions.

The first version of the Know Risk questionnaire addressed four theses: Who are you?; Do you feel safe?; What do you risk?; What would you do?.

The section "Who are you?" collected information on municipality of residence of the respondents, their age, gender, place of birth, nationality, level of education, work activity, civil status, family unit composition, and information on the building where the respondents lived.

The section "Do you feel safe?" is the perception dimension. It was designed to provide indications of how interviewees view earthquakes and its effects. The main indicators pertaining to the risk have been tested in previous research (Crescimbene et al. 2013, 2014, 2015a, b, 2016). The section was designed to be able to derive perceived risk by using the indicators of Hazard, Exposure and Vulnerability.

The section "What do you risk?" is the knowledge dimension. It contained indicators of key concepts on risk. These notions are important to evaluate the success of KnowRISK project, and to understand if, at the end of the project actions, the level of knowledge has increased. Respondents were asked to rate their level of information on earthquakes and indicate the sources of such information. They were also asked if they participated at risk reduction initiatives and, if so, the level of their involvement. This section also included a comparison between the perceived probabilities of occurrence of an earthquake with respect to other natural hazards. In addition, to test the level of knowledge, we solicited a comparison between the Magnitude and the Intensity of an earthquake defined as mild or strong. The respondents were asked where, in case of an earthquake, they would feel safer; what are the key concepts to define risk and if they are able to recognize the structural/non-structural elements of buildings. A question about what damage they expect due to non-structural elements closed the session.

The section "What would you do?" is the practice dimension. It addresses preventive measures that respondents are willing to take. In addition they are asked to indicate the possible difficulties in taking actions, what they think are the basic steps to protect themselves, and what might persuade to adopt them.

35.4.1 The First Test of the KnowRisk-Q V1

The first test of the KnowRISK Questionnaire version 1.0 was executed from March 29, 2016 to May 12, 2016 in three schools (one each) of the Lombardia (N = 127), Lazio (N = 24) and Liguria regions (N = 14), for a total of 165 students. The sample consisted of 74 males (44.8%) and 91 females (55.2%), with average age of 14.26 years and modal value of 13 years. In all schools, the questionnaire was administered before the intervention actions of the project were carried out.

35.4.2 Data Analysis of the KnowRisk −Q V1

Perception- The risk perception area is composed by 71 differential-method-based items and uses a seven-points rate scales, these items are related to eight indicators (see Table 35.1).

For example, as shown in Table 35.2, the hazard perception score is obtained summing the scores of the ten scales (unexpected-expected; slight-intense; little-big and so on) and dividing the result for 10 (the scale number).

With this procedure for each indicator, we can calculate the mean adding the obtained score of each scale and dividing for the number of the scales of the indicator (Fig. 35.1).

Knowledge- The risk knowledge section had several indicators, in the following we show the information that can be derived.

Table 35.1 KnowRisk questionnaire V1 – indicators of perception

Indicators	#Scales	Mean
Hazard	10	2.90
House vulnerability	6	2.76
School vulnerability	6	3.09
What will happen in house	10	2.90
What will happen in school	10	3.31
Social perception	7	3.37
Exposure	7	4.11
Phenomenon description	15	4.18
Total perception	71	3.40

Table 35.2 Perception dimension and related differential-method items: If you imagine an earthquake in the area where you live, how would you describe it?

Unexpected	1	2	3	4	5	6	7	Expected
Slight	1	2	3	4	5	6	7	Intense
Little	1	2	3	4	5	6	7	Big
Faraway (geographic)	1	2	3	4	5	6	7	Near
Predictable	1	2	3	4	5	6	7	Unpredictable
Short	1	2	3	4	5	6	7	Long
Moderate	1	2	3	4	5	6	7	Violent
Slow	1	2	3	4	5	6	7	Quick
Innocuous	1	2	3	4	5	6	7	Dangerous
Far in time	1	2	3	4	5	6	7	Near in time

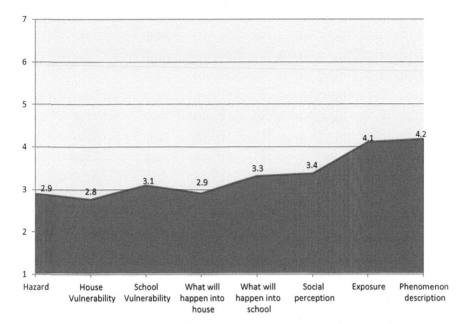

Fig. 35.1 Perception dimension- the mean scores calculated for the indicators of the risk perception (N = 165)

For the question: "how much do you know about earthquake risk?"; data show that the mean is 2.46 in a range of points of 1–4 (from "not at all" 1, to "a lot" 4). The most frequently used sources to inform themselves are: television 25.7%; school 18.5%; newspaper 12.3%; web and social media 12.1%; parents and friends 9.9%; books 8.8%; radio 6.8%; the rest regards other initiatives and campaigns regional 2.4%; municipals 2.2%; provincials 1.3%. Natural phenomena considered more likely than earthquakes were: fires 39.9%; floods 30.2%; landslides 17.2%; the residual 12.7% is distributed in whirlwinds (8.2%), tsunami (2.6%) and volcanic eruptions (1.9%).

With regard to participation in initiatives for the seismic risk reduction, 10.9% of the respondents were involved in the past, while 89.1% were never involved. But the 10.9% that participated indicated a good level of involvement with a mean of 5.06 out of 7.

We asked students to describe an earthquake moderate and intense considering its Magnitude, Intensity and Effects. the question was "Should you define an earthquake "moderate" (or severe) when its Magnitude/Intensity/Consequence is. . . .

Figure 35.2 shows the comparison between the means of scores of Magnitude, Intensity and Effects for earthquake that is considered moderate (blue) or severe (red) in an equivalent scale by 1 to 7. The differences between Moderate and Severe earthquakes are of 3.23 for Intensity, 3.45 points for Magnitude and 3.25 for Effects. Absolute scores indicate that an earthquake is considered strong when its magnitude is around 5.0, while it is considered mild with a magnitude around 1.6.

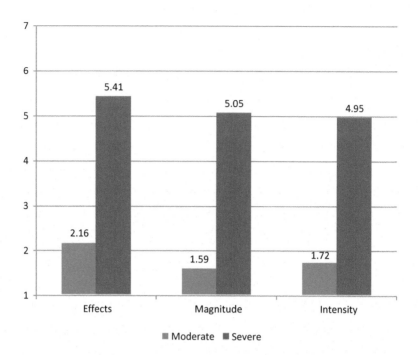

Fig. 35.2 Knowledge dimension- comparison between means of the scores of hazard, intensity and effects for earthquake that is considered moderate (blue) or severe (red) in a equivalent scale by 1–7

Figure 35.3 shows the percentages of answers to the question: "In your opinion, how important are the following concepts to define the risk?

The last two questions of are dedicated to evaluate the knowledge about the structural and non-structural elements of buildings. The first question is: "Could you recognize the structural and non-structural elements of a building?". Figure 35.4 shows the answers given by students.

The last question is: "In your opinion, what consequences can result from non-structural elements during an earthquake?". Figure 35.5 shows the Students' answers to the possible consequences that may result from the non-structural elements of buildings.

Practice- The area dedicated to assess intentions of the interviewees to adopt measures to reduce risk, opens with the question: "Do you plan to take preventive measures against a mild earthquake?". 57.6% of respondents answered "yes" to this question, while 38.8% say "no" and 3.6% "don't know".

Figure 35.6 shows the responses to the question: "What preventive measures would you take?"

We also asked students to indicate "What difficulties you may encounter in making these protective measures?" and "What might happen to you if you do not?". The answers are summarized in Tables 35.3 and 35.4.

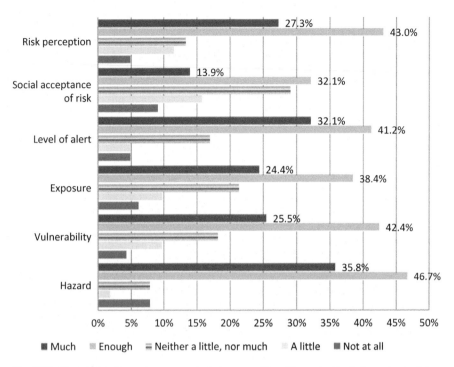

Fig. 35.3 Knowledge dimension- concepts that are considered important to define seismic risk

Then, respondents were asked to indicate, according to them, which would be the measures to be taken to protect against minor earthquakes? The answer most respondents indicated is "to fix the furniture and book shelves" with the 24.5%, but the majority of students, 30%, did not respond.

The answers to the last question, "What could convince you to take measures to defend yourself from earthquakes?", are shows in Fig. 35.7.

35.4.3 The Second Test of the KnowRisk-Q V1.2

The second test of the KnowRisk Questionnaire version 1.2 was executed from November 11, 2016 to February 24, 2017 in three regions Emilia-Romagna (N = 11), Liguria (N = 239) and Toscana (N = 8) for a total of 258 students and some adults. The sample was composed of 124 males (48.1%) and 134 females (51.9%), with average age of 16.3 years and modal value of 13 years. In all the cases, the test was performed before the training sessions of the project.

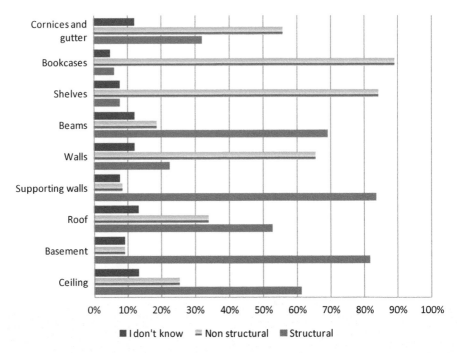

Fig. 35.4 Knowledge dimension- distinction between structural and non-structural elements of buildings

35.4.4 Data Analysis of the KnowRisk-Q V1.2

The results from this survey were similar to those shown above. The only change from the previous version of questionnaire regards the attempt to reduce the number of questions and items of the three areas considered. For these reason we do not report, in this paper, the data elaboration of the second version of the questionnaire.

35.5 Discussion

Considering that these are the results of a test, only some comments are appropriate. Scores of all the indicators considered (hazard, house and school vulnerability, what will happen in the house and the school, social perception, exposure, earthquake phenomenon) for the perception dimension seems to be too low compared to the level of actual risk. We can state this even without making any comparison with the actual risk because the average of all these indicators is always below 4 or just above (Exposure = 4.11; Phenomenon = 4.18; see Fig. 35.1), which are very close to the median values of the scales.

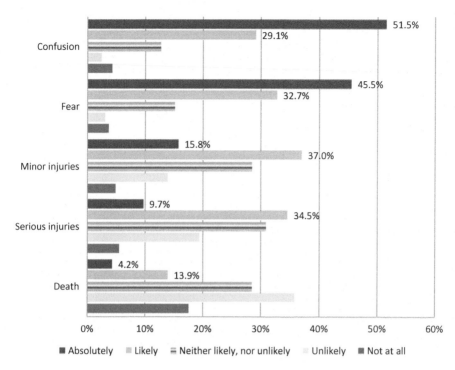

Fig. 35.5 Knowledge dimension- consequences that may result from the non-structural elements of buildings

On the knowledge dimension respondents seem to have a medium level of information 2.46 of 4.0 (between little and enough). The most commonly medium used to be informed on earthquake is television (25.7%), followed by school (18.5%). Respondents seem to know the difference between a strong and a mild earthquake (see Fig. 35.2). Knowledge of hazard (35.8%) and alert levels (32.1%) are considered the most important concepts to reduce risk (see Fig. 35.3). Most students properly recognize structural and non-structural elements of a building (see Fig. 35.4), but only a small part of them think that non-structural elements can definitely cause minor (15.8%) or serious (9.7%) injuries or death (4.2%) (see Fig. 35.5).

The practice dimension inquires respondents' intentions to adopt measures to reduce risk. Here 57.6% of respondents answered "yes" to the question: "Do you plan to take preventive measures against a mild earthquake?". But on the measures to be taken there are very different answers and perhaps ideas on what to do are not clear (see Fig. 35.6). The answers in Table 35.3 indicate that the economics is among the most important obstacle towards the adoption of measures to reduce risk. This is despite the fact that more than 43% think that they could get hurt if they did not take preventative actions (Table 35.4). With respect to the measures to be taken to protect against minor earthquakes, the answer of most of the respondents (24.5%) is "to fix

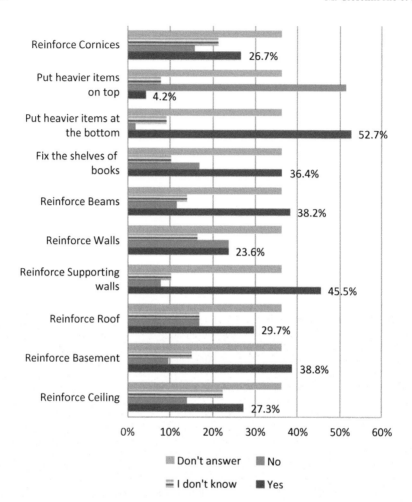

Fig. 35.6 Practice dimension- answers to the question: what preventive measures would you take?

Table 35.3 Practice dimension: what difficulties you may...?

What difficulties...?	%
No difficulty	6.0%
Economic	32.1%
It would take too much time	15.6%
I would find it hard to convince my family/cohabitants	10.6%
It would be too difficult	7.8%
No answer	28.0%

Table 35.4 Practice dimension: What might happen to you if you do not?

What might happen…?	%
I might get hurt	43.1%
I could die	12.0%
Nothing	3.0%
I do not know	5.4%
No answer	36.5%

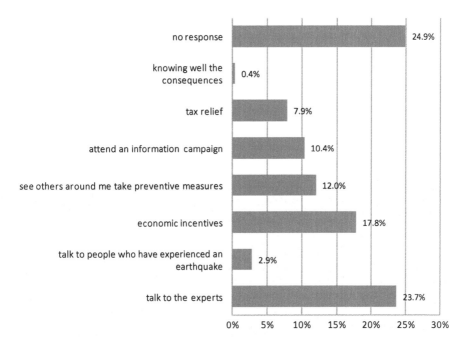

Fig. 35.7 Practice dimension: "What could convince you to take measures to defend yourself from earthquakes?"

the furniture and book shelves", but the majority of students (30%), did not respond, showing a lack of knowledge. The answers to the question "What would convince you to take measures to defend yourself from earthquakes?" indicate that talk with an expert could be most effective.

35.6 Guidelines for the Creation of a Common Questionnaire

From the tests described above, some useful guideline for the construction of a single shared questionnaire to be used in all the study areas (Italy, Portugal, and Iceland) could be derived:

(a) it is possible to build a common questionnaire to assess the project even if the partners are from different countries with different culture, building tradition and past experiences.
(b) the questionnaire should consider both qualitative and quantitative aspects. In this regard it is helpful to use Likert scales and methods such as the semantic differential that consider both aspects and allow quantitative scoring which are amenable to sophisticated statistical data analysis.
(c) the three areas that we consider important to evaluate, considering the changes in thought, behaviour and actions that the project actions could induce in the participations are: perception; knowledge; and practice, meant as the intention to act to reduce the risk.

Acknowledgements This study was co-financed by the European Commission's Humanitarian Aid and Civil Protection (Grant agreement ECHO.A.5 (2015) 3916812).

References

Chaiken S, Trope Y (1999) Dual-process theories in social psychology. Guilford, New York
Covello VT, Merkhoher MW (2013) Risk assessment methods: approaches for assessing health and environmental risks. Springer US, https://books.google.it/books?id=4pS1BwAAQBAJ
Crescimbene M, La Longa F, Camassi R, Pino NA (2013) Seismic risk perception test. Geophys Res Abstr 15
Crescimbene M, La Longa F, Camassi R, Pino NA, Peruzza L (2014) What's the seismic risk perception in Italy? Eng Geol Soc Territ 7:69–75
Crescimbene M, La Longa F, Camassi R, Pino NA (2015a) The seismic risk perception questionnaire. Geol Soc Lond Spec Publ 419:69–77
Crescimbene M, La Longa F, Camassi R, Pino NA, Peruzza L, Pessina V, Cerbara L, Crescimbene C (2015b) INGV EOS8 geoethics for society: general aspects and case studies in geosciences
Crescimbene M, La Longa F, Peruzza L, Pessina V (2016) The seismic risk perception in Italy compared to some hazard, exposure and vulnerability indicators. International Conference of Urban Risk, Lisbon (Portugal) 30 June – 2 July, 2016
Fischhoff B, Slovic P, Lichtenstein S, Read S, Combs B (1978) How safe is safe enough? A psychometric study of attitudes towards technological risks and benefits. Policy Sci 9(2):127–152
Infanti J, Sixsmith J, Barry MM, Núñez-Córdoba J, Oroviogoicoechea-Ortega C, Guillén-Grima F (2013) A literature review on effective risk communication for the prevention and control of communicable diseases in Europe. Stockholm, ECDC
Likert R (1932) Technique for the measure of attitudes. Arch Psychol 22(140):5–55
Osgood CE, Suci G, Tannenbaum P (1957) The measurement of meaning. University of Illinois Press, Urbana
Slovic P (1987) Perception of risk. Science 236:280–285
Slovic P (1992) Perceptions of risk: reflections on the psychometric paradigm. In: Krimsky S, Golding D (eds) Social theories of risk. Praeger, Westport/London, pp 117–152
Slovic P (2000) The perception of risk. Earthscan Publications, Ltd, London
Slovic P, Weber E (2002) Perception of risk posed by extreme events. Conference on risk management strategies in an Uncertain World, Palisades, New York, April 12–13, 2002
Wachinger G, Renn O (2010) Risk perception and natural hazards. CapHaz-Net WP3 Report, DIALOGIK Non-Profit Institute for Communication and Cooperative Research, Stuttgart. Aavailable at: http://caphaz-net.org/outcomes-results/CapHaz-Net_WP3_Risk-Perception.pdf

Chapter 36
Awareness on Seismic Risk: How Can Augmented Reality Help?

Danilo Reitano, Susanna Falsaperla, Gemma Musacchio, and Riccardo Merenda

Abstract To communicate the importance of knowing the risk of non-structural damage caused by earthquakes, we developed applications based on Augmented Reality (AR) features. These applications run on mobile devices, such as tablets and smartphones, by using their video camera and other on-board sensors, such as GPS, accelerometer, and gyrocompass, from which AR users do take advantage. Combined with a specifically designed exhibit, our AR applications can contribute to increase the common awareness on seismic risk, providing useful information on how to have safer homes in case of an earthquake. Building codes do not take into account non-structural elements, leaving communities at risk of injuries, blocking escapes and even causing deaths. In this framework, the personal preparedness is of paramount importance. The development of our AR applications is supported by the European project KnowRISK (Know your city, Reduce seISmic risK through non-structural elements).

Keywords Non-structural damage · Earthquake hazard · Augmented reality · Risk reduction · Dissemination

---

D. Reitano (✉) · S. Falsaperla
Istituto Nazionale di Geofisica e Vulcanologia, Osservatorio Etneo, Catania, Italy
e-mail: danilo.reitano@ingv.it; susanna.falsaperla@ingv.it

G. Musacchio
Istituto Nazionale di Geofisica e Vulcanologia, Roma, Italy
e-mail: gemma.musacchio@ingv.it

R. Merenda
Università di Catania, Catania, DIEEI, Catania, Italy

36.1 Introduction

The collapse of buildings is the major reason of victims from earthquakes. The societal impacts of earthquakes, however, is not limited to the collapse. Injury, loss of property and of functionality from earthquakes can also occur in resilient buildings in form of non-structural damage. Furniture, partitions, balconies, hydraulic system are only a few examples of non-structural elements of buildings the damage of which is still grossly underestimated. Indeed, even the effects of a small (low-magnitude), superficial earthquake can lead to high costs for the lack of simple (and low-cost) solutions that can prevent non-structural damage. Moreover, in medium-to-high magnitude earthquakes moderate-heavy damage (D2–D3), when non-structural elements start to play a major role, usually occurs over a wide area. Communities are scarcely aware of the relevance of this kind of damage and, therefore, they are not prepared (Crescimbene et al. 2018). Often there is a lack of effective communication tools specifically tailored for the public engagement in its own safety.

The European project KnowRISK (Know your city, Reduce seISmic risK through non-structural elements; Grant agreement ECHO/SUB/2015/718655/PREV28) brings together engineers, seismologists, architects and sociologists. Focusing on non-structural damage, the goal of the project is twofold. First, it explores solutions to reduce non-structural damage from earthquakes in pilot areas of Portugal, Iceland, and Italy. The second goal is the promotion of preventive action, with a public engagement approach (Musacchio et al. 2018). Indeed, seismic risk reduction calls for preparedness not only in terms of countermeasures for building construction and reinforcement. It also requires effective scientific outreach activity to convey useful information to people living in regions prone to earthquakes, increasing their personal capability to cope with hazard and being prepared.

A key issue for a successful outreach activity is the choice of tools that can prove to be simple and effective. We propose Augmented Reality applications as they fulfil both requirements. In the following, we describe the prototype of the KnowRISK exhibit, an interactive poster that was specifically designed for our scientific dissemination purposes.

36.2 Augmented Reality

Unlike reality, which is the state of things as they actually exist (Oxford University Press 2017), Augmented Reality (AR) enriches the real world with digital information by using a cutting-edge technology. It operates overlaying real-time images (coming from a video camera) with virtual elements, such as 3D models, pictures, and videos. Virtual and multimedia elements are superimposed using different information layers in order to obtain a unique view for users' eyes. Elements that can "increase" reality can be viewed through a mobile device, such as a smartphone or through a tablet with a video camera. Point Of Interest (POI) can also be added to the real world using other on-board sensors, such as the internal GPS device, nowadays present in every new-generation mobile phone. The process to create AR is based on the real-time capture of images from any device (typically on-board cameras) and GPS location; software generates layers full of virtual items, such as image contents, queries to a web page or to a database, etc. AR also gathers a wide variety of user's experiences. It is possible to distinguish three main categories of AR tools (Augment Web Site 2016):

"Augmented Reality 3D viewers, like Augment, allow to place life-size 3D models in your environment with or without the use of trackers. Augmented Reality browsers enrich your camera feed with contextual information. For example, you can point your smartphone at a building to display its history or estimated value. Augmented Reality games create immersive gaming experiences, like shooting games with zombies walking in your own bedroom!"

The best results of AR require the development of a complex software (generally one or more APPs – an APP is typically a small, specialized program downloaded onto mobile devices), working with image processing and computer graphics. Most of the useful data can be directly derived from real-time and/or offline images. For example, imagine a user who needs to know how many restaurants are located around his/her own actual position. In this example, the results of the search with AR mark the GPS coordinates of restaurants extracted from internal data to the user's device (Fig. 36.1).

Throughout the process of graphics overlay, images can be added to or can even remove/hide parts of the real environment. Optical and video AR technologies are both under development (a tablet screen vs optical glasses) to better increase user's perception of reality.

Fig. 36.1 A typical example of AR from Nokia-Microsoft APP (HERE City Lens) using internal GPS and compass

36.3 Software: The Wikitude Experience

Developed by an Austrian company (Wikitude GmbH), Wikitude was the first AR APP distributed worldwide. Wikitude (2016; http://www.wikitude.com/products/studio/) is an impressive tool to create AR software. It can visualize data of the real environment through a camera and includes an Image Recognition Tool and a 3D model. Furthermore, Wikitude distributes a social framework called Wikitude Studio, with which developers have the opportunity to make exciting AR experiences.

36.3.1 KnowRISK APP

Our KnowRISK APP was built using the Wikitude Software Development Kit (SDK) for Android. This exploits web technology (i.e., HTML, JavaScript, CSS) and allows the developer to write software within a multi-platform system. Architect elements are the basic building blocks; an Application Programming Interface (API) is also available to connect the APP to the most used operating systems (e.g., Windows, Linux, Android). The interaction between the Wikitude SDK and the KnowRISK APP was obtained by adding a special "view" called ARchitectView user interface to our platform.

Three main different modules were implemented in order to obtain the results described in Sect. 36.4: Image recognition, Video Overlay, and Location Based Service.

36.3.1.1 Image Recognition

Based on the Computer Vision concept, the Image recognition is one of the most powerful tools available inside the Wikitude framework. When the Image recognition tool is active, the AR APP is ready to display new contents on layers that overlay to the real world. In addition, the user can navigate and interact with local or remote data provided by the application. This interactivity is based on three main methods to acquire data:

- Target image: If a target image (or multiple different images) is recognized by the viewfinder, then an animation runs (for example, a new layer with images, videos or HTML content will appear);
- Target collection: The user will access an archive containing all the images the tracker can recognize;
- Client tracker: By using the live camera image of the smartphone, it detects the targets stored in the related target collection.

36.3.1.2 Video Overlay

In addition to images, text, and HTML content, it is possible to add videos with the help of Wikitude libraries. The Video Overlay module can display a video on any image recognition target as well as at any geolocation. We use this tool to "activate" the exhibit described in the following showcase. Like any other drawable element, it is possible to position, scale, rotate and change the opacity of the video. As with all other resources, the video can be loaded both locally, from the application bundle, and remotely, from any server.

36.3.1.3 Location Based Service

The Location-Based Service (LBS) is a software-level service that uses location data to control features inside Wikitude APPs. Enabling the geolocation permission, the APP is able to run an animation or anything the developer chooses, depending on his/her real position. In this way, images produced by the live camera will be "augmented" with new elements introduced during the development of the APP. The user will also be able to access multiple contents depending only on his/her own actual position.

36.4 The Showcase of ScienzAperta

We tested in 2016 our KnowRISK APP along with a prototype of interactive poster during "ScienzAperta", an open-door scientific, outreach event that INGV yearly organizes in Italy for schools and the public (Falsaperla et al. 2016). We designed an "animated" exhibit according to the working scheme of Fig. 36.3. We divided our poster into three different sections, containing target (static) images (Fig. 36.2). These target images were the "virtual button" to activate our AR application. It is worth noting that AR might apply to all senses, not just sight. For example, our APP included sounds: a target image associated with a seismogram (at the top of our poster, in Sect. 36.1, see Fig. 36.2) allowed visitors to experience the "sound" of an earthquake by using a proper frequency shift.

The exhibit was open to the visitors at INGV Catania, Italy, during the 5-day-long event, from 16 to 20 May 2016. Part of the images of the poster in Fig. 36.2 were frames of videos depicting visible effects inside and outside a building anchored to a shaking table. Our APP allowed visitors to see what happens during a shaking-table test, which simulates different scenarios based on non-destructive earthquakes. When a target image came into focus, the software recognized it and played the video on a tablet (Fig. 36.3).

The total number of persons who visited the exhibit was ~600. As each group of visitors was large (between 12 and 18 people – Fig. 36.4), we decided to transfer the

Fig. 36.2 Scientific content and layout of the poster

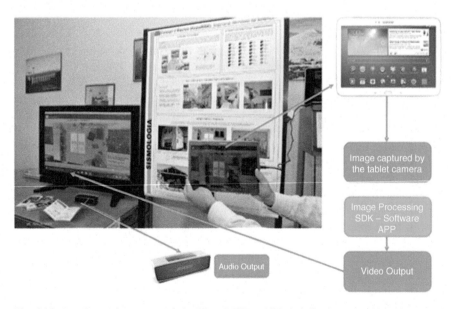

Fig. 36.3 The interactive poster of the KnowRISK exhibit based on our working scheme of augmented reality

Fig. 36.4 Presentation of the KnowRISK exhibit at Catania during "ScienzAperta" 2016

data from the tablet to a bigger monitor to improve the vision (Figs. 36.3 and 36.4). An audio-surround device was also associated with the videos to make effects more impressive (Fig. 36.3). Eventually, the poster itself "suggested" solutions to improve safety in buildings, by switching various images on the screen; for example, before and after the application of simple solutions to fix a problem (e.g., fall of objects, heavy furniture).

36.5 Discussion and Conclusions

In case of high-magnitude earthquakes, the non-structural damage can potentially strike a region of vast extension in addition to the collapse of buildings in the mesoseismal area. This can heavily affect the return time to normal life, also with repercussions on the resumption of economic activity. Yet, even low-magnitude earthquakes with shallow foci can yield high costs due to non-structural damage (Azzaro et al. 2018). In this case, the mesoseismal region will be smaller, but effects may be comparatively costly.

There are many solutions to prevent the possible tumbling and falling of objects and furniture. Some of them have low costs (a few Euros only), and their application does not require technical skills. Simple examples are steel brackets to fasten tall furniture to the wall, latches on cupboard doors and drawers, etc. The adoption of these solutions along with others that require specialized staff – for example, the anchorage of chimneys or utility systems, such as gas, water, and power– is of paramount importance in regions potentially prone to damaging earthquakes (e.g., EQC Earthquake Commission 2016).

We propose an interactive exhibit combined with AR software (KnowRISK APP), which allow the user to focus on non-structural elements inside and outside buildings. The KnowRISK APP provides information in form of sounds, video clips, and animated snapshots. The interactivity offers an easy engagement of the user.

Italian pupils and students tested the first prototype of the KnowRISK exhibit at INGV Catania in May 2016. A heartening measure of the success of the exhibit was the wide participation of children, who were also asked to discuss the safety of their home and, in particular, of their bedroom. The exhibit made them aware of the potential danger of heavy furnishings (e.g., sport cups) above their bed or close to doors, causing injury or hindering escape in case of fall.

The majority of visitors was aware of the concept of Virtual Reality while they were not familiar with AR. The launching of the game "Pokémon go" in summer 2016 gave to AR worldwide visibility and resonance. Hardware and software have evolved fast ever since. This also fosters our efforts towards new developments of AR APPs, which can contribute to seismic risk reduction.

Acknowledgements This study was co-financed by the European Commission's Humanitarian Aid and Civil Protection (Grant agreement ECHO/SUB/2015/718655/PREV28). We acknowledge Carlos Sousa Oliveira, Mário Lopes, and all the KnowRISK partners for the fruitful discussions and support.

References

Augment Web Site (2016) http://www.augment.com
Azzaro R, D'Amico S, Langer H, Meroni F, Rupakhety R, Squarcina T, Tusa G, Tuvè T (2018) From seismic input to damage scenario: an example for the pilot area of Mt. Etna volcano (Italy) in the KnowRISK project. Submitted to this volume
Crescimbene M, Pino NA, Musacchio G (2018) Between perception and knowledge: the construction of the Italian questionnaire to assess the KnowRISK project actions. Submitted to this volume
EQC Earthquake Commission (2016) http://www.eqc.govt.nz
Falsaperla S, Reitano D, Merenda R, Benbachir M (2016) Augmented reality applications as dissemination tools for the mitigation of non-structural damage from earthquakes. Oral presentation at the 2nd general meeting KnowRISK, Catania, Italy, 15–17 December 2016. Vol. 33 MISCELLANEA INGV, ISSN 2039-6651, 39–40. http://istituto.ingv.it/l-ingv/produzione-scientifica/miscellanea-ingv/, http://hdl.handle.net/2122/10428
Musacchio G, Falsaperla S, Solarino S, Piangiamore GL, Crescimbene M, Pino NA, Eva E, Reitano D, Manzoli F, Fabbri M, Butturi M, Accardo M (2018) KnowRISK on Seismic Risk Communication: The Set-Up of a Participatory Strategy- Italy Case Study. Submitted to this volume
Oxford University Press (2017) https://en.oxforddictionaries.com/definition/reality
Wikitude Web Site (2016) http://www.wikitude.com

Chapter 37
Development of a Common (European) Tool to Assess Earthquake Risk Communication

Stephen Platt, Gemma Musacchio, Massimo Crescimbene,
Nicola Alessandro Pino, Delta S. Silva, Mónica A. Ferreira,
Carlos S. Oliveira, Mário Lopes, and Rajesh Rupakhety

Abstract This paper reports work on an on-going EC project called KnowRISK aimed at reducing the seismic risk from non-structural elements in buildings. Specifically it reports work on the development of a European tool to assess the effectiveness of risk communication interventions and awareness raising training with middle and high school children in case study areas in Portugal, Italy and Iceland. It describes the difficulties research teams faced in agreeing a theoretical

---

S. Platt (✉)
Cambridge Architectural Research Ltd, Cambridge, UK
e-mail: steve.platt@carltd.com

G. Musacchio
Istituto Nazionale di Geofisica e Vulcanologia, Roma, Italy
e-mail: gemma.musacchio@ingv.it

M. Crescimbene
Istituto Nazionale di Geofisica e Vulcanologia, Centro Nazionale Terremoti, Roma, Italy
e-mail: massimo.crescimbene@ingv.it

N. A. Pino
Istituto Nazionale di Geofisica e Vulcanologia, Osservatorio Vesuviano, Neaples, Italy
e-mail: alessandro.pino@ingv.it

D. S. Silva
Laboratório Nacional de Engenharia Civil, Lisbon (LNEC), Lisboa, Portugal
e-mail: delta@lnec.pt

M. A. Ferreira · C. S. Oliveira · M. Lopes
Instituto Superior Técnico, Lisboa, Department of Civil Engineering, Architecture and Georesources, CEris, Lisboa, Portugal
e-mail: monicaf@civil.ist.utl.pt; csoliv@civil.ist.utl.pt; mariolopes@tecnico.ulisboa.pt

R. Rupakhety
Faculty of Civil and Environmental Engineering, School of Engineering and Natural Sciences, University of Iceland, Reykjavík, Iceland

Director of Research, Earthquake Engineering Research Centre (EERC), University of Iceland, Selfoss, Iceland
e-mail: rajesh@hi.is

framework and in devising the survey tool. Although they all agreed it was essential to have a common survey if the findings from the research were to be compared across the three countries, one year into the two-year project two of the teams were moving in different directions. This was significant since some of the pre-intervention surveys had already been conducted. Both theoretical frameworks had merit and each of the questionnaires were capable of assessing the efficacy of the training. However, they were in no way comparable. Finally the paper details how these difficulties were resolved and a common questionnaire was devised that embodied virtues from both surveys. This was then applied in all three countries to provide comparable data, the findings from which will be reported elsewhere.

Keywords Non-structural earthquake risk · Risk communication · Assessment tool · Children's attitudes

37.1 Introduction

37.1.1 The KnowRISK Project

This paper reports work on an on-going EC project called KnowRISK aimed at reducing the seismic risk from non-structural elements in buildings. The KnowRISK consortium consists of four partner organisations from three European Countries (Portugal, Italy and Iceland) – the Instituto Superior Técnico, Lisbon (IST), Laboratório Nacional de Engenharia Civil, Lisbon (LNEC), Istituto Nazionale di Geofisica e Vulcanologia (INGV) and Earthquake Engineering Research Centre, Iceland (EERC). Specifically the paper describes the development of a survey tool to assess the impact of the KnowRISK risk communication strategy in middle-high schools in pilot-areas in Italy, Portugal and Iceland. The survey is part of a communication exercise aimed at engaging students and their social surrounding in active citizenship to foster risk reduction. The survey tool will evaluate children's risk awareness and knowledge of seismic protection both before and after the training and awareness raising initiative in schools. The overall aim is to develop a tool that might be used in other European countries.

As with all EC projects one of the key aims is to foster collaboration and understanding between scientists from different countries. However, this is not always easy. Professionals come with their own ideas and experience. This was the case on this KnowRISK study where the Italian team introduced a questionnaire they had used on previous projects that the Portuguese team found impossible to adapt to the different characteristics of the Portuguese pilot-area. The Portuguese assessment strategy was, therefore, designed in conjunction with planning the intervention in the two schools (see Table 37.1) rather than by adopting a ready-made framework and questionnaire.

This paper is an analysis of the two questionnaires in order to produce a common one.

Table 37.1 Planned interventions in schools

	Date	Schools	Place	Children	Institute	Version
Italy	Apr 2016 – May 2017	5	La Spezia and Laveno Mombello	714	INGV	Italian
Portugal	Oct 2017 – Mar 2017	2	Lisbon	117	LNEC	Portuguese
Iceland	March 2017	1	Selfoss	60	EERC	Common

37.2 Pilot Areas: Knowledge, Attitude and Practice

Seismic hazard is different in Portugal, Italy and Iceland. There are differences in the relevance of seismic hazard compared to other hazards; in the frequency of earthquakes, the time span from last damaging event, type of damage, level of implementation of protective measures, cultural attitudes towards prevention and previous actions to raise awareness that needed to be taken into account in setting up a communication strategy. This was the rationale behind the choice of pilot areas. Lisbon is an area were perception of seismic risk is likely to be low; Italy encompasses a wide range of seismic hazard, level of perception and implementation of preventative action; Iceland as the area were although prevention was at the best compared to the other two areas, non-structural element are still widely underestimated.

37.2.1 The Portuguese Case Study

The Portuguese case study differs from Italian and Icelandic cases in terms of the infrequency of earthquakes and the low-level disaster experience. For the inhabitants of Lisbon, seismic risk is something distant and something they do not think about in their daily lives. Given this, the school children in our target-group for risk communication, are likely to be less aware of seismic risk, including the risk of non-structural damage and injury, than children in Italy and Iceland.

Risk communication in the Lisbon pilot-area covered two secondary schools – Rainha D. Leonor and Padre António Vieira – and includes approximately 120 students of the 7th and 8th grades aged 12–16 years old. The intervention was structured into a set of activities included as part of the curriculum "education for citizenship". The programme of KnowRISK intervention was the same in both schools and brought students and scientists together on a regular basis, over a period of 2–4 months.

The risk communication assessment research strategy for Portugal was designed as a quasi-experimental (Bryman 2011) where one group, the experimental group, was subject to the intervention, and a second group, the control group was not. The Portuguese questionnaire was applied to both experimental group and control group before the start of the KnowRISK intervention (T0) and after the intervention (T1). The experimental group comprised 117 students from Rainha D. Leonor and Padre

António Vieira Schools and the control group a 100 students from a similar age profile from the neighbouring Eugénio dos Santos School.

During the research it became clear that the assessment could not rely solely on a questionnaire. It was felt necessary to add qualitative data that could add to and verify the survey results. A "KnowRISK notebook" was given to students at the start of the intervention and students were asked to make notes. After the intervention, these notebooks were collected and subjected to content analysis. In addition, focus groups are planned for the end of the school year.

37.2.2 The Italian Case Study

In general terms Italy is a country with recurrent earthquakes but a low level of prevention. Two pilot-areas were chosen – the Mt Etna volcano region and Northern Italy. They were selected based on two criteria: (i) areas affected by the most common non-structural vulnerability; (ii) areas where it was possible to have a high range of target public.

In Mt Etna pilot area the focus was on the lower eastern flank of the volcano where recent earthquakes, associated with moderate shaking, had caused non-structural damage. The Northern Italy Pilot area was chosen to implement communication in regions where PGA was expected to be lower than 0.15g (G.U. n.108 del 11/05/2006); these being seismic zones where strong earthquakes are rare but non-structural damage, for example during the 2012 Emilia earthquake, can be widespread and cause anxiety for people living all over Northern Italy.

The Italian action applied the questionnaire to a sample of target group before the start of the communication intervention (T0), and an ex-post survey at the end of the process (T1). This design was meant to assess the initial condition of the students at T0, in order to evaluate the impact of the project action and to acquire important information for future interventions.

In recent years, the Italian team had analysed seismic risk perception in Italy, by employing a web-based questionnaire in 2014 to collect responses from a sample of over 8500 people. Statistical analysis of this dataset showed good reliability of the indicators of hazard, exposure and vulnerability.

After revision to adapt the format for telephone interview, the questionnaire was applied to a sample of the Italian population, consisting of 4000 people, using the Computer Assisted Telephone Interview (CATI) technique. There were small differences between the two findings, likely to be due to self-selection in the web-based survey ie. These respondents may have been more willing to respond and have therefore been more sensitive to the subject matter. Therefore, a higher risk assessment might be expected in the web-based survey than the telephone interview CATI survey,

The results relating to the hazard perception in the CATI survey suggest that only 6% and 17% of people respectively living in higher and lower hazard areas have adequate perception. Overall, 61% of the entire population admit to being only "slightly" or "not at all informed" about earthquakes and less than 5% have ever participated in risk reduction initiatives.

37.2.3 The Icelandic Case Study

The Icelandic case study focuses on the South Iceland Seismic Zone (SISZ), which contains the largest agricultural region in Iceland, and is seismically very active, generating frequent tectonic and volcanic earthquakes. Several small towns or villages, schools, medical centres, industrial plants, geothermal and hydropower plants, and several major bridges are within this area. Three moderate to strong earthquakes have occurred in SISZ in the recent past. These are the June 2000 South Iceland earthquakes (Sigbjörnsson and Olafsson 2004) and the May 2008 Ölfus Earthquake (Sigbjörnsson et al. 2009). Most of the damages caused by these recent earthquakes were non-structural (see, for example, Bessason and Bjarnasson 2016). Akason et al. (2006) conducted a study on perceived and observed residential safety during the June 2000 earthquakes. According to their analysis, the victims of the June 2000 South Iceland earthquakes, who were inside their residences during the earthquakes, generally found themselves in significant, deadly danger, at least in the epicentral areas, mostly due to different loose or improperly fastened household items. The authors claim that the low number of physical injury caused by the earthquake was due to the fortunate timing of the earthquake, when many people were outside of their houses. The study concluded that the following factors had a positive impact on residential safety.

- the instructions to the general public regarding earthquake safety measures disseminated by the National Civil Defence in the years and decades preceding the June 2000 earthquakes
- the significantly high level of the victims' knowledge and awareness of these instructions before the earthquakes
- many victims being able to apply these instructions during the earthquake to move to some kind of safe spots inside their dwellings

During the June 2000 earthquakes, many people in the epicentral area reported having difficulty or finding it impossible to move to a safe place inside their dwellings (Sigbjörnsson et al. 2017). The potential danger due to movement of household objects was reduced because many residents were outside of their dwellings during the earthquakes. Had this not been the case, more people could have been injured. Therefore, great importance needs to be placed on pre-earthquake arrangements for loose household articles (e.g., fastening down bookshelves, closets, loose and "poorly" fastened articles, etc.), as this will offer safety to at least those who are unable to move during an earthquake exposure.

From this recent experience, it is evident that most of the residential buildings in SISZ can sustain significant ground shaking without major structural damage. In this context, in contrary to areas where buildings are prone to collapse, it is important to advise the residents to stay calm and maintain balance during an earthquake rather than trying to run out of the house, which could be more hazardous. This scenario was carefully considered in designing the questionnaire.

Experience from past earthquakes in SISZ have shown that damage to room heating radiators and subsequent leakage of hot and cold water caused major damage in some houses. In some cases, damage to electric circuits in the houses was hazardous. These observations were also considered in designing the questionnaire. Apart from these issues, the study area in Iceland differs from those in Italy in Portugal, in terms of building typology and their seismic vulnerability. These differences have been accommodated in the questionnaire.

37.3 Theoretical Framework

37.3.1 Italian Framework

The Italian theoretical framework is structured on a distinction between three dimensions: perception (what people think about earthquake risk), awareness (what they know and understand about earthquakes) and behaviour (their propensity to take preventative action). These, together with information about respondents and their home and school, formed the four-part structure of the Italian questionnaire.

It is widely assumed that residents' low risk awareness is among the main causes of an insufficient level of preparedness and an inadequate response to disasters (Grothmann and Reusswig 2006; Miceli et al. 2008; Terpstra et al. 2009; Maidl and Buchecker 2015). And a willingness to adopt precautionary measures is positively related in many cases with the level of risk awareness (Neuwirth et al. 2000; Floyd et al. 2000; Scolobig et al. 2012). This assumption not only formed a part of the Italian theoretical framework, but also of the whole KnowRISK project.

Within the social sciences the term 'risk perception' is used to represent the bundle of feelings, attitudes, knowledge and experience that influence how people consider the seriousness and acceptability of risks (Slovic 1987). Yet risks cannot be 'perceived' by the human senses. The mental models that people use to judge risks (such as heuristics and imagery) are internalized through social learning and are constantly moderated (reinforced, modified, amplified or attenuated) by media reports, peer influences and other communication processes (Morgan et al. 2001). There are two main approaches to the study of risk perception (Renn 2008). The realist approach assumes that there is an outside objective world and aims 'to bring perception as close as possible to the objective risk of an event'(Rosa 1998, 2008). Changing perception involves simply better information and a greater understanding of the risk. In contrast, constructivists argue that risk perception is not objective but is subjective and socially constructed (Jasanoff 1998).

To reconcile this difference the Italian team constructed the seismic risk perception questionnaire to measure risk perception taking into account seismic hazard, vulnerability and exposure (Crescimbene et al. 2014, 2015a). This questionnaire had

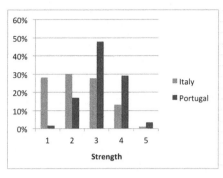 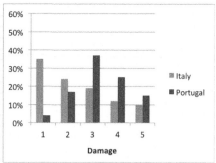

Fig. 37.1 Comparison of preliminary results for question "How would you describe an earthquake in your neighbourhood" La Spezia, Italy (n = 72) and Lisbon, Portugal (n = 117)

been used previously in various projects and was incorporated into the KnowRISK questionnaire (Crescimbene et al. 2015b).

Questions about people's level of awareness and knowledge were constructed considering the specific goals of the KnowRISK project, specifically about damage from non-structural elements and some basic concepts about risk from mild and strong earthquakes. The final part of the questionnaire focuses on behaviour to reduce risk with particular reference to non-structural risk mitigation. This part also investigates barriers to potential mitigation measures to be taken.

One of the key theoretical considerations in the Italian questionnaire was the distinction between perception, awareness and behaviour. Even if aware of the risk, people do not necessarily change their behaviour or adopt protective measures. The Italian team argued that the reason for inaction laid in people's feelings, attitudes and beliefs ie. their perception of risk. In order to induce a real change in behaviour, an effective communication strategy should be designed that both increases people's knowledge and awareness and also addresses their perception of risk.

The Italian questionnaire made extensive use of the semantic differential (64% of all questions). Semantic differential scales are a simple way of obtaining data on people's attitudes that are cross-cultural and work equally well with children and adults. Importantly they allow subjective issues to be quantified, and make qualitative judgments amenable to statistical analysis. The method uses bipolar Likert scales (typically 5 or 7 point scales) using contrasting adjectives, for example strong and weak (Likert 1932). Osgood et al. (1957) championed the use of semantic differential and it has been used in a wide variety of studies of risk perception, for example Slovic and Weber (2002) (Fig. 37.1). Osgood et al. (1957) performed a factor analysis on their data and found that three dimensions, that they labelled evaluation, potency, and activity, accounted for most of the co-variation in ratings. This approach was adopted by Semantic Risk Questionnaire used in previous studies (Crescimbene et al. 2014, 2015a) and formed the basis of the Italian KnowRISK questionnaire.

37.3.2 The Portuguese Framework

The Portuguese theoretical framework is largely based the Precaution Adoption Process (PAP) as elaborated by Weinstein (1988). Weinstein began with the deceptively simple question, "When will people act to protect themselves from harm?" He was critical of cost-benefit theories that these theories assume that people weigh the expected benefits of a precaution against its costs and adopt the precaution if the balance appears favourable. His contention was that the adoption of protective behaviours is a process of development over time of beliefs and intentions that lead to action.

Other ideas also influenced the Portuguese team's thinking about how people appraise or perceive threats. The Protection Motivation Theory (Rogers 1983; Mullis et al. 1990) describes adaptive and maladaptive coping with a threat as a result of two appraisal processes – a process of threat appraisal and a process of coping appraisal, in which the behavioural options to diminish the threat are evaluated (Boer and Seydel 1996). The intention to protect one self depends upon four perceptions: the severity of a threat (in this case an earthquake); the probability of its occurrence; the efficacy of preventive behaviour and confidence in one's ability to take preventive action (Rogers 1975).

Lindell (1994) developed the Protective Action Decision Model (PADM) based on research into people's responses to environmental hazards and disasters to help explain the decision to take preventative action. In their revised model Lindell and Perry (2012) identify three core factors: the threat, alternative protective action, and social stakeholders (for example scientists and government authorities) that, they argue, determine how people respond to a threat.

All these models share the assumption that decision to act protectively is influenced by the following factors:

- Individuals' beliefs concerning their own susceptibility towards risk (perceived susceptibility);
- The severity they assign to a certain event (perceived severity);
- Individuals' beliefs concerning efficacy of protective measures (perceived efficacy);
- The demands of knowledge, effort, time and money that actions imply (perceived costs).

In short, *an individuals' willingness to act protectively is based on the double evaluation they make about their own susceptibility towards a certain risk and the efficacy they attribute to the alternatives of protection* (Silva 2017 p 2).

The Portuguese team realised that it was unrealistic to expect that their intervention in schools would immediately induce the adoption of non-structural seismic protective measures. Nevertheless, they argue that, "there are series of actions that individuals may take which indicate proactivity. The act of looking for more information after hearing about risk; the act of discussing with others on the subject; the act of influencing parents to take precautions are examples of the above-

Fig. 37.2 Stages in the precaution adoption process (After Weinstein 1988)

mentioned proactivity that should not disregarded in our assessment procedure. These actions will be taken as signals of progression" (Silva 2017 p 1).

The adoption of protective behaviours is a process, made of advances and retreats, where individuals pass through several stages until they decide to initiate new behaviours. As shown in Fig. 37.2, Weinstein (op cit.) depicts perceived susceptibility to harm into four stages.

37.4 Assessment of the Italian and Portuguese Questionnaires

37.4.1 Comparison of Theoretical Frameworks

As described above the Italian and Portuguese questionnaires had similar but subtly different theoretical frameworks – the Italian questionnaire put greater emphasis on trying to measure what people unconsciously do and the Italian team felt it was important, in the context of this project, to distinguish between rational, emotional and intuitive understanding and behaviour, (Crescimbene et al. 2017).

Some people distinguish between awareness and perception, as in knowing about or having opinion about something (for example EC 2016). For some however, the difference is less clear-cut and in English these words are synonymous. The Italian team made this distinction in their theoretical framework and in the subheadings of their questionnaire. The Portuguese team did not make this distinction.

37.4.2 Method of Assessment

An initial assessment was made of the Italian and Portuguese questionnaires. Working with the native language versions of the two questionnaires, together with some data from the pre-intervention surveys, the task was to judge how well the questions met the principal aim of assessing children's risk awareness and to what extent the two surveys were comparable. The main difference was in the length of the two surveys. The Italian questionnaire was much longer (164 distinct fields in the Italian questionnaire compared to 65 in the Portuguese version), and in the extensive use of the semantic differential in the Italian version. In the judgement of the assessor the Italian questionnaire would have met the objectives of the KnowRISK project if it omitted about half the questions.

The assessment procedure involved drawing up a correspondence table comparing the two questionnaires and in analysing some of the key questions. The questionnaires also differed in the way questions were constructed. The Portuguese questionnaire posed direct questions about risk awareness and behaviour, while the Italian questionnaire tended to use more indirect questioning, using semantic differential scales with polar opposite adjectives, mixing relevant and non-relevant attributes from which they could infer perception and awareness. Both approaches have merit and in the judgement of the independent consultant, both would achieve the project objectives of measuring awareness, knowledge and action.

37.4.3 The Italian Questionnaire

The data set used to test the Italian questionnaire (72 answers by 12–13 year-old children living in La Spezia, Liguria) was a subset of the complete data collected (Crescimbene et al. 2017). The area where the schools are located lies in a moderate seismic hazard area. For this reduced data set, there was little difference in risk awareness according to building structure, age, and number of stories or level of maintenance. For example, children living in masonry buildings perceive themselves to be at the same risk as children living in reinforced concrete buildings. If the awareness raising intervention worked, one would expect this to change.

In general, the children underestimated the risk. They made little distinction between the impact of mild and strong earthquakes, and those living in older masonry buildings did not perceive themselves to be at greater risk. The research hypothesis is that the planned intervention will increase children's awareness and understanding of non-structural risk and it was though that it might have been interesting at this stage to test whether the questionnaire is likely to measure change between T0 and T1 by comparing the responses in this low risk area with responses from children in a higher risk area.

The Italian questionnaire, in the way it grouped questions, distinguished between the respondents' attitudes and feelings about earthquakes, and other hazards, and their level of knowledge. However, there was ambiguity about whether some of the questions were trying to measure opinion or knowledge. For example, why were questions about objects falling not considered to be about perception rather than awareness or knowledge? In fact the distinction between attitude and knowledge may not be this clear-cut in the minds of the respondents.

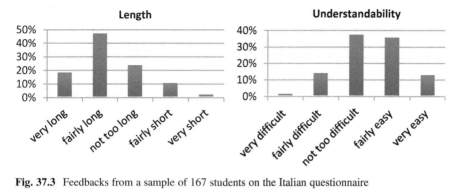

Fig. 37.3 Feedbacks from a sample of 167 students on the Italian questionnaire

The last section of the survey asked children direct questions about the probability of their families taking preventative action. These are important questions but respondents would be tired by the time they got to this stage in the questionnaire.

Feedback on the questionnaire was sought from a sample of 167 students. Although there was concern about the length of the questionnaire (66% found it long or fairly long) and the use of words that might be unfamiliar to children, the results showed that most of the students found it fairly easy to understand and complete (Fig. 37.3).

The analysis made some detailed observations about date ranges and the phrasing of some questions to make them more comprehensible to children. Finally the assessment suggested how the questionnaire might be shortened, for example by omitting questions not specifically relevant to children, for example questions about work place and community, and the duplication of questions for both weak and strong earthquakes.

37.4.4 The Portuguese Questionnaire

The Portuguese questionnaire was devised specifically for this project and did not build on the Italian version. It could have adopted more semantic differential scale questions from the Italian version. Instead it asked direct questions about the respondents' awareness, knowledge and intended actions.

The data set used to test the Portuguese survey was for 117 11–14 year-old children (plus 2 15–16 year-olds) living in or near Lisbon This area, although high risk, has not experienced a major earthquake disaster since the 1755 Great Lisbon Earthquake. Eight-nine percent of respondents had never experienced an earthquake but 80% knew that there have been destructive earthquakes in the past in Portugal. To the question had they received information about earthquake risk, 7% answered rarely or never.

Asked if people could do anything to protect themselves and minimize the damage caused by an earthquake, 59% responded with yes, 3% with no and 19% with don't know. If the intervention is successful, the percentage of children responding yes should rise. Asked about the possibility of an earthquake in Lisbon 52% responded that it is unlikely, 41% that it is likely and 7% did not know. It will be interesting to see if these percentages change after the intervention.

One set of questions used a 4-point semantic differential scale to ask if an earthquake occurred in Lisbon how weak/strong; harmless/damaging it would be. Nineteen percent responded weak and 33% responded strong; 18% responded harmless and 40% responded damaging. It will be interesting to see if these percentages change after the intervention. It was presumed that the omission of a mid-point in the scale was intentional but it would, however, make comparison easier if this question used a 5 point-scale.

The questionnaire also used 5-point Likert scales for five questions: to ask about the consequences of an earthquake in Lisbon, the child's neighbourhood and their own home. In general children seemed to underestimate the risk in their own home and neighbourhood. It would have allowed a useful comparison if this question had been included in the Italian version and if both used 5-point scales (Table 37.2).

The Portuguese survey asked a number of simple yes/no/don't know questions that might usefully have be incorporated in the Italian version. For example, 61% of children think precautions should be taken in Lisbon but 33% don't know. The key questions in the Portuguese survey asked about 10 measures that might reduce non-structural risk, asking if the child knew about them and how useful they thought they would be in saving lives and avoiding damage.

37.4.5 Comparison of the Questionnaires

There was a correspondence, although imperfect, between the two questionnaires on 24 questions. For example Fig. 37.1 shows that children in Lisbon (a place with very infrequent but devastating earthquakes) consider a possible future earthquake in the home area to be stronger and more damaging than do children in La Spezia, Italy (a region of moderate seismic hazard).

37.4.6 Alternative Course of Action

There seemed to be three alternative courses of action.

1. Do nothing and accept the discrepancy between the two surveys and report them independently.

Table 37.2 Portuguese survey: pre-intervention results

	Lisbon (%)	Home area (%)	Home (%)
Little or no damage	28	49	44
High damage	57	36	30
Don't know	15	15	25

This had the merit of not involving the partners in additional work and would have allowed each team to proceed with their own programme of intervention and assessment.

2. Bring the two questionnaires into line by adding questions to both

This would have involved adding about 60 questions from the Italian questionnaire to the Portuguese version, and by adding 39 questions from the Portuguese version to the Italian. It was judged that this was unrealistic at this halfway stage in the project.

3. Devise a short common questionnaire embodying the virtues of both.

It was still possible even at this late stage to produce a common version, adopting the best parts of both. The questionnaire might allow a few specific questions relevant to the particular case, for example the infrequency of events in Portugal compared to Italy, but by and large the questionnaire would be common to all three countries. This final alternative had considerable appeal and was in fact adopted by the partners. However, the problem remained that the bulk of the pre-intervention survey had already been conducted in Portugal and so the Portuguese version, and not the common version, would need to be used in these schools.

The Portuguese team adopted course of action 1. The KnowRISK intervention in the schools was on going and the T0 questionnaire had already been applied. A change in the questionnaire instrument in the middle of the process would have derailed the adopted assessment methodology. However, they decided to pre-test the common questionnaire (February 23, 2017) on a small group of Portuguese school children and teachers who were visiting Instituto Superior Técnico (IST).

The Italian team adopted a mixed strategy of continuing to use the Italian questionnaire is schools already surveyed and the new common questionnaire in schools not previously surveyed. The Icelandic team opted for the new common questionnaire with some minor modifications to suit their particular circumstances.

37.5 Design of the 'Common' Survey Tool

The main aim of the questionnaire was to measure children's awareness of non-structural earthquake risk before and after awareness raising interventions. This meant that the questions had to be capable of measuring change in attitude and intended behaviour and also needed to be clear simple and comprehensible to children.

The design of the common questionnaire built on the Italian and Portuguese theoretical frameworks and on the two existing surveys. This meant using the existing questions as a basis for the proposed version. It also meant understanding the rationale underpinning the design of the existing surveys, ie the Italian reference frame and the Portuguese theoretical framework, and embodying this understanding in the common version.

The survey was structured following the Portuguese theoretical framework that envisaged stages of increasing understanding of risk, from lack of awareness through to taking preventative action. Following the Italian reference frame the common questionnaire adopted the notion of perception and awareness, of feeling and knowing, in the way the questions were framed. However, the questionnaire did not flag the distinction in terms of headings or divisions in the way the Italian version had. As in the Italian version the semantic differential and Likert scale questions were used extensively although their number was reduced. The common questionnaire adopted a number of direct questions from the Portuguese questionnaire. It also added a couple of questions from a recent risk perception survey in Nepal (NSET 2017)).

Following suggestions by the Portuguese team, as well as a focus on non-structural risk, questions about structural risk and building collapse were included as were questions about risk at home and school. However, the distinction between the impact of mild and strong earthquakes in the Italian version was removed.

The questionnaire used a few words as possible and avoided using additional descriptions and qualifying sentences wherever possible. It asked questions a child of 11–13 might be capable of answering and used language they were likely to understand. The approach taken was to help inform respondents as well as gather data.

Efforts were made to make the questionnaire as short as possible while ensuring it covered both structural and non-structural risk at home and at school and collected sufficient data to assess children's risk awareness before and after the training intervention. Google Docs was used to host the survey. Google Docs had been used by the Italian questionnaire. It imposes a discipline on questionnaire design and it could be printed and used in paper form or used to collect data online.

37.5.1 Piloting and Modifications to Questionnaire

37.5.1.1 Piloting in Portugal

The Portuguese team tested version the questionnaire with fifty one 15–16 year-old children and four teachers. They found that the teachers were more demanding than students. In particular, the teachers wanted to label the intermediate levels of the semantic differential scales. In contrast, the students did not give too much importance to the meaning of the levels and were able to use the scales without difficulty. It was decided to adopt the feedback from the Italian team to use 7-point scales rather than 5-point since 7-point scales are generally used for semantic differential and never have intermediate labels. Whereas 5-point Likert scales often have labels on intermediate points.

The Portuguese team also had a number of detailed comments. They thought a number of multiple-choice questions lacked a "don't know" category. Maintenance of the home and school they thought should include the category "reasonable". Built to building code should be replaced with "earthquake resistant design". People asked what type of earthquake was implied in the question, "What would happen in your home if there was an earthquake?" In the opinion of the Portuguese team, because the focus of the project is on non-structural damage, the size of earthquake should be a specified as "mild/moderate earthquake". The option "never found" should be included in the question "What is the main way you find out about earthquake risk?" These changes were made.

37.5.2 Piloting in Italy

The Italian INGV team suggested that the various questions about the child's attitude to earthquakes could usefully be divided into two parts, subtitled "What do you think about earthquakes and "How would you describe an earthquake in your home area?" And they thought it would be helpful to turn the question "Do you know about earthquakes in your region?" into a Likert scale rather than a simple yes/no answer.

They also had a number of detailed suggestions. They felt it was wrong to offer the option "don't know" for two of the questions: "What is the strongest earthquake you have experienced?" and "Number of floors in the home". They wanted to add, "Keep calm and wait the end of shaking" to the list of possible preventative actions. Finally they thought printing from the Google doc worked fine.

37.5.3 Piloting in Iceland

The EERC team discussed and trialled the questionnaire internally with colleagues. They suggested the following minor changes. In the questions about construction of the home and school they wanted to add an option of "timber" since this is a common construction material in Iceland. And in question about number of floors in the home, they wanted to split the value "2–5" into "2–3" and "4–5" since most buildings in South Iceland have either 1 floor or 2–3 floors, whilst higher buildings are very uncommon.

They also had a more substantive suggestion. They took issue with the preventative action, "Run out of the house". In Iceland, they argued, it might be safer to stay calm until the quake is over instead of running during the shaking. This is because there is very little chance of building collapse during earthquakes in their study area. On the other hand, and from their own experience, of vulnerable buildings in Nepal, "get out of the house" might be the best advice in some areas with masonry buildings including in Italy and Portugal.

The problem they saw was that including this question in the survey implied this action was recommended, which is not the case in Iceland. A more relevant action in their study area would be "Stay calm and try to keep your balance". It was decided to include this proposed action, but to retain "Run out of the house since it was felt to be important to test if the intervention eliminates any tendency to see this action as desirable".

The other difference the EERC team saw in their study area was related to the action "Move beds from windows". In the last three earthquakes, they argued, there was no evidence of window glass getting broken, even in areas where PGA was as high as 0.9g. This is mainly because the buildings are made with shear walls, are very stiff, and consequently the interstory drift is very low. They would therefore have preferred not to have this question in the survey in Iceland, as it seemed unnecessary. It was, however, decided to retain this action for the same reason as the previous suggestion i.e. to see if the intervention affects the children's deep understanding of the true nature of the risk.

Finally the EERC team suggested adding two other important preventative measures: "Knowledge of where and how to turn off cold and hot water supply" because damage to room heating ovens and subsequently flooding was a major problem in past earthquakes, and "Knowledge of where electricity main switch is and how to turn it off", to prevent electric shocks and potential fires. Both were added.

37.6 Discussion and Conclusion

The main conclusions from this analysis may be summarised as follows.

1. Differences between the two case study countries or differences in culture go some way to explain the difference in approach and content between the Portuguese and Italian approaches. The differences also due to legitimate differences between the two teams about how best to design a survey to meet the objectives of the project.
2. It is helpful on pan European projects that involve collaboration between different research teams, to have an independent consultant help develop a common agreed approach.
3. A key aspect of the research design was to use the same assessment tool, ie. the common questionnaire, both before and after the training/awareness raising interventions. However, the time interval between the two interventions was relatively short and there was a high probability that memory answering the

first survey would have influenced the second. If, however, one accepts that the assessment procedure is actually an integral part of the awareness raising exercise then this is not as much of a problem as it first seems. Nevertheless in the teams Iceland and Portugal advanced the idea of using a control group in the second T1 survey, who would attend the training but not complete the first questionnaire.
4. The tool is only able to measure an increase in awareness and understanding since the time interval between the intervention and the subsequent intervention is too short to measure change in behaviour and an improvement in preventative action. Nor is the subject group, children aged 11–15, capable of taking effective action, either in their own homes or at school. However, the questionnaire did seek to measure avowed intention to take action, either taking to friends and parents about the risk and the need to take action.
5. The project had a limited time span of 2 years. It is therefore impossible to thoroughly assess the effectiveness of the communication. The tools the project is building, including the questionnaire, will therefore need to be modified to ensure better performance. The ex-ante and ex-post questionnaire, that might be used to assess future communication actions, should ideally be conducted over a much wider time range, several months after the end of communication intervention.

Acknowledgements This study was financed by the European Commission's Humanitarian Aid and Civil Protection (Grant agreement ECHO.A.5 (2015) 3916812) – KnowRISK (Know your city, Reduce seISmic risK through non-structural elements).

References

Akason J, Olafsson S, Sigbjörnsson R (2006) Perception and observation of residential safety during earthquake exposure: a case study. Saf Sci 44:919–933
Bessason B, Bjarnasson J (2016) Seismic vulnerability of low-rise residential buildings based on damage data from three earthquakes (Mw 6.5, 6.5, 6.3). Eng Struct 111:64–79
Boer H, Seydel E (1996) Protection motivation theory. In: Connor M, Norman P (eds) Predicting health behavior. Open University Press, Buckingham
Bryman A (2011) Social research methods. Oxford Press, London
Crescimbene M, La Longa F, Camassi R, Alessandro Pino N, Peruzza L (2014) What's the seismic risk perception in Italy? Eng Geol Soc Territ 7:69–75
Crescimbene M, La Longa F, Camassi R, Pino N (2015a) The seismic risk perception questionnaire. Geol Soc Lond Spec Publ 419:69–77
Crescimbene M, La Longa F, Camassi R, Pino N, Peruzza L, Pessina V, Cerbara L, Crescimbene C (2015b) INGV. *EOS8 geoethics for society*: general aspects and case studies in geosciences, Istituto Nazionale di Geofisica e Vulcanologia, Rome
Crescimbene M, Pino N, Musacchio G (2017) *Reference frame: KnowRisk questionnaire* (ITA) Internal document, Istituto Nazionale di Geofisica e Vulcanologia, Rome
EC (2016) Perception and awareness about transparency of state aid. Special Eurobarometer 448 – Wave EB85.3 – TNS opinion & social
Floyd D, Prentice-Dunn S, Rogers R (2000) A meta-analysis of research on protection motivation theory. J Appl Soc Psychol 30:407–429
Grothmann T, Reusswig F (2006) People at risk of flooding: why some residents take precautionary action while others do not. Nat Hazards 38:101–120

Likert R (1932) Technique for the measure of attitudes. Arch Psychol 22(140):5–55
Lindell M (1994) Perceived characteristics of environmental hazards. Int J Mass Emerg Dis 12:303–326
Lindell M, Perry R (2012) The protective action decision model: theoretical modifications and additional evidence. Risk Anal 32(4):616–632
Maidl E, Buchecker M (2015) Raising risk preparedness by flood risk communication. Nat Haz Earth Syst Sci 15:1577–1595
Miceli R, Sotgiu I, Settanni M (2008) Disaster preparedness and perception of flood risk: a study in an alpine valley in Italy. J Environ Psychol 28:164–173
Neuwirth K, Dunwoody S, Griffin R (2000) Protection motivation and risk communication. Risk Anal 20:721–734
NSET (2017) Risk perception survey in Bhimeshwor municipality. National Society for Earthquake Technology-Nepal (NSET) 16WCEE conference Chile 9 January 2017
Osgood C, Suci G, Tannenbaum P (1957) The measurement of meaning. University of Illinois Press, Urbana
Rogers R (1975) A protection motivation theory of fear appeals and attitude change. J Psychol 91:93–11
Scolobig A, De Marchi B, Borga M (2012) The missing link between flood risk awareness and preparedness: findings from case studies in an alpine region. Nat Haz 63(2):499–520
Sigbjörnsson R, Ólafsson S (2004) On the South Iceland earthquakes in June 2000: strong-motion effects and damage. Boll Geofis Teor Appl 45(3):131–152
Sigbjörnsson R, Snæbjörnsson J, Higgins S, Halldórsson B, Ólafsson S (2009) A note on the M6.3 earthquake in Iceland on 29 May 2008 at 15:45 UTC. Bull Earthq Eng 7(1):113–126
Sigbjörnsson R, Ragnarsdottir S, Rupakhety R (2017) Is perception of earthquakes gender dependent. In: Rupakhety R, Olafsson S (eds) Earthquake engineering and structural dynamics in memory of Prof Ragnar Sigbjörnsson: selected topics. Springer, Cham
Silva D (2017) Theoretical framework for risk communication impact assessment. LNEC. KnowRisk Project Task 4 Action D3
Slovic P, Weber E. (2002) Perception of risk posed by extreme events. In: Conference on risk management strategies in an uncertain world, Palisades, New York, April 12–13, 2002
Terpstra T, Lindell M, Gutteling J (2009) Does communicating (flood) risk affect (flood) risk perceptions? Results of a quasi-experimental study. Risk Anal 29:1141–1155
Weinstein N (1988) The precaution adoption process. Health Psychol 7(4):355–386

Printed by Printforce, the Netherlands